D1446489

AIRPLANE
PERFORMANCE
STABILITY
AND
CONTROL

AIRPLANE
PERFORMANCE
STABILITY
AND
CONTROL

BY

COURTLAND D. PERKINS

Professor and Chairman, Aeronautical Engineering Department
Princeton University

ROBERT E. HAGE

Senior Engineer, Preliminary Design Department
Boeing Airplane Company

JOHN WILEY & SONS, INC.
NEW YORK · LONDON · SYDNEY

PRINTED IN THE UNITED STATES OF AMERICA

PREFACE

The purpose of this book is to present those elements of applied aerodynamics that bear directly on the problem of airplane design. Following conventional aeronautical engineering practice, we have divided the book into two major parts, the first dealing with the problems of airplane design for performance, and the second involving the design of the airplane for adequate stability and control characteristics or, more popularly, its flying qualities.

The book has been written to meet the needs of the practicing aeronautical engineer, as well as the requirements of class instruction. The text is based largely on material developed by us while working together in the Aerodynamics Branch of the United States Air Force Matériel Command at Wright Field, during the years of World War II. During 1944 we used a preliminary copy of the manuscript while teaching officers in the Air Force Institute of Technology at Wright Field. Since that time we have expanded the text at Princeton University and at the Boeing Airplane Company in Seattle. All the material has been used in undergraduate and graduate course instruction at Princeton University during the past three years.

In the first part of the book, we have presented the classical methods of estimating airplane performance, substantiated by modern airplane data within security limitations. In this section the estimation of airplane performance in the subsonic, transonic, and supersonic speed ranges is discussed, as well as the effect on airplane performance of the new reaction-type power plants. In the second part of the book, we have dealt rather fully with the comparatively new art of designing the airplane for adequate flying qualities. In this section only subsonic flight is discussed. Several chapters dealing with transonic and supersonic stability and control have been deleted as premature.

Many references to publications of the National Advisory Committee for Aeronautics have been used in the preparation of this book, for which we express our appreciation to the committee. We also wish to acknowledge the work of Miss Alice I. Dethman, Mrs. Richard C. Jones, and Messrs. R. D. Fitzsimmons and H. W. Withington of the Boeing Airplane Company for their assistance in the preparation of the material for Part 1; and Messrs. D. C. Hazen, P. F. Sheridan, and G. W. Brooks of Princeton University for their assistance in the preparation of Part 2. Special acknowledgement is made to Mr. S. D.

Hage, director of the Propulsion Development Unit of the Boeing
Airplane Company, for his invaluable assistance during the preparation of the chapter on propulsion.   Finally, we should like to thank all
our students, of both the Air Force Institute of Technology and
Princeton University, for their patience during the periods when the
rough edges of the manuscript were being smoothed off, as well as for
their efforts toward supplying the correct answers to the many
problems.

COURTLAND D. PERKINS
ROBERT E. HAGE

Princeton, N. J.
September 1949

CONTENTS

Part 1 AIRPLANE PERFORMANCE

Part 2 AIRPLANE STABILITY AND CONTROL

Part 1

AIRPLANE PERFORMANCE

Part 1

AIRPLANE PERFORMANCE

CHAPTER 1

INTRODUCTION

1-1 The Airplane as a Rigid Body

The airplane, considered as a rigid body flying through the airspace, moves along paths that are determined by the airplane's inertia characteristics, the attraction from the Earth's gravitational field, the propulsive forces generated by its power plant, and the aerodynamic forces and moments created on it because of the reaction between it and the air through which it moves. The forces and moments created on the airplane are functions of the velocity of the airplane, the density of the air through which it flies, the geometry of the airplane, and finally the angle that the relative wind makes with the airplane. The major purpose of this book is to introduce the reader to the theoretical and empirical means for estimating these forces and to study the characteristics of the airplane's motion along the resulting flight paths.

The paths along which the airplane can fly in the airspace are limited only by the aerodynamic characteristics of the airplane, its propulsive system, and the structural strength of the airframe. These limitations, if found, indicate the maximum performance and maneuverability of the airplane. If the airplane is to realize maximum utility, it must be safely controllable by the pilot to these limits without exceeding his strength and without requiring acrobatic ability on his part. The design problems involved are those of predicting the performance of the airplane and of providing it with satisfactory stability and control characteristics or flying qualities. The first part of this volume will deal with the problems of airplane performance, and the second part will deal with the problem of stability and control.

If the airplane is to fly steadily along any arbitrary flight path within its aerodynamic or strength limitations, the forces acting on it must be in static equilibrium if the path is a straight one, and in dynamic equilibrium if the flight path is curved or accelerated in any way. In order to discuss these equilibriums, it is first necessary to adopt some nomenclature and symbols. To this end it is convenient to replace and represent the airplane by a set of mutually perpendicular axes with

3

their origin at the airplane's center of gravity, referred to throughout this book as the airplane's c.g. This axis system is shown in Figure 1-1 and is a right-hand axis system with the positive X and Z axes in the plane of symmetry and with the X axis out the nose of the airplane usually taken pointing into the wind. The Z axis is perpendicular to the X axis, positive downwards, and the positive Y axis is out the right wing perpendicular to the plane of symmetry.

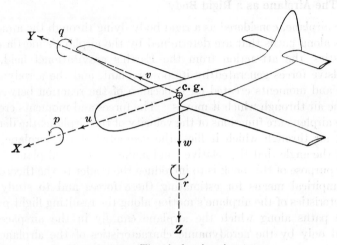

FIGURE 1-1. The airplane's axis system.

The components of the forces and moments acting on the airplane, and the components of the airplane's motion referred to this axis system, are as follows:

AXIS	FORCE ALONG	MOMENT ABOUT	LINEAR VEL.	ANG. DISP.	ANG. VEL.	INERTIA
X	F_x	L	u	ϕ	p	I_x
Y	F_y	M	v	θ	q	I_y
Z	F_z	N	w	ψ	r	I_z

1-2 The Airplane as a Dynamic System

As the airplane's motion in space can be completely defined only if six velocity components are given, the airplane is considered to be a dynamic system in six degrees of freedom. For equilibrium along any straight unaccelerated flight path, the equation of statics applied to each degree of freedom must be satisfied.

$$\Sigma F_x = 0 \qquad \Sigma L = 0$$
$$\Sigma F_y = 0 \qquad \Sigma M = 0 \qquad\qquad (1\text{-}1)$$
$$\Sigma F_z = 0 \qquad \Sigma N = 0$$

The airplane's plane of symmetry that cuts the airplane into two symmetric halves (Figure 1–2) contains components of motion only

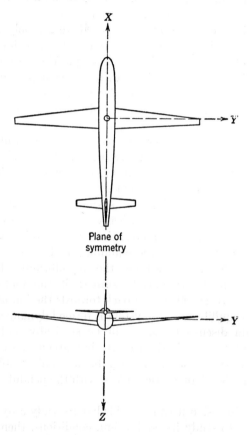

FIGURE 1–2. The airplane's plane of symmetry.

along the X and Z axes and about the Y axis. The other three components of the airplane's motion lie outside the plane of symmetry. The symmetric degrees of freedom comprise what is usually referred to as the airplane's longitudinal motion, and the asymmetric degrees of freedom comprise the airplane's lateral motion.

In order to investigate the motion of any dynamic system, it is necessary first to establish that the system can be brought into a condition of equilibrium. Beyond this, it is necessary to determine the stability characteristics of the equilibrium. For the equilibrium to be statically stable, a disturbance of the equilibrium must create forces or moments within the system that tend to start the motion back towards the equilibrium. For example, consider the simple pendulum (Figure 1–3). The pendulum is free only to rotate about the pivot and therefore is a dynamic system in one degree of freedom. There are only two positions at which the pendulum can be brought into a condition of static equilibrium, namely, positions A and B as shown in Figure 1–2. There is, however, a profound difference in these equilibriums, for if once in balance at A, the moment created by any disturbances will tend to start the pendulum moving away from A, but if in balance at B, the moment created by a disturbance will tend to start the pendulum back towards B. The equilibrium at A is termed statically unstable; the equilibrium at B is termed statically stable.

Equilibrium
A

B
Equilibrium

FIGURE   1–3. The single-degree-of-freedom pendulum.

Once the dynamic system is brought into equilibrium and made statically stable, it is necessary to examine the characteristics of its motion following any disturbance of the equilibrium. If the subsequent motion finally restores the equilibrium, the system is termed dynamically stable; if the motion, although starting the system back towards the initial equilibrium, never restores equilibrium, the system is termed dynamically unstable. The pendulum discussed above will be statically and dynamically unstable about the equilibrium at A, but about the equilibrium at B, although statically stable, it may be neutrally stable in the dynamic sense if the pivot is frictionless, with the pendulum swinging in a vacuum.

From this discussion it can be said that to study a dynamic system it is necessary to study its equilibrium conditions, then establish the static stability of the equilibrium and the dynamic stability or the response characteristics of the system to disturbances. The magnitude of static and dynamic stability required of any dynamic system can be dictated only after the characteristics of the response are known.

1-3 Equilibrium Conditions

The problem of satisfying the equilibrium equations (1–1) is a very simple one for the asymmetric degrees of freedom. A quick look at

Figure 1–2 will demonstrate that for motion in the plane of symmetry the asymmetric degrees of freedom will be in equilibrium because of the airplane's symmetry alone and constitute no particular design problem. The equilibrium equations for the longitudinal degrees of freedom are not as easily dealt with and must be considered in more detail.

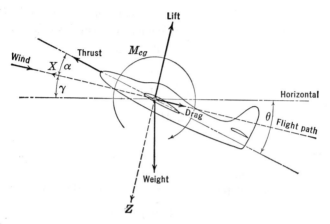

FIGURE 1–4.   Forces and moment in plane of symmetry.

For symmetric flight along a straight path, the airplane's axis system is taken as indicated in Figure 1–4. In this figure the X axis is arbitrarily taken as always pointing into the wind with the Z axis perpendicular to it in the plane of symmetry. The major forces acting on the airplane are shown in this figure as all acting through the airplane's center of gravity (c.g.). All the moments acting on the airplane are lumped into one couple, M_{cg}. The inclination of the flight path to the horizontal is called γ, and the angle between the flight path and the airplane's reference axis is α. The displacement of the reference axis from the horizontal is the airplane's pitch angle, θ.

Summation of the forces along the X and Z axes and the total moments acting about the Y axis yields the equations of static equilibrium for the airplane in straight symmetric flight.

$$\Sigma F_x = T \cos \alpha - D - W \sin \gamma = 0 \qquad (1\text{–}2)$$

$$\Sigma F_z = W \cos \gamma - L - T \sin \alpha = 0 \qquad (1\text{–}3)$$

$$\Sigma M = \Sigma M_{cg} = 0 \qquad (1\text{–}4)$$

The assumption is made that the angle of attack is always a relatively small angle, and therefore that little error is involved in assuming

$\cos \alpha = 1$ and $\sin \alpha = 0$. Under this assumption, the equilibrium equations reduce to the following:

$$T - D = W \sin \gamma \tag{1-5}$$

$$L = W \cos \gamma \tag{1-6}$$

$$\Sigma M_{cg} = 0 \tag{1-7}$$

The velocity of climb, v_c (feet per second), or rate of climb, R/C, is equal to $v \sin \gamma$; therefore the static equations can be written:

$$v_c = \frac{Tv - Dv}{W} \tag{1-8}$$

$$L = W \cos \gamma \tag{1-9}$$

$$\Sigma M_{cg} = 0 \tag{1-10}$$

If equation (1-9) is expressed in terms of the lift coefficient, $C_L = L/qS$, where L is the lift in pounds, q the dynamic pressure in pounds per square foot, and S the wing area in square feet, the airplane's velocity in feet per second can be expressed as follows:

$$v = \sqrt{\frac{2 \cos \gamma \, W/S}{\rho C_L}} \tag{1-11}$$

For all flight paths whose angle is small, $\cos \gamma$ can be assumed equal to unity, and the airplane's velocity for a given wing loading and altitude is purely a function of the lift coefficient. It is well known that the lift coefficient is a function of the airplane's angle of attack through the lift curve slope; therefore the airplane's speed along shallow paths is purely a function of the angle of attack or lift coefficient. In order for the airplane's speed to be controlled, then, the pilot must be able to control the equilibrium angle of attack or lift coefficient.

If equation (1-8) is examined in detail, it can be seen that whether the airplane goes up or down at a given speed depends on the differences between the thrust and drag at this speed. If the thrust just equals the drag ($T = D$), then the rate of climb will be zero and the airplane will be in level flight. Equation (1-8), if expressed in terms of the drag coefficient, $C_D = D/qS$, can be stated as follows:

$$R/C \text{ or } v_c = \frac{Tv - C_D \frac{1}{2}\rho v^3 S}{W} \tag{1-12}$$

The airplane's drag coefficient is a function of the lift coefficient, so that once the airplane's lift coefficient (or angle of attack) is specified everything in equation (1–12) is known except the thrust, T. The thrust in general is a function of the airplane's speed and throttle setting. For simplicity, it will be assumed that the thrust is only a function of throttle setting. Under this condition, the airplane's speed is determined by the value of the equilibrium lift coefficient, while its rate of climb or descent is regulated principally through the throttle control. For very large angles of climb or descent, this simple picture can no longer be used.

The solution of the first two equations of statics, (1–5) and (1–6), yields the major characteristics of the airplane's performance along unaccelerated flight paths. The solutions for the airplane's perform-ance characteristics, such as maximum speed, rate of climb, time to climb, range, and take-off, are all predicated from estimates of the variation of the lift, drag, and thrust forces, as functions of angle of attack, altitude, and throttle setting. Once these variables are known, the total performance of the airplane can be estimated. Methods for estimating the lift and drag forces acting on any airplane are discussed in Chapter 2, and methods for estimating the thrust output of different types of power plants are discussed in Chapter 3. Finally, in Chapter 4, the over-all performance of the airplane is evaluated.

The solutions for the so-called performance equations, (1–8) and (1–9), state that, for a given thrust and drag variation with angle of attack or lift coefficient, the airplane will fly at a definite speed and climb at a definite rate. All this assumes that the airplane can be brought into equilibrium at any arbitrary angle of attack by some means provided the pilot. This condition must be formally satisfied by requiring the summation of moments about the Y axis to equal zero at a given angle of attack or lift coefficient by equation (1–10). The airplane must be designed to allow control over the equilibrium lift coefficient, through-out the usable angle of attack range of the wing. Furthermore, it is necessary to design the airplane to permit the pilot to hold the equi-librium lift coefficient, without the application of excessive forces to the controls, and without exceeding the pilot's strength or reaction abilities. These problems are those of airplane stability and control and are tied in very closely to the third static equation, (1–10). The problem of airplane stability and control is that of designing the air-plane to be not only flyable, but also to be safe, easy, and pleasant to fly. This design problem is sometimes referred to as that of providing the airplane with adequate "flying qualities." The United States Air Force and the United States Navy both have recently published specifications setting definite minimum flying quality standards. These specifications require the airplane designer to provide basic stability and control characteristics in his design that will usually result in meeting the requirement of a safe, nicely flying airplane.

In order to complete the study of the airplane's equilibrium, it is necessary to look at the equation of pitching moments more carefully. The major forces and moments operating on the airplane can be thought of as acting as shown in Figure 1-5.

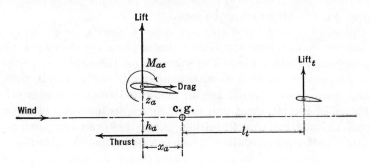

FIGURE 1-5. Summation of moments about center of gravity.

Summation of moments about the airplane's center of gravity yields the equation for static equilibrium in pitch.

$$M_{cg} = Lx_a + Dz_a + Th_a + M_{ac} - L_t l_t \qquad (1\text{-}13)$$

In coefficient form with $M_{cg} = C_{m_{cg}} q S c$, where c is the wing mean aerodynamic chord in feet and S_t, the area of the horizontal tail in square feet, this equation is:

$$C_{m_{cg}} = C_L \frac{x_a}{c} + C_D \frac{z_a}{c} + T_c \frac{h_a}{c} + C_{m_{ac}} - C_{L_t} \frac{S_t}{S} \frac{l_t}{c} \frac{q_t}{q} \qquad (1\text{-}14)$$

As the drag coefficient and tail lift coefficient, C_{L_t}, are functions of C_L, it can be seen that the lift coefficient is the major variable involved, and the pitching moment coefficient is a function of C_L. The equilibrium can be established if the components of the airplane are proportioned to allow $C_{m_{cg}} = 0$ at some useful lift coefficient.

It is through this equation in pitch that control over the airplane's lift coefficient is given the pilot. If the pilot is given control over the constant terms of equation (1-14), he will be able to change the equilibrium at will and therefore his speed; and if supplied with a throttle control, the pilot will be able to control the airplane's flight path angle within its performance limits.

The equilibrium in the longitudinal degrees of freedom is therefore established only if the very important degree of freedom in pitch is satisfied by careful design.

It is repeated again that the symmetry of the airplane eliminates as a design problem the equilibrium in the asymmetric degrees of freedom.

1–4 Static Stability Conditions

The static stability of the three longitudinal equations of equilibrium can be handled simultaneously by considering the result of a change in the airplane's angle of attack from the equilibrium condition. Increasing the airplane's angle of attack will throw all three equilibrium equations immediately out of balance. Through equation (1–5) there will be an increase in drag which will start to slow the airplane down, and through equation (1–6) the increase in lift will start to curve the flight path upwards. If the increase in lift coefficient produces a nose-down pitching moment through equation (1–7), the airplane's angle of attack will tend to be reduced, which will tend to restore the speed, straighten the flight path, and restore the original equilibrium conditions. If, however, the increase in lift coefficient produces a nose-up moment about the airplane's center of gravity through (1–7), the airplane will tend to diverge further and further away from the original equilibrium. The condition for static stability of the longitudinal degrees of freedom therefore depends on the manner in which the airplane's pitching moment varies with lift coefficient. If the slope of a plot of C_m versus C_L is negative, then the equilibrium will be statically stable; if positive, the equilibrium will be unstable. The longitudinal static stability then is tied up with the very important equation of pitch, and the design conditions required to satisfy the static stability are one of the major problems of airplane design. The slope of the pitching moment curve, C_m versus C_L, expressed as the derivative dC_m/dC_L has come to be the criterion of static longitudinal stability.

Although the airplane is automatically in equilibrium in the asymmetric degrees of freedom when the flight path is in the plane of symmetry, these equilibriums are not necessarily stable. Stability of the airplane in the lateral degree of freedom is somewhat confusing at first because mere changes in airplane heading or displacements in roll will create no forces tending to restore the airplane to the neutral heading or to eliminate the roll. The lateral degrees of freedom are sensitive only to the direction of the relative wind to the plane of symmetry. This is called the angle of sideslip, β; the airplane is designed to resist the development of sideslip during all normal flight maneuvers.

The ability of the airplane to create yawing moments that tend to eliminate any sideslip that might develop for one reason or other is called directional stability or, in some cases, weathercock stability. It is essentially the problem of static stability of the degree of freedom about the Z axis and must be carefully designed, for it is extremely important to the flying qualities of the airplane.

The second type of static lateral stability is associated with the rolling moments produced by sideslip. Although in the strict sense of the word this is not a stability, it nevertheless has a bearing on the airplane's flying qualities. The rolling moment created because of sideslip is usually referred to as a dihedral effect and is considered stable when the rolling moments are such as to raise the wing in the direction of the sideslip.

The major problems of equilibrium and static stability can be stated as those pertaining to the moments about the three axes and their variation with angles of attack and angle of sideslip. The problems of static equilibrium, stability, and control, for the longitudinal and lateral degrees of freedom, are discussed in Chapters 5 through 9.

1-5 Airplane Dynamics

The requirements of static and dynamic stability for any dynamic system arise from the characteristics of the system's response to disturbances or to its controls. If a dynamic system responds to some disturbance very slowly with respect to the time required for the person controlling the system to apply corrective control, static and dynamic stability may not be required at all. An example of this type of system is the boat or the large, lighter-than-air dirigible. Both of these examples are usually statically unstable, but the divergent motion involved is very slow and, as a matter of fact, is useful in improving the control characteristics.

The high-speed airplane, on the other hand, responds very rapidly to disturbances, especially with regard to angle of attack. If the airplane is allowed any large unstable margin, the divergent response to disturbances will be so fast that the airplane could easily fail structurally before the pilot could react to the motion at all. For this reason some static stability is required in the major airplane degrees of freedom. The nature of the response of the airplane to disturbances and to its controls is discussed at length in Chapters 10 and 11.

1-6 Airplane Control

The flight path of the airplane can be controlled within the limitations of its aerodynamic characteristics and structural strength, through control over the equilibrium angle of attack, α, angle of sideslip, β, angle of bank, ϕ, and the output of the power plant. These controls are the elevator, the rudder, the aileron, and the throttle.

The design of the elevator, rudder, and aileron for lightness, effectiveness, and harmony of action is a major design problem. The controls must be made effective enough to allow the airplane to realize the maximum utility, and at the same time light enough so that maneuvering the airplane will not tax the pilot's strength, yet never so light that the pilot with very little effort can inadvertently maneuver the airplane past its structural design limits. These problems become more difficult as airplanes become larger and faster.

CHAPTER 2

DRAG ESTIMATION

2-1 Drag Aerodynamics

The study of drag and lift aerodynamics, or the nature of the forces on an aerodynamic body moving relative to a real fluid, may be somewhat simplified by analyzing these forces, first with respect to a special frictionless incompressible fluid—sometimes defined as a perfect or ideal fluid—then with respect to a viscous fluid, and finally with respect to a compressible fluid. Drag forces at relatively low speeds on an aerodynamic surface, as in Figure 2–1, are fundamentally the result of the horizontal components of the normal and tangential forces trans-

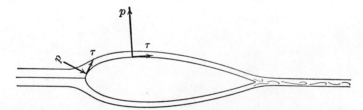

FIGURE 2–1.   Normal and tangential forces on an aerodynamic surface.

mitted from the air to the body. The pressure or normal forces can be determined from the streamline pattern about the body and the use of Bernoulli's principle as applied in an ideal fluid. The friction or tangential forces are the result of the effects of viscosity within the boundary layer. To the rear of the body an additional pressure drag results because of separation in the formation of a turbulent wake. The drag resulting from the pressure variation over and behind the surface is sometimes defined as pressure or form drag; that due to the shear forces in the boundary layer is usually called skin-friction drag. For an airplane wing the summation of the two types is referred to as profile drag.

Dimensional Analysis

Before proceeding with a detailed discussion of airflow in an ideal, viscous, and compressible fluid, an introduction to dimensional analysis

14

of aerodynamic forces should indicate fundamental parameters on which these forces should depend and permit an orderly presentation of them. In considering the units of various physical quantities, it is customary to divide them into two groups, fundamental and derived. The fundamental units are mass, length, and time, and all physical quantities have dimensions that are derived from a combination of these three fundamental units. For example, the coefficient of viscosity in a viscous fluid has dimensions of M/LT. In Figure 2–2 a differential force acting on one of two adjacent layers of a moving viscous fluid

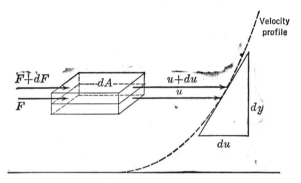

FIGURE 2–2. Shear stresses between layers of a moving fluid (viscosity).

sets up a shear stress between the layers which is a function of the viscosity of and the distance between layers. This shear stress is directly proportional to the rate of change of velocity along y or, in general, is inversely proportional to the slope of the velocity profile at a given point.

$$\tau = \frac{dF}{dA} \sim \frac{du}{dy} = \text{Constant}\,\frac{du}{dy} = \mu\,\frac{du}{dy}$$

The coefficient of viscosity, μ, is the constant that relates the linear relationship between this force and the rate of shear, and has the dimensions of pound-seconds per square foot, or slugs per foot-second, or, in terms of the fundamental units, M/LT.

$$\mu = \frac{\tau}{du/dy} = \frac{\text{Shear stress}}{\text{Velocity/Length}} = \frac{\tau}{v/L}$$

$$= \frac{M/LT^2}{L/TL} = \frac{M}{LT}$$

The degree of viscosity is practically independent of pressure. Viscosity varies widely with temperature, depending on the cohesion and intermolecular characteristics of the fluid. In a gas, where cohesion between molecules is negligible and exchange of molecules between moving layers is exhibited, μ is approximately proportional to the $\frac{3}{4}$ power of the absolute temperature. For liquids where cohesion holds the molecules together, the viscosity decreases as the temperature increases. A term combining both the viscous and density properties of a fluid is defined as kinematic viscosity,

$$\nu = \frac{\mu}{\rho}$$

and has the dimensions L^2/T or square feet per second.

Various combinations of physical quantities define the nature or types of forces that act on any body immersed in air. The existence of these forces as defined below determines the nature of the relative flow between the air and the body.

$$F_p = \text{Pressure force} = pA = pL^2 \tag{2–1}$$

$$F_I = \text{Inertia force} = ma = \rho L^3 \frac{L}{T^2} = \rho \frac{L^2}{T^2} L^2 = \rho v^2 L^2 \tag{2–2}$$

$$F_\mu = \text{Viscosity force} = \mu \frac{du}{dy} A = \mu \frac{v}{L} L^2 = \mu v L \tag{2–3}$$

$$F_E = \text{Elastic force} = EA = EL^2 = \rho a^2 L^2 \tag{2–4}$$

Pressure force is similar to inertia force as the pressure on any surface is a function of density and the second power of velocity. Furthermore, by Bernoulli's law for an ideal fluid, it can be shown that p and $\frac{1}{2}\rho v^2$ are interchangeable. Two other forces, gravity and surface tension, are important only if the body is floating in a relatively high-density fluid such as water. In this case the buoyant force on the body is dependent on the gravity force, and the surface tension is dependent on the free surface that separates the two fluids of different densities. In the study of aerodynamics the elastic force is important only at speeds high enough to cause an appreciable compression of air. The effects of compressibility will be presented in more detail later. In the following discussion the relative importance of various combinations of pressure, inertia, viscous, and elastic forces will be demonstrated.

One of the most important principles in the study of dimensional analysis is dimensional homogeneity. This means that expressions of physical relationships involving physical quantities must have in each term of the equation the same units in order that the equation may have physical significance. The use of this principle is twofold. It may be utilized to derive the nature of fundamental physical equations. It may be used to find an error in a derived physical equation by a check of the units of each term of the equation. To illustrate the first use of this mathematical tool, consider the nature of the resultant force existing on an aerodynamic body. First consider an ideal fluid. By inspection of the nature of the fluid and the body, it may be reasoned that the force is dependent on the density of the fluid, the size of the body, and the relative velocity between the two.

$$F = K\rho^a v^b l^c$$

where K is a constant of proportionality and a, b, and c are exponents to be determined. Expressing the same equation in terms of fundamental and derived units,

$$\frac{ML}{T^2} = K\left(\frac{M}{L^3}\right)^a \left(\frac{L}{T}\right)^b L^c$$

Since the dimensions on both sides of the equation must be identical in order that it have physical significance,

$$M = M^a$$

and

$$a = 1$$

$$T^{-2} = T^{-b}$$

and

$$b = 2$$

$$L = L^{-3a}L^b L^c$$

and by the exponential law for the product of two quantities,

$$1 = -3a + b + c$$

$$c = 2$$

The general equation may now be written

$$F = K\rho v^2 l^2 \tag{2–5}$$

Since lift and drag are two components of the resultant force, the equations for lift and drag of an airplane or airplane wing may be written in the following manner:

$$\text{Lift} = L = C_L \tfrac{1}{2}\rho v^2 S = C_L qS \tag{2–6}$$

$$\text{Drag} = D = C_D \tfrac{1}{2}\rho v^2 S = C_D qS \tag{2–7}$$

where C_L and C_D are lift and drag coefficients, respectively, and the area term in equation (2-5) is taken arbitrarily as the wing area. In Figure 2-3 the conventions related to an airfoil are indicated. The chord, c, is usually a convenient reference axis joining the trailing edge of the airfoil with a point on the leading edge. The angle of attack, α, is the angle between the chord and the remote relative air velocity, v. Lift and drag are components of the resultant force, F, and act perpendicular and parallel to the remote air velocity or relative motion. The moment about some point on the chord is indicated in Figure 2-3

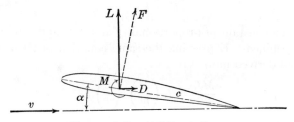

FIGURE 2-3.   Airfoil notation.

and is the product of the resultant force and a length or lever arm. To make the moment equation dimensionally correct a length is arbitrarily selected as the mean chord of a wing or the chord, c, of an airfoil. Therefore

$$M = C_m q S c \qquad (2\text{-}8)$$

where C_m is defined as a moment coefficient. The nomenclature in Figure 2-3 is positive as indicated.

For an ideal fluid, the preceding analysis shows that the nondimensional coefficients, C_L, C_D, and C_m, are functions of the shape and attitude of the body under consideration. For a real fluid other variables affect the nature of these forces and moments, and if equations (2-6), (2-7), and (2-8) are still to apply, then the coefficients must be functions of additional variables not previously considered. To illustrate, the lift force in a real fluid is a function of the attitude, size, and shape of the body, the relative velocity, and the fluid properties.

Lift $= f$ (Shape, Size, Attitude, Velocity, Fluid properties)   (2-9)

However, density is not the only fluid property that affects the nature of the force. Viscous, elastic, and turbulent properties are also important. Also, in addition to the shape and attitude of the body, the surface texture or roughness has some effect on the force. For a given shape of airfoil, then, it can be demonstrated by dimensional analysis that

$$\text{Lift} = \rho v^2 l^2 f\left(\alpha, \frac{\rho v l}{\mu}, \frac{v}{a}, \text{Surface roughness, Air turbulence}\right) \quad (2\text{–}10)$$

where

$$\frac{\rho v l}{\mu} = \text{Reynolds number} = R$$

and

$$\frac{v}{a} = \text{Mach number} = M$$

Reynolds number and Mach number are two very important dimensionless quantities associated with flow in a viscous and compressible fluid.

The effect of surface roughness and air turbulence is usually small and is sometimes considered as a modification of Reynolds number. To demonstrate the derivation of Reynolds number by dimensional analysis, assume that lift is some function of density, velocity, length, and viscosity.

$$\text{Lift} = K\rho^a v^b l^c \mu^d$$

Writing the force equation in terms of the fundamental units,

$$\frac{ML}{T^2} = K\left(\frac{M}{L^3}\right)^a \left(\frac{L}{T}\right)^b L^c \left(\frac{M}{LT}\right)^d$$

$$M = M^a M^d \qquad \text{or} \qquad 1 = a + d$$

$$L = L^{-3a} L^b L^c L^{-d} \qquad \text{or} \qquad 1 = -3a + b + c - d$$

$$T^{-2} = T^{-b} T^{-d} \qquad \text{or} \qquad 2 = b + d$$

Since the above three equations contain four unknowns, it is necessary to express them in terms of the other three. Since viscosity for a fluid such as air is the least important of the four variables, a, b, and c are expressed in terms of the exponent of viscosity, d.

$$a = 1 - d$$

$$b = 2 - d$$

$$c = 2 - d$$

The lift equation now becomes

$$\text{Lift} = K\rho^{1-d} v^{2-d} l^{2-d} \mu^d = K\rho v^2 l^2 \left(\frac{\mu}{\rho v l}\right)^d$$

or

$$\text{Lift} = \rho v^2 l^2 f(K,R)$$

From equation (2–6)

$$C_L = \frac{L}{qS}$$

where

$$C_L = f\,(\alpha,\,\text{Shape},\,R)$$

In this case the lift coefficient will be a function of Reynolds number which is a ratio of inertia to viscous forces. To illustrate, equation (2–2) divided by equation (2–3) yields

$$\frac{F_I}{F_\mu} = \frac{\rho v^2 L^2}{\mu v L} = \frac{\rho v l}{\mu} = R$$

In a similar manner, the ratio of inertia to elastic forces, equations (2–2) and (2–4), shows

$$\frac{F_I}{F_E} = \frac{\rho v^2 L^2}{\rho a^2 L^2} = \frac{v^2}{a^2} = M^2$$

If the speed of sound, a, is included as one of the variables, dimensional analysis will indicate that

$$C_L,\ C_D,\ \text{or}\ C_m = f\,(\alpha,\,\text{Shape},\,R,\,M) \qquad (2\text{–}11)$$

Now, in order to systematize force and moment aerodynamic data obtained in many wind tunnels on different sizes of models of the same prototype and at several air speeds and attitudes under different conditions of air density and viscosity, it is customary to plot α, C_D, and C_m versus C_L as illustrated in Figure 2–4.

It should be noted in Figure 2–4 that Reynolds number and Mach number are specified for the test data because the non-dimensional coefficients, C_D, C_L, and C_m, are still functions of these two variables, as indicated by equation (2–11). In order that the test data be reliable, it is important that the test model have not only geometrical but also kinematic similarity to the prototype. The independent force ratios of Reynolds number and Mach number must be the same in the prototype and model. Because the cost of testing a full-scale model in a wind tunnel is prohibitive, test data from the majority of wind tunnels must be extrapolated to full-scale values of Reynolds number to be reliable. Since the speed of wind tunnels can usually duplicate the estimated speed of the prototype, test Reynolds numbers are sometimes increased in magnitude by pressurizing tunnels to several times atmospheric conditions. For high-speed airplanes designed to fly near and above the speed of sound, specially designed high-speed tunnels duplicate actual Mach numbers expected in flight by the prototype airplane. A few of these tunnels are also pressurized, permitting the attainment of full-scale Reynolds numbers.

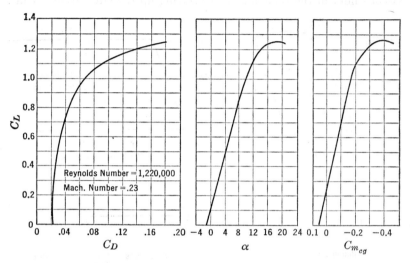

FIGURE 2–4. Aerodynamic data for a typical airplane. From GALCIT Report 298, "Wind Tunnel Tests on a .068 Scale Model of the Boeing Model XB–29 Bomber."

Potential Flow

From classical hydrodynamic theory for flow in an ideal fluid, or potential flow, it can be shown that with no frictional drag or formation of a turbulent wake, regardless of the shape of the body, the pressure drag components about a body such as in Figure 2–1 add up to a net sum of zero drag. It follows then that the assumption of an ideal fluid is not rigorous for explaining the existence of finite pressure or frictional drag forces on aerodynamic bodies. Fortunately, however, the existence of frictional forces within the boundary layer of a real fluid has little or no effect on the net lift of most streamline bodies. Where separation is not extreme, the assumption of an ideal fluid in developing the theory of lift forces on airplane wings is entirely rigorous. This assumption greatly simplifies the mathematical treatment of the theory of lift for both two- and three-dimensional flow. A third type of drag, resulting from the formation of this lift, is called induced drag and may easily be derived from the potential flow theory.

From the potential flow theory it can be shown that with no circulation, Γ, about an aerodynamic body the lift is equal to zero.   With a finite value of circulation a lift force results which in turn produces an induced drag force.   From the classical theory it was impossible to explain the formation of this circulation without the assumption of a viscous fluid in the boundary layer setting up the circulation.   Until

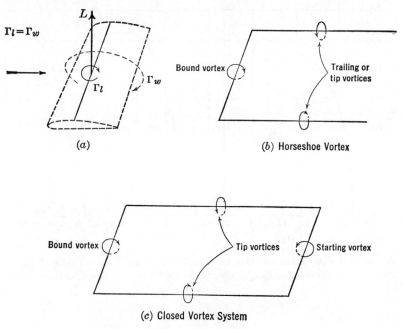

(a)

(b) Horseshoe Vortex

(c) Closed Vortex System

FIGURE 2-5.   Prandtl vortex systems.

the concept of a viscous fluid was developed, it was reasoned that no lift existed on aerodynamic bodies because no flow pattern could be set up in an ideal fluid that would produce this lift.   From studies by Prandtl in the early part of the twentieth century, it was shown that the lift per unit span is directly proportional to circulation and that the wing can be replaced by an hypothetical bound vortex system about the wing and trailing vortices off each wing tip.   The downwash produced by a system of so-called bound and trailing vortices permits a flexible pattern with which it is mathematically possible to calculate any type of lift or downwash distribution desired.   (See Figure 2–5.)

Induced Drag

Now, in order to investigate the effects of the downwash distribution and the lift and drag characteristics, consider first a two-dimensional

section of a wing at some position along the span (Figure 2–6). In
front of the wing (Figure 2–6a) the bound vortex creates a strong
upwash which increases in magnitude as the flow approaches the wing.
At the wing the induced downwash velocity, w, is produced by the
trailing vortices extending only in the downstream direction. Far

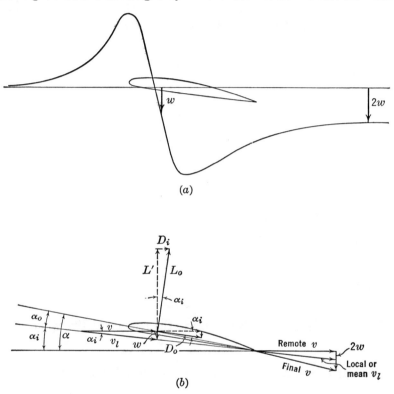

FIGURE 2–6. Two- and three-dimensional airfoil geometry and forces.

behind the wing the theoretical induced velocity is the result of vortices
extending upstream and downstream to infinity and is twice the
magnitude of the local downwash, w. Actually, the trailing vortices
are dissipated as a result of the effects of viscosity, and the downwash
will eventually disappear completely. From a position on the wing
(Figure 2–6b) the direction of the local velocity of the air is then the
mean between the remote free-stream direction and the final direction
of the air. Also, as far as the wing is concerned, the local lift, L_o, is
measured perpendicular to the local velocity, v_l; and the local drag,
D_o, is measured parallel to v_l. The magnitudes of both quantities are

dependent on the effective angle of attack, α_o, measured between the chord and the direction of the local velocity. The induced angle of attack, α_i, is measured between the remote and local relative wind velocities. In stability and control notation, where downwash at the tail is important, this angle is defined as ϵ. The component of the local lift parallel to the remote velocity is defined as the induced drag, D_i. From the geometry of Figure 2–6 it is seen that

$$\tan \alpha_i = \frac{w}{v} = \frac{D_i}{L'} = \frac{C_{D_i}}{C_L'}$$

and

$$\cos \alpha_i = \frac{v}{v_l} = \frac{L'}{L_o} = \frac{C_L'}{C_{L_o}}$$

Since from theory and experiment

$$w \ll v$$

then $\tan \alpha_i$ or $\sin \alpha_i \cong \alpha_i$ in radians, and $\cos \alpha_i \cong 1$. Summing force components perpendicular to the remote relative velocity,

$$L = L_o \cos \alpha_i - D_o \sin \alpha_i$$

Since $D_o \ll L_o$ and $\sin \alpha_i \ll \cos \alpha_i$

$$L \cong L_o \cong L'$$

or, in coefficient form,

$$C_L \cong C_{L_o}$$

Summing force components parallel to the remote relative wind,

$$D = D_o \cos \alpha_i + L_o \sin \alpha_i$$

and if $\cos \alpha_i \cong 1$, $L \cong L_o$, and $\sin \alpha_i \cong \alpha_i$ in radians,

$$D = D_o + L\alpha_i = D_o + D_i \tag{2–12}$$

and

$$C_D = C_{D_o} + C_L\alpha_i = C_{D_o} + C_{D_i} \tag{2–13}$$

Also

$$\alpha = \alpha_o + \alpha_i \tag{2–14}$$

To summarize the results of the above equations, the effect of finite span is to introduce vortices which induce a downwash velocity along the span. At any position along the wing span, the vector addition of the downwash and remote velocity results in a local mean velocity and reduced angle of attack. To maintain the same effective angle of attack or lifting force, with a finite span, the total angle of attack, α, must be increased by an amount equal to the induced angle of attack.

The resultant lifting force on the wing is rotated backwards, creating a component of drag, D_i, parallel to the remote velocity. The planform characteristic of span or aspect ratio is instrumental then in introducing an additional drag increment on the wing.

The determination of the magnitude of the induced drag for the entire wing depends on the distribution of downwash across the span. If the lift distribution over a wing is a semiellipse, with major axis equal to the span and one-half the minor axis equal to the maximum local lift and located at the midspan, Prandtl discovered that the downwash is constant along the span, and for a given condition of lift, span, and velocity, the induced drag is a minimum possible value. This elliptic lift distribution occurs on an untwisted wing of elliptical planform. For the constant downwash condition, therefore, equations (2–13) and (2–14) apply not only to one wing section but to the entire wing. To determine the magnitude of α_i and C_{D_i}, consider a finite wing of constant lift, resulting from a given mass of air per second, m', deflected downward with a velocity equal to $2w$.

$$L = m'2w$$

Since

$$\frac{w}{v} \cong \alpha_i \text{ or } \epsilon$$

$$\alpha_i = \frac{L}{2vm'} = \frac{L}{2v\rho A'v}$$

Prandtl also discovered that the hypothetical cross-sectional area, A', of the mass flow of air that is deflected downward is equivalent to a circular area of diameter equal to the wing span.

$$A' = \frac{\pi b^2}{4}$$

Therefore

$$\alpha_i = \frac{C_L\frac{1}{2}\rho v^2 S}{2v^2\rho \dfrac{\pi b^2}{4}} = \frac{C_L S}{\pi b^2}$$

By defining aspect ratio, A, as

$$A = \frac{b^2}{S}$$

then
$$\alpha_i \text{ in radians} = \frac{C_L}{\pi A}$$

Also, since

$$C_{Di} = C_L \alpha_i$$

$$C_{Di} = \frac{C_L{}^2}{\pi A}$$

In terms of induced drag

$$D_i = C_{Di} qS$$

and since

$$C_L = \frac{L}{qS}$$

$$D_i = \frac{L^2 qS}{q^2 S^2 \pi A} = \frac{L^2}{\pi q b^2} \tag{2–15}$$

In the preceding analysis it must be remembered that the product of m' and $2w$ is a constant for all aspect ratios in order that the assumption of a constant lift is not invalidated. From the above relationships it can be seen that α_i and C_{Di} approach zero as the aspect ratio approaches infinity. However, m' also approaches infinity as the downwash velocity approaches zero, and the product of the two may still be considered constant. For the condition of a wing with elliptical lift distribution and constant downwash along the span, equations (2–14) and (2–13) may now be written

$$\alpha \text{ in radians} = \alpha_o + \frac{C_L}{\pi A} \tag{2–16}$$

and

$$C_D = C_{Do} + \frac{C_L{}^2}{\pi A} \tag{2–17}$$

Both α_o and C_{Do} refer to wing characteristics for infinite aspect ratio and depend only on the profile shape of the airfoil section. C_{Do} is defined as the wing profile drag coefficient and, as previously discussed, is composed of the pressure drag and skin-friction drag on the wing.

In general, airfoil characteristics are presented for an infinite aspect ratio condition as illustrated in Figure 2–7. Angles of attack and drag coefficients for any value of lift coefficient can then be easily calculated by the relations set up in equations (2–16) and (2–17).

In all the preceding theoretical developments it is noticed that lift coefficient rather than angle of attack is taken as the basic parameter. Since one is directly proportional to the other, this choice is immaterial and is only done for convenience. For example, in order to determine the effect of a change in aspect ratio on wing characteristics, it is convenient to fix a value of lift coefficient and let the wing rotate to a new

angle of attack.   In Figure 2-8 the increment differences in angle of
attack and drag coefficient for two different aspect ratios are illustrated.

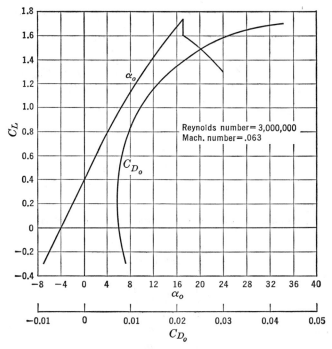

FIGURE 2-7. Wing characteristics for infinite aspect ratio. From NACA TR 669,
"Airfoil Section Data Obtained in the NACA Variable-density Tunnel as Affected
by Support Interferences and Other Corrections," by Eastman N. Jacobs and Ira
H. Abbott.

The mathematical presentation follows as derived from equations
(2-16) and (2-17).

$$\alpha_1 - \alpha_2 \text{ in degrees} = \alpha_{i1} - \alpha_{i2} = \frac{57.3 C_L}{\pi}\left(\frac{1}{A_1} - \frac{1}{A_2}\right) \quad (2\text{-}18)$$

$$C_{D1} - C_{D2} = C_{Di1} - C_{Di2} = \frac{C_L{}^2}{\pi}\left(\frac{1}{A_1} - \frac{1}{A_2}\right) \quad (2\text{-}19)$$

To find the slope of the lift curve for any aspect ratio in terms of that
for infinite aspect ratio the following relationships are derived:

$$\alpha \text{ in radians} = \alpha_o + \frac{C_L}{\pi A}$$

$$\frac{d\alpha}{dC_L} = \frac{d\alpha_o}{dC_L} + \frac{1}{\pi A}$$

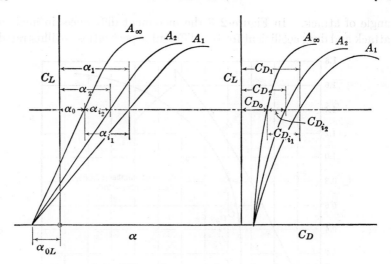

FIGURE 2–8.   Effect of aspect ratio on wing characteristics.

Let
$$a = \frac{dC_L}{d\alpha} \quad \text{and} \quad a_o = \frac{dC_L}{d\alpha_o}$$

Then
$$\frac{1}{a} = \frac{1}{a_o} + \frac{1}{\pi A}$$

from which
$$a = \frac{a_o}{1 + (a_o/\pi A)} \tag{2-20}$$

$$a_o = \frac{a}{1 - (a/\pi A)} \tag{2-21}$$

In the preceding discussions of induced angle and induced drag it should again be noted that the results apply only to wings with elliptical lift distributions. For the practical case of rectangular planforms these results are very slightly modified, as will be shown later. In Figure 2–9a is presented a solution of equation (2–20) in which a_o is given as slope per degree. In Figure 2–9b the end-plate effect of the fuselage and vertical tail on the horizontal tail is presented in the form of a correction factor, r. The slope of the lift curve of the horizontal tail is defined as a_t.

Flow of a Viscous Fluid

In order to explain some of the resistance properties or drag phenomena for aerodynamic bodies it is necessary to introduce the property

of fluid viscosity into the discussion.   A real fluid as it exists in nature
is viscous and compressible.   Although air may be easily compressed
when confined by external forces, the nature of its unrestricted flow
about aerodynamic bodies is such that except for very high speeds
approaching the speed of sound it can be treated as an incompressible

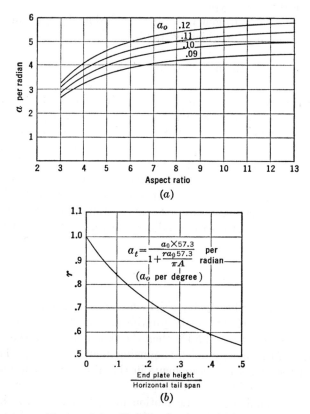

FIGURE 2–9.   Variation in slope of lift curve.

fluid.   Since the analysis of both viscous and compressible fluid motion
taken together is very complex and beyond the scope of this book, a
separate discussion of each of these flows will follow.   Later the effects
of boundary-layer flow on the formation of shock waves will be briefly
presented.

Resistance to motion is induced by shear or friction forces between
fluid particles.   The effects of viscosity on the character of fluid flow
were first reported by the English physicist, Osborne Reynolds, during
the latter part of the nineteenth century.   His experiments indicated

that at a certain critical speed the resistance of a tube or pipe to the passage of a given fluid suddenly changes from a function of the first power of the relative velocity to a function of the second power of the velocity. Furthermore, this change in the law of resistance occurs simultaneously with a change in the character of the flow from a laminar to a turbulent type.

Reynolds was able to generalize his conclusions by introducing the dimensionless coefficient which bears his name,

$$\text{Reynolds number} = R = \frac{\rho v l}{\mu}$$

above a certain value of which the flow in the tube remains turbulent. As has been previously shown this dimensionless number is the ratio of inertia to viscous forces. The nature of the flow about aerodynamic bodies is governed to some extent by the value of Reynolds number. At a small value of Reynolds number viscous forces predominate, and at a large value inertia forces predominate. A discussion of the character of the shear stresses in the two regions of flow will further indicate the reason for the change in the resistance properties of the fluid.

The shear stress between two adjacent layers of fluid in a laminar flow can be expressed as

$$\tau \sim \frac{du}{dy} \tag{2-22}$$

For turbulent flow the fluid particles are not restrained to parallel layers but move in chaotic or fluctuating fashion from layer to layer. The average speed of the fluid is then the time average of the velocity fluctuations of the many particles bouncing in all directions. The time average of the pulsating components in the direction of the motion is the speed of the fluid. The time average of the components of the pulsations perpendicular to the motion must be equal to zero. The shear stress between two layers for this case is not dependent on the viscosity of the fluid but instead on the momentum exchange between layers. Reynolds showed that this shear stress is proportional to the second power of the rate of change of velocity with respect to the distance between moving layers.

$$\tau \sim \left(\frac{du}{dy}\right)^2 \tag{2-23}$$

In 1904 Prandtl introduced the now famous boundary-layer theory which describes the influence of viscosity on the character of the fluid flow adjacent to a solid surface. The starting point in the physical conception of the boundary layer is that a condition of no slip exists between an infinitesimal-thickness layer of the viscous fluid and the boundary surface.   In other words, no relative motion exists because the molecules actually touching the surface adhere to the surface because of the viscous properties of the fluid.   Furthermore, a diagram showing the velocity variation or velocity profile perpendicular

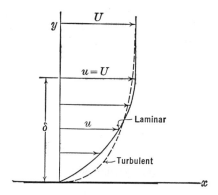

FIGURE 2–10.   Boundary layer notation.

to the fluid motion indicates a speed of zero at the boundary surface, increasing in magnitude as the distance from the surface increases. Figure 2–10 illustrates standard boundary-layer notation and profiles. The thickness of the layer, δ, is the perpendicular distance from the surface at which point the velocity has just reached the free-stream velocity, U, of the undisturbed fluid.   The thickness of this layer is always very small, before separation occurs, and is dependent on the nature of the fluid and the boundary surface over which it forms.

From Figure 2–10 it can be seen that the velocity gradient, du/dy, at the fixed surface or wall is greater for the turbulent boundary layer than for a laminar one.   Also the skin friction is a function of $(du/dy)^2$ instead of du/dy as in the case of a laminar boundary layer.   The skin friction, therefore, is much larger for a turbulent boundary layer. The problem of reducing skin-friction drag is then one of maintaining a laminar boundary layer.

It can be reasoned that for both laminar and turbulent flow and for reasonably smooth and streamline bodies the viscous properties of a real fluid as they affect the nature of the motion are confined to a very thin surface near the boundary.   Outside this boundary layer the fluid may be considered as ideal or perfect.   It is this concept that simplifies the prediction of the velocities and paths of air particles and the corresponding pressures about areodynamic bodies. The nature of the flow about a cylinder shown in Figure 2–11, based on this assumption, is exact only until the fluid breaks away or separates from the surface on the rear portion of the body. (See Figure 2–11a.)

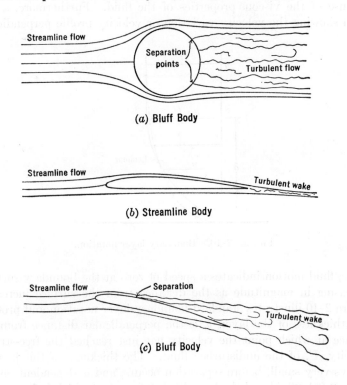

FIGURE 2–11. Flow of a real fluid around bluff and streamline bodies.

For streamline bodies such as an airfoil with the flow nearly parallel to the longitudinal axis of the body (Figure 2–11b) the prediction of force characteristics, based on calculated pressure distributions for an ideal fluid, agrees very closely with experimental results for test speeds less than sonic values. However, this same airfoil, when inclined to the relative wind (Figure 2–11c) so that separation occurs, may be considered a bluff body, and the flow characteristics for an ideal and real fluid are the same only up to the point of separation. Thus, it may be seen that the boundary layer in adhering to or breaking away from the surface of a body has an important effect on the flow characteristics and resulting forces on the body.

In order to investigate further the properties of separation consider the nature of the boundary layer along the surface of the cylinder in Figure 2–11a. Figure 2–12 shows selected boundary-layer profiles along the surface of the cylinder with the y scale greatly expanded and perpendicular to the surface at each point. If a particle of air is traced around the contour of the cylinder, the velocity of the particle reaches a maximum at the top where the streamline spacing is a minimum. The pressures and velocities along the surface outside the boundary layer change according to Bernoulli's law. One of the conclusions of the boundary-layer theory is that at any given point on the surface of the cylinder the pressure remains a constant throughout the depth of the boundary layer even though the velocity changes.

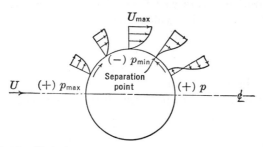

FIGURE 2–12.   Variation in boundary layer around a circular cylinder.

Bernoulli's equation no longer applies for this case because the flow is not steady and energy is being subtracted from the flow because of the internal work of friction. Now a particle of air, in traveling from the front stagnation point to the top of the cylinder, is assisted by a favorable pressure gradient. If the flow were ideal, the maximum kinetic energy reserve at the top would be just sufficient to carry the particle against an unfavorable pressure gradient to the rear stagnation point where its velocity would again be zero. However, for the flow of a real fluid the particle of air is slowed down because of viscous effects and comes to rest before the rear stagnation point is reached. At this point the particles just behind pile up and cause a breakaway or separation of the flow from the surface. Figure 2–12 shows the nature of the boundary-layer profile at the separation point. Beyond this point a reversal of flow exists at the surface. The turbulent wake to the rear is very unstable and breaks up into oscillating waves as eddies form because of the reverse flow behind the separation point. The fluctuating local velocities in the wake are damped out as a result of viscosity and represent a loss in energy which manifests itself into an increase in drag on the body over that which exists in a perfect or ideal fluid.

The degree of separation on an aerodynamic body is largely dependent on the magnitude of the unfavorable pressure gradient to the rear of the point of minimum pressure or maximum surface velocity. If the pressure gradient, dp/dx, along the surface from this point is equal to or less than zero, then no separation exists. If the pressure gradient is gradual, the separation occurs so near the rear of the body that only a very small turbulent wake is produced and the boundary layer is extremely thin everywhere. The body may then be considered as streamline, and the perfect and real-fluid motion and corresponding pressure distributions are nearly identical. The drag of such a body is small and arises mostly from skin friction. If the pressure gradient is high, separation occurs well forward of the rear stagnation point and a turbulent wake exists which alters the potential-flow picture and pressure distribution, resulting in an increase in drag. If this flow phenomenon exists, the body is considered as bluff, and drag forces as predicted from a thin boundary layer in an ideal fluid no longer hold. Lift forces for both streamline and bluff bodies may be predicted with reasonable accuracy from pressure distributions obtained from a perfect-fluid analysis.

Parasite Drag

It has now been shown that the resistance of aerodynamic bodies immersed in a real fluid is the result of three types of drag. The resistance resulting from the pressure variation over the surface of the body is defined as pressure drag, and that due to the shear stresses in the boundary layer is defined as frictional drag. A third type of drag associated with the lift on a wing is defined as induced drag. It is customary to call the sum of the first two types parasite drag, D_p, such that

$$D_p = D - D_i \qquad (2\text{–}24)$$

For an airplane, D_p may also be written in the familiar coefficient form

$$D_p = C_{D_p} q S \qquad (2\text{–}25)$$

where C_{D_p} is the parasite drag coefficient and is based on wing area.

In expressing C_{D_p} for component parts of the airplane or for aerodynamic bodies as illustrated in Figure 2–11 it is convenient to base the coefficient on total projected frontal area or surface area, whichever is more representative of the surface which contributes the greatest to the drag. Such an area is sometimes referred to as the proper area for the particular shape of body, and the corresponding drag coefficient is defined as the proper drag coefficient, C_{D_π}. For example, the drag on a flat plate parallel to the airstream is largely frictional drag, and the proper area contributing most to this drag is the total wetted or surface area.   In coefficient form

$$C_F = C_{D_\pi} = \frac{\text{Skin friction}}{qA_W} \qquad (2\text{-}26)$$

where A_W is the wetted surface exposed to the air.   For the same plate perpendicular to the airstream, pressure drag is largely predominant, and the proper area contributing most to this drag is the frontal area.

$$C_{D_\pi} = 1.28 = \frac{\text{Force}}{qA_F} \qquad (2\text{-}27)$$

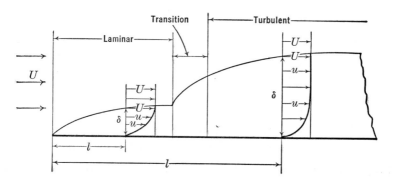

FIGURE 2-13.   Boundary layers on a flat plate.

For a sphere the proper area is usually chosen as the projected frontal area.   For a streamline body of revolution either the frontal area or the surface area may be used, depending on the fineness ratio.   In any case a proper area is conveniently chosen so that the proper drag coefficient is both a qualitative and quantitative measure of the relative magnitude of drag change of a given type of body because of change in size.

It is also recalled from previous discussions that the drag properties of an aerodynamic body are closely associated with the flow of the air in the boundary layer as affecting frictional drag and separation. Consider first the frictional drag associated with the boundary layer on one side of a smooth flat plate aligned with the airstream (Figure 2-13).   Referring to the boundary-layer notation and the shear stress relationships between layers of fluid in a laminar and turbulent flow it is apparent that the frictional drag along the plate is a function of the type of flow and shape of the velocity profile in the flow. The boundary layer starts from zero thickness at the edge of the plate and increases in thickness in the downstream direction as the air is decelerated by the resistance of the plate. The resistance of the plate at any position along the plate is obtained by summing up the resistance between successive layers of the air throughout the depth of the boundary layer.

The determination of the thickness of the laminar boundary layer and the corresponding skin-friction coefficient within this region was developed by Prandtl and Blasius through an application of the shear stress relationship at the surface of the plate

$$\tau_0 = \mu \left(\frac{du}{dy}\right)_0 \qquad\qquad (2\text{–}28)$$

The thickness of this layer is usually expressed in terms of Reynolds number along the plate

$$\frac{\delta}{l} = \frac{5.20}{\sqrt{R_l}} \qquad\qquad (2\text{–}29)$$

where $R_l = \dfrac{\rho v l}{\mu}$, and

l = length of plate from leading to trailing edge or to the point where the thickness is measured.

Equation (2–29) leads to the conclusions that δ/l approaches zero as R_l approaches infinity, so that the flow over a plate with a very thin laminar boundary layer approaches the flow characteristics of a perfect fluid. If, in equation (2–28), μ approaches zero and $(du/dy)_0$ approaches infinity as in a perfect fluid, then

$$\tau_0 = 0 \times \infty$$

which is an indeterminate form and was shown by Prandtl to be proportional to $1/\sqrt{R_l}$. In other words, even though R_l becomes very large, a finite skin friction exists. For flow with a very large Reynolds number, corresponding to real fluids with very small viscosity, a finite drag force exists. Equation (2–26) as evaluated by Blasius for a laminar condition can then be written as

$$C_F = \frac{\text{Skin friction}}{q A_W} = \frac{1.328}{\sqrt{R_l}} \qquad\qquad (2\text{–}30)$$

Unfortunately these relatively simple expressions for boundary-layer phenomena are not used extensively by the aeronautical engineer because the flow over most airplane surfaces in flight is largely turbulent. At a critical R_l approximately between 100,000 and 3,000,000 a transition to turbulent flow in the boundary layer occurs. The shear stress in this type of flow is dependent on the momentum exchange between layers and is much greater than that for laminar flow. Since the average velocity of the air near the surface is increased by the exchange of higher-velocity particles from the free stream for the ones already retarded at the surface, the velocity profile differs from that of the laminar profile in that it is fuller near the surface and flatter away from it. In addition, the turbulent boundary layer is thicker and increases more rapidly with the length of the surface. (See Figure 2–13.) The mathematical analysis of the characteristics of the turbulent boundary layer is too complex for development in most textbooks. Prandtl has shown that the approximate thickness of such a turbulent layer on a smooth plate is given by

$$\frac{\delta}{l} = \frac{.37}{R_l{}^{.2}} \qquad (2\text{–}31)$$

where l is the length of plate from the beginning of the transition. The corresponding skin-friction coefficient may be expressed in the following form

$$C_F = \frac{.455}{(\log_{10} R_l)^{2.58}} \qquad (2\text{–}32)$$

A plot of C_F versus R_l for laminar, turbulent, and transition flows is given in Figure 2–14.

The critical R_l at which transition occurs is somewhat variable, depending on the turbulence level in the free stream and the roughness of the plate leading edge. An equation suggested by Prandtl which sometimes satisfies experimental results in the transition region and which determines the critical R_l follows:

$$C_F = \frac{.455}{(\log_{10} R_l)^{2.58}} - \frac{1700}{R_l} \qquad (2\text{–}33)$$

This equation establishes the critical R_l at approximately 500,000 (see Figure 2–14), although recent experimental data indicate transitions at considerably higher values. Since the critical $R_{l_{cr}}$

$$R_{l_{cr}} = \frac{\rho v l}{\mu}$$

and is a constant for a given plate and turbulence level, the transition moves forward as the velocity is increased. For streamline bodies such as airfoils at low angles of attack the profile drag varies considerably with $R_{l_{cr}}$. To maintain the lowest profile drag, laminar flow must be maintained as long as possible. Since any roughness of the surface itself also decreases the $R_{l_{cr}}$, attempts have been made to maintain smooth wings by careful manufacture and surface finish. However, at normal-flight Reynolds numbers, even the smoothest wings can maintain laminar flow over only a portion of the surface, as the deflection of the wing in the loaded condition produces local waves in

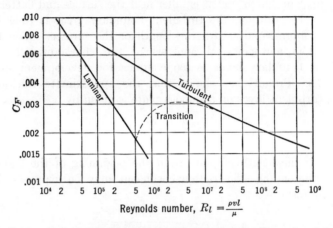

FIGURE 2–14.   Drag coefficient for smooth flat plates.   Reproduced by permission from *Elementary Fluid Mechanics*, First Edition, by J. K. Vennard, published by John Wiley & Sons, Inc., 1940.

the surface. The most effective method of obtaining laminar flow is by partial removal of the boundary layer by suction through slots in the wing surface. However, this process requires power and additional weight for ducting, and the net gain is dependent on the refinement with which these additional design requirements are accomplished.

The preceding boundary-layer picture applies also over the curved surface of a bluff body. In the following discussion some important characteristics of the magnitude of the pressure drag behind a bluff body are explained by the application of the fundamentals of boundary-layer theory and separation phenomena. As an example consider the pressure drag and separation on a circular cylinder or sphere with two types of boundary layer (Figure 2–15). For the case of the laminar boundary layer, separation occurs early, creating a large turbulent wake and high pressure drag. For the turbulent boundary-layer condition, however, separation occurs at a more rearward position on the periphery of the sphere, causing a smaller wake and drag. In this case the increase in momentum of the particles of air adjacent to the surface in a turbulent boundary layer permits the air to travel farther

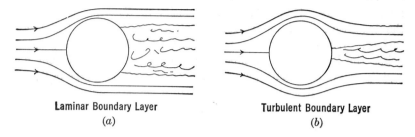

Laminar Boundary Layer
(a)

Turbulent Boundary Layer
(b)

FIGURE 2–15.   Laminar and turbulent flow about spheres.

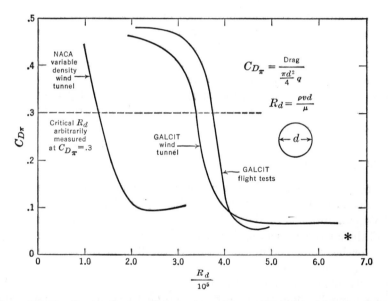

FIGURE 2–16. Proper drag coefficients for spheres. Reproduced by permission from *Aerodynamics of the Airplane*, by C. B. Millikan, published by John Wiley & Sons, Inc., 1941.

around the surface before coming to rest and promoting separation. Now consider the effect of $R_{l_{cr}}$ on this phenomenon.   (See Figure 2–16.)

Here again the same fundamental principles that control separation along a flat plate determine when these two different types of flow exist. For a given airstream and sphere the transition from laminar to turbulent flow in the boundary layer occurs at a constant critical Reynolds number, $R_{l_{cr}} = \rho v l/\mu$, where l is measured along the surface of the sphere downstream from the front stagnation point. As long as the transition occurs downstream of the laminar separation point, where the flow is already turbulent, only a small change in C_{D_π} with Reynolds number occurs. However, as the velocity or $R_d = \rho v d/\mu$ increases, the transition point moves forward or upstream along the sphere surface. When the transition moves forward of the laminar separation point, the flow in the boundary layer then becomes turbulent and rapidly moves the separation point associated with this type of flow to a much more rearward position (Figure 2–15b), creating a large decrease in drag. As the Reynolds number increases still further, the C_{D_π} remains approximately constant. Since $R_{l_{cr}}$ depends also on airstream turbulence and surface roughness, changes in these variables affect the point at which the large change in drag occurs. Spheres with a given surface texture are often used in wind tunnels to measure the degree of airstream turbulence by noting the Reynolds number at which this phenomenon occurs. For streamline bodies this phenomenon is of lesser practical importance to the aeronautical engineer because of the very low Reynolds number associated with the change. For bluff bodies, however, the range of $R_{l_{cr}}$ is within the normal test speeds, and therefore great care should be taken in extrapolating data through this region.

Flow of a Compressible Fluid

In the preceding discussions of aerodynamic phenomena, it is assumed that the air through which the airplane flies is incompressible. This assumption is valid only if the relative speed between the airplane and the air is considerably below the speed of sound. As the airplane approaches sonic speed, the nature of the flow of the air around the body approximates that of a compressible fluid. Military aircraft are now flying at sonic and supersonic speeds. A discussion of the flow of a compressible fluid is a necessary forerunner to a clear understanding of the nature of the moments and forces of the airplane at these high speeds. The following remarks are intended for the reader as only a review of some of the fundamental high-speed concepts contained in much greater detail in the literature.

Density variations may occur in all fluids because of the compression of the fluid and consequent change in space between molecules. This compressible property of air has an appreciable effect on its flow properties at high speeds. The reciprocal of the modulus of elasticity of a fluid is a direct measure of its compressibility. From Hooke's law,

$$\frac{\text{Stress}}{\text{Strain}} = \frac{F/A}{-\Delta\bar{v}/\bar{v}} = \frac{\Delta p}{-\Delta\bar{v}/\bar{v}} = \frac{dp}{-d\bar{v}/\bar{v}} = E, \text{ the bulk modulus} \qquad (2\text{--}34)$$

$$\text{Compressibility,} \quad = \frac{1}{E} = \frac{-d\bar{v}/\bar{v}}{dp} = +\frac{d\rho/\rho}{dp} \qquad (2\text{--}35)$$

which represents the specific change in volume or density per unit change in pressure. The negative sign represents a decrease in volume with increase in pressure. The positive sign represents an increase in density with increase in pressure.

Pressure disturbances move in waves from point to point. The speed of propagation of such waves is dependent on the elastic properties of the fluid. Fluids that are easily compressible, with corresponding low value of E, transmit pressures with a lesser speed than those which are difficult to compress. The speed of propagation of small-pressure waves is expressed* by

$$a = \sqrt{\frac{dp}{d\rho}} \qquad (2\text{--}36)$$

and is equal to the local speed of sound in a fluid.

For gases sound waves move by a series of adiabatic compressions and rarefactions. For this change of state of the gas,

$$p = \text{Constant } \rho^\gamma$$

$$\text{and} \qquad a = \sqrt{\frac{\gamma p}{\rho}} = \sqrt{\gamma g R T} \quad (R \text{ for air} = 53.3 \text{ ft/°F abs}) \qquad (2\text{--}37)$$

$$\text{and for air,} \qquad\qquad a = 49.1\sqrt{T} \qquad (2\text{--}38)$$

and is equal to the speed of sound in feet per second if T is in degrees Fahrenheit absolute. The speed of sound varies linearly from approximately 761 miles per hour at sea level to approximately 662 miles per hour at 35,000 feet standard altitude. Values of the physical properties of the air at different altitudes are tabulated in the standard atmosphere tables in the Appendix.

Within the subsonic flight region inertia and viscous forces largely determine the behavior of flight. As the speed of an aerodynamic body approaches the speed of sound, the density of the air is no longer constant because of the elastic properties of the air and consequent

---

* *Introduction to Aerodynamics of a Compressible Fluid*, by Liepmann and Puckett, page 21, John Wiley & Sons, 1947.

compression of the air as it approaches the body. As the body approaches the speed of sound, the pressure disturbances that are propagated by the body and which travel at a speed equal to the speed of sound are set up closer and closer to the body. At or near a Mach number of unity a compression normal shock wave is set up in front of the body. Behind this wave the flow is very turbulent and represents a loss in energy or an increase in the drag coefficient of the body. In passing through a normal shock wave the flow is reduced in total pressure and velocity and increased in static pressure, density, and temperature. Behind such a wave, which is normal to the direction of relative motion, the flow is reduced from a Mach number greater than one to a value less than one. As the Mach number increases above unity, the shock wave attaches itself to the nose of the body and trails back on either side, forming an oblique wave. Since the energy loss in this type of wave is less than in a normal wave, the drag coefficient decreases. The flow component normal to the oblique wave behaves in the same manner as the flow through a normal shock wave. The component parallel to the wave front remains unchanged. In Figure 2–17b, $M_2 < M_1$ but may be less than, more than, or equal to one. Minor compression and expansion waves may also be propagated on critical points on the body where the local air speed exceeds the speed of sound, because of a change in the direction of flow. In Figure 2–17a the Mach angle, β, is a reference by which the speed of the body may be calculated, as the Mach waves form the

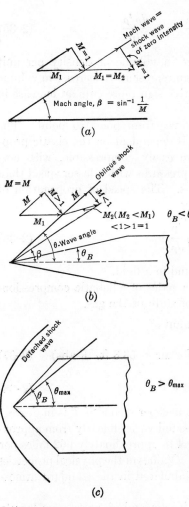

FIGURE 2–17. Supersonic flow notation.

locus of a tangent to the spherical waves propagated by the leading edge.

$$\beta = \sin^{-1}\frac{a}{v} = \sin^{-1}\frac{1}{M} \qquad (2\text{-}39)$$

For this infinitely thin body the component normal to the wave front is always equal to a Mach number of one. A Mach wave is a shock wave of zero intensity, as no change in direction or magnitude of flow occurs through the wave. In Figure 2-17b the wave angle, θ, is always more than the Mach angle in the region of the body. If, as in Figure 2-17c, the deflection angle of the body, θ_B, is larger than the maximum

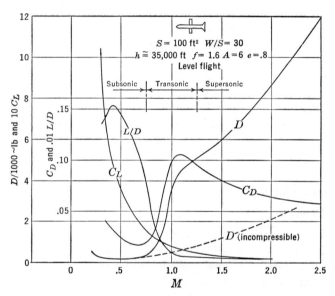

FIGURE 2-18. Variation in lift and drag with Mach number.

angle, θ_M, through which the flow can be turned and remain supersonic, a head wave or detached shock wave is formed ahead of the body, becoming more oblique on either side. As the free-stream Mach number is increased, the critical value of θ_B for which a detached wave is formed is increased. At a Mach number of one even a very thin leading edge will produce a detached head wave.

The drag characteristics of a typical aerodynamic body are presented in Figure 2-18. The difference between the drag for a compressible and an incompressible fluid is presented. In Figure 2-18 is also presented the lift coefficient, drag coefficient, and L/D characteristics of this body as a function of Mach number. From these curves it can be seen that the performance of aircraft is dependent on the intensity of the shock waves. If the shock waves could be avoided, the thrust or power requirements necessary to maintain level flight at a given speed would be greatly reduced. The designer's problem then becomes one of trying to reduce the intensity of shock phenomena as it affects the airplane drag. At supersonic speeds this can be accomplished by designing sharp leading edges which increase the obliquity of the shock waves. The stability of the laminar boundary layer on aerodynamic surfaces has an appreciable effect on the formation of shock waves. If the boundary layer becomes turbulent, the separation behind the shock wave may be less severe than in the case of a laminar boundary layer. In wind-tunnel tests of aerodynamic bodies at a high speed it is therefore important that comparable Reynolds numbers as well as Mach numbers be approximated for the flight test article. In Figure 2–18 is presented the aerodynamic characteristics of a complete airplane designed for subsonic speeds. In the following section the aerodynamic characteristics of the wing and other component parts of the airplane are described in detail. These data are useful to the engineer in his attempt to estimate airplane performance, stability, and control at speeds in the subsonic, transonic, and supersonic regions of flight.

2–2 Aerodynamic Data

Section Characteristics

Airfoil characteristics for infinite aspect ratio as illustrated in Figure 2–7 are defined as section characteristics. In order to distinguish the two-dimensional coefficients of a particular section profile from the average section characteristics of a complete wing, the following nomenclature has been generally adopted for most of the important airfoil characteristics upon which an airplane designer's choice of an airfoil is made (Figure 2–19). The section characteristics listed in Figure 2–19 are dependent only on the profile shapes of the sections and are independent of planform characteristics of the wing. Three-dimensional or average section characteristics for the complete wing are dependent on planform characteristics and will be discussed in detail later in this chapter. Aerodynamic characteristics of the NACA 2412 and 65_1–212 airfoil sections from tests in the two-dimensional tunnel at Langley Field are presented in Figure 2–20. Most of the parameters appearing in Figures 2–19 and 2–20 have been previously discussed. The optimum lift coefficient, $c_{l_{opt}}$, is defined as the lift coefficient for minimum drag coefficient, $c_{d_{min}}$, and is approximately the same as the design lift coefficient, c_{l_i}. Moment coefficients and the aerodynamic center, a.c., require further discussion.

The general form of the moment equation has been presented as

$$M = C_m q S c$$

For an average airfoil section of a wing, c is usually defined as the mean aerodynamic chord, m.a.c., and represents an average chord which, when multiplied by the average section moment coefficient, dynamic

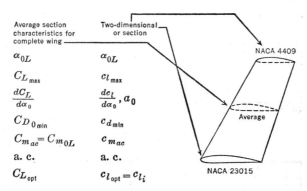

Average section characteristics for complete wing	Two-dimensional or section
α_{0L}	α_{0L}
$C_{L\,max}$	$c_{l\,max}$
$\dfrac{dC_L}{d\alpha_0}$	$\dfrac{dc_l}{d\alpha_0}, a_0$
$C_{D\,0\,min}$	$c_{d\,min}$
$C_{m\,ac} = C_{m\,0L}$	$c_{m\,ac}$
a. c.	a. c.
$C_{L\,opt}$	$c_{l\,opt} = c_{l_i}$

NACA 4409

Average

NACA 23015

FIGURE 2-19.   Airfoil section notation.

pressure, and wing area, gives the moment for the entire wing. The value of moment coefficient depends on the airfoil attitude and the location along the chord about which the corresponding moment is taken. Hydrodynamic theory shows that for a particular position on an airfoil section the corresponding moment coefficient is a constant, independent of lift coefficient. This point is defined as the aerodynamic center, a.c., and the moment coefficient about this point is denoted as $C_{m_{ac}}$. The aerodynamic center usually lies very close to the chord and from 22 to 26 per cent of the chord from the leading edge. The magnitude of the moment or moment coefficient about a point on the chord is important for stress analysis and stability and control calculations. For a given airfoil section it depends on the lift coefficient and center of pressure location. Center of pressure, c.p., is the distance from the leading edge to a point on the chord through which the resultant of all the pressure forces on the airfoil section is assumed to act. Consider first an average airfoil section at a given angle of attack

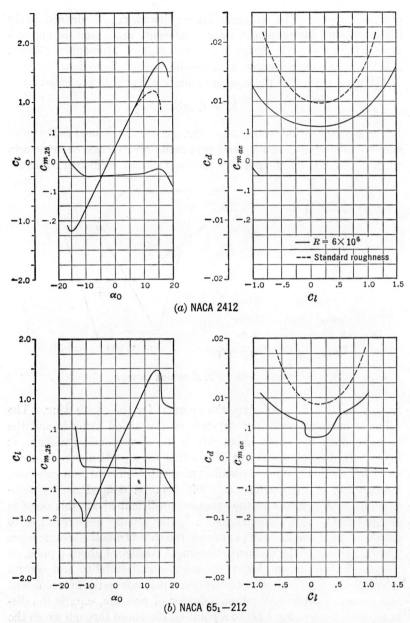

(a) NACA 2412

(b) NACA 65₁—212

Figure 2–20. Typical section characteristics. From NACA WR L–560, "Summary of Airfoil Data," by Ira H. Abbott, Albert E. von Doenhoff, and Louis S. Stivers, Jr.

(Figure 2–21).　Neglecting the small contribution of the drag force
and taking moments about a point n' distance from the leading edge

$$-M_n = N(\text{c.p.} - n') \cong L(\text{c.p.} - n')$$

Dividing by qSc,

$$\frac{-M_n}{qSc} = \frac{L}{qS}\left(\frac{\text{c.p.} - n'}{c}\right)$$

and

$$-C_{m_n} = C_L(C_p - n) \qquad (2\text{–}40)$$

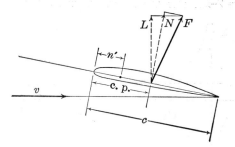

FIGURE 2–21.　Moment and center of pressure.

where C_p and n are per cent distances along the chord, $\text{c.p.}/c = C_p$,
and $n'/c = n$.
Solving for C_p,

$$C_p = n - \frac{C_{m_n}}{C_L} \qquad (2\text{–}41)$$

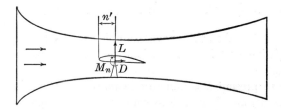

FIGURE 2–22.　Measurement of forces and moments in wind tunnel.

Now, if an airfoil is tested in a wind tunnel, lift and drag forces and
moments are transmitted to the balance through a supporting trunnion
located at some point along the chord of the airfoil (Figure 2–22).
The values of M_n, L, and D are recorded for several angles of attack of
the airfoil.　After converting these values to coefficient form, equation
(2–41) can be used to calculate the magnitude of C_p for several values

of C_L. Figure 2–23 is a typical graphical representation of the relationship between C_p and C_L for a cambered airfoil section. It is important to note that a straight-line relationship holds when these experimental data are plotted as in Figure 2–23b. This means that equation (2–41) can be written as

$$C_p = K - \frac{K'}{C_L} = \text{a.c.} - \frac{C_{mac}}{C_L} \qquad (2\text{–}42)$$

where the two constants K and K' are respectively the intercept and slope in Figure 2–23b and are defined as the a.c. and C_{mac}. Experi-

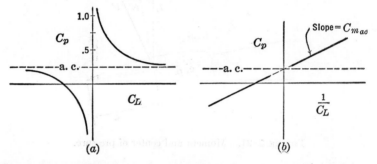

FIGURE 2–23. Relationship between center of pressure and lift coefficient.

mental evidence thus bears out hydrodynamic theory and proves that the C_{mac} is independent of C_L. Furthermore,

$$C_{mac} = C_{m_n} \text{ at zero } C_L = C_{m0L} = \text{Constant} \qquad (2\text{–}43)$$

also $M_{ac} = M_{0L} = $ Constant for a given dynamic pressure

For the airfoil tested in the above example the absolute magnitude of C_{mac} is negative and is characteristic of conventional airfoils with positive camber.

In order to present a complete explanation of the relationships in equation (2–43) and Figure 2–23 consider the lift coefficient distribution about a symmetrical and cambered airfoil section for different values of lift coefficient (Figure 2–24). At zero lift coefficient the C_{mac} for the symmetrical section is obviously zero. For the cambered section (Figure 2–24b) the C_{mac} is not zero and is the result of the couple produced by the basic curved shape of the airfoil. For a finite lift coefficient the center of pressure for the symmetrical section (Figure 2–24c) remains at the aerodynamic center. However, in the case of the cambered section, as indicated in Figure 2–24d, the center of pressure moves off the trailing edge as the lift coefficient approaches zero. It must be noted that the actual forces remain on the airfoil (Figure 2–24d), but this mathematical conception indicates a net lift coefficient that is equal to the difference between the values on the upper and lower surfaces and so located that the moment coefficient it produces about the aerodynamic center is equivalent to C_{mac}.

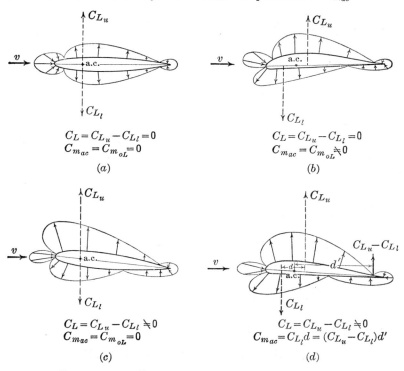

FIGURE 2–24. C_{mac} for symmetrical and cambered airfoils.

The basic conceptions presented in the preceding discussion lead to a convenient system for designating moment and force coefficients on an airfoil section (Figure 2–25). In Figure 2–25a, it can be seen that as the angle of attack changes the magnitudes of the force coefficients and the position of the center of pressure change. By replacing the product of lift coefficient and the distance to the aerodynamic center by the C_{mac} (Figure 2–25b), which is a constant for all angles of attack, aerodynamic calculations are simplified. For example, in calculating the wing moment coefficient contribution about some point on the airplane for several different angles of attack, the magnitude of the force coefficients in Figure 2–25b change, but the lever arms and

C_{mac} remain constant; whereas in Figure 2–25a both the force coefficients and lever arms change with angle of attack.

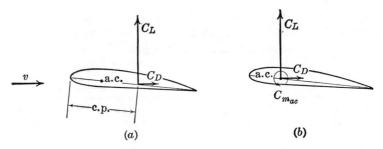

FIGURE 2–25.   Airfoil notation.

By equating equations 2–41 and 2–42 a useful expression for determining the aerodynamic center from typical airfoil data may be derived.

$$\text{a.c.} - \frac{C_{mac}}{C_L} = n - \frac{C_{mn}}{C_L}$$

$$\text{a.c.} = n - \left(\frac{C_{mn} - C_{mac}}{C_L}\right) = n - \left(\frac{C_{mn} - C_{m0L}}{C_L}\right)$$

$$\text{a.c.} = n - \frac{dC_{mn}}{dC_L} \tag{2-44}$$

For example, in Figure 2–20a

$$\text{a.c. of the NACA 2412} \cong .25 - .01 \cong .24$$

In Table 2, Appendix, are presented some of the fundamental section characteristics of a representative group of modern airfoils developed by the National Advisory Committee for Aeronautics. The first popular airfoils of this family are designated by numbers having 4 digits, such as the NACA 2412 section presented in Figure 2–20a. The geometry of this series and subsequent ones is defined in terms of the mean line and thickness distribution about this line (Figure 2–26). The abscissas, ordinates, and slopes of the mean line are designated as x_c, y_c, and $\tan \theta$ respectively. Ordinates of the section are always laid out perpendicular to the mean line. To avoid a geometric layout perpendicular to the point on the mean line the coordinates of a point on the upper surface, x_u, y_u, can be expressed by the following relations, where y_t is the ordinate of the symmetrical thickness distribution at a chordwise position, x.

$$x_u = x_c - y_t \sin \theta$$
$$y_u = y_c + y_t \cos \theta \tag{2-45}$$

Similarly, the corresponding expressions for the coordinates of the lower surface are

$$x_l = x_c + y_t \sin \theta$$
$$y_l = y_c - y_t \cos \theta \tag{2-46}$$

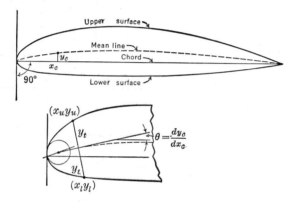

FIGURE 2-26. Geometry of airfoil sections. From NACA TR 460, "The Characteristics of 78 Related Airfoil Sections from Tests in the Variable-density Wind Tunnel," by Eastman N. Jacobs, Kenneth E. Ward, and Robert M. Pinkerton.

The center of the leading-edge radius is usually obtained by drawing a line through the end of the chord with slope equal to the slope of the mean line at .5 per cent of the chord and laying off the leading-edge radius along this line.

The mean line of the 4-digit series and explanation of the numbering system for a typical section are illustrated in Figure 2-27. The coordinates of this airfoil are also listed in Figure 2-27.

The basic thickness distribution for both the 4- and 5-digit series is presented in Figure 2-28. The position of the maximum thickness is always 30 per cent of the chord for both series. For the basic thickness function the ratio of maximum thickness to chord, t/c, is 20 per cent. For any other thickness ratio, such as the NACA 2412 section shown in Figure 2-27, a direct ratio of the basic thickness distribution is taken, e.g., 12/20. The leading-edge radius varies as the square of the thickness ratio.

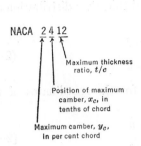

NACA 2412		
$x_l,\ x_u$	y_u	y_l
0	0	0
1.25	2.15	−1.65
2.5	2.99	−2.27
5.0	4.13	−3.01
7.5	4.96	−3.46
10	5.63	−3.75
15	6.61	−4.10
20	7.26	−4.23
25	7.67	−4.22
30	7.88	−4.12
40	7.80	−3.80
50	7.24	−3.34
60	6.36	−2.76
70	5.18	−2.14
80	3.75	−1.50
90	2.08	−.82
95	1.14	−.48
100	0	0
L. E. radius 1.58		
Slope of radius through L. E. .10		

FIGURE 2–27. NACA 4-digit series notation. From NACA TR 460, "The Characteristics of 78 Related Airfoil Sections from Tests in the Variable-density Wind Tunnel," by Eastman N. Jacobs, Kenneth E. Ward, and Robert M. Pinkerton.

A slight modification to the 4-digit series is obtained by varying the leading-edge radius and position of maximum thickness. This series is designated by two numbers added to the 4-digit notation.

$$4312 - 34$$

LEADING EDGE RADIUS

0—Sharp leading edge
3—¼ Normal leading edge
6—Normal leading edge
9—Three times normal
        leading edge

POSITION OF MAXIMUM THICKNESS IN TENTHS OF CHORD

The mean line of the 5-digit series and explanation of the numbering system for a typical section are illustrated in Figure 2–29.

The latest sections tested by the NACA are designed to decrease the skin-friction drag by increasing the relative extent of the laminar boundary layer. This is accomplished on airfoils designed to have a favorable pressure gradient over a large percentage of the chord, tested under controlled conditions of low airstream turbulence. Successive attempts to design this type of airfoil led to families of airfoils designated NACA 1 to 7 series sections, of which the most widely used are the 6 series airfoils, a typical example of which is shown in Figure 2–30·

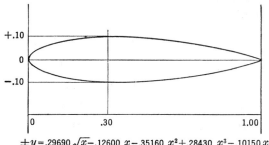

$$\pm y = .29690 \sqrt{x} - .12600 \ x - .35160 \ x^2 + .28430 \ x^3 - .10150 \ x^4$$

FIGURE 2–28. Basic thickness distribution for NACA 4- and 5-digit series. From NACA TR 460, "The Characteristics of 78 Related Airfoil Sections from Tests in the Variable-density Wind Tunnel," by Eastman N. Jacobs, Kenneth E. Ward, and Robert M. Pinkerton.

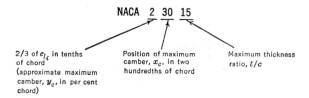

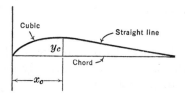

FIGURE 2–29. NACA 5-digit series notation. From NACA TR 460, "The Characteristics of 78 Related Airfoil Sections from Tests in the Variable-density Wind Tunnel," by Eastman N. Jacobs, Kenneth E. Ward, and Robert M. Pinkerton.

Data for the thickness distributions of the 1 and 6 series airfoils are similar to the NACA 4-digit series and are individually presented in NACA WR L–560. The mean lines usually used with the NACA 6 series airfoils produce a uniform chordwise normal force loading from the leading edge to $x/c = a$ and a linearly decreasing load from this point to the trailing edge. Equation (2–47) is a simplification of the general equation for mean-lines ordinates for $a = 1$.

$$y_c = -\frac{c_{l_i}}{4\pi}[(1 - x)\ln(1 - x) + x\ln x] \qquad (2\text{–}47)$$

These data are presented in NACA WR L–560 for c_{l_i} equal to unity. Corresponding data for other design lift coefficients may be calculated by direct ratios of the values for c_{l_i} equal to unity. These mean lines are combined with thickness forms in the usual manner previously described.

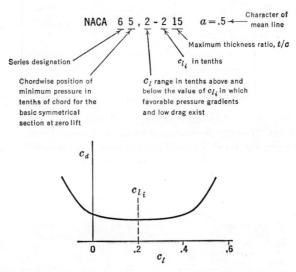

FIGURE 2–30.   NACA 6-digit series notation.   From NACA WR L–560, "Summary of Airfoil Data," by Ira H. Abbott, Albert E. von Doenhoff, and Louis S. Stivers, Jr.

Airfoils having a thickness distribution obtained by linearly increasing or decreasing the ordinates of a previously derived thickness distribution are designated as follows:

$$65\ (215)\ 318 \qquad a = .5$$

In this example the numbers in the parentheses refer to the low-drag range and thickness of the originally derived section illustrated in Figure 2–30. The remaining numbers retain the same definitions as previously discussed. The more recent NACA sections are derived from slightly modified thickness families. These airfoils are differentiated from the originally derived sections by writing as a subscript the number indicating the low-drag range. For example, the airfoil in Figure 2–30 would be modified in designation as follows:

$$65_2\text{-}215$$

A knowledge of the pressure distributions about the airfoils previously described is important to the designer for several reasons. The structural design of the wing depends on the magnitude and location of the normal and chord forces and resulting moments. The high-speed characteristics of the airfoil section depend on the pressure distribution. Finally the pressure distribution has a large effect on the boundary-layer flow, which in turn affects airfoil characteristics. The theoretical calculation of a pressure distribution about an airfoil, although very accurate, is often too lengthy to permit the designer to compute such a distribution for more than a very few airfoils at one time. However, pressure distribution data either from theoretical calculation or from test data on a limited number of basic mean line shapes and thickness distributions can be used to calculate rapidly pressure distributions for a large number of related airfoil sections.

To illustrate, the pressure coefficient, P, can be expressed in terms of the ratio of the local velocity, v, at any station along the airfoil to the free-stream velocity, v_0, by use of Bernoulli's equation.

$$p - p_0 = q_0 - q$$

$$\frac{p - p_0}{q_0} = P = 1 - \frac{q}{q_0} = 1 - \frac{v^2}{v_0^2} \qquad (2\text{-}48)$$

or

$$\left(\frac{v}{v_0}\right)^2 = 1 - P = S$$

Values of S for a particular station on a particular section can be derived from data given for basic thickness distributions and mean-line shapes by the following equation:

$$S = \left(\frac{v}{v_0}\right)^2 = \left(\frac{v}{v_0} \pm \frac{\Delta v}{v_0} \pm \frac{\Delta v_a}{v_0}\right)^2 \qquad (2\text{-}49)$$

where

1. v/v_0 is the velocity ratio distribution over the basic thickness form at zero angle of attack and is illustrated in Figure 2-31 for the NACA 0012 and NACA 65, 2-016 sections. For any other thickness ratio the values of v/v_0 may be obtained approximately by linear scaling of the corresponding values of v/v_0 for the nearest thickness ratio.

$$\left(\frac{v}{v_0}\right)_{t_2} = \left[\left(\frac{v}{v_0}\right)_{t_1} - 1\right]\frac{t_2}{t_1} + 1 \qquad (2\text{-}50)$$

2. $\Delta v/v_0$ corresponds to the design load distribution (difference in pressure between upper and lower surface) of the mean line and is

illustrated in Figure 2–32 for an NACA 64 mean line and NACA $a = .5$ mean line. For any other camber or c_{l_i}, $\Delta v/v_0$ may be obtained by a direct ratio. The resultant pressure ratio, P_R, is the difference between the local upper-surface and lower-surface pressure coefficients.

3. $\Delta v_a/v_0$ corresponds to the additional load distribution associated with angle of attack, values of which are presented for the given airfoils in Figure 2–31 for an additional lift coefficient of unity. These values are approximately independent of thickness ratio. Pressure distributions are usually desired for some lift coefficient, c_l, not corresponding to c_{l_i}. In this case the ratio $\Delta v_a/v_0$ may be obtained by multiplying the tabulated value by a factor, $f(\alpha)$, where

$$f(\alpha) = c_l - c_{l_i} \tag{2–51}$$

To illustrate the use of equation (2–49) consider the following example. Find the pressure coefficient, S, at the station $x = .25$ on the upper and lower surfaces of the NACA 2413 section at a lift coefficient of .5. From Figure 2–31

$$\left(\frac{v}{v_0}\right)_{t/c=.12} = 1.174$$

From equation (2–50)

$$\left(\frac{v}{v_0}\right)_{t/c=.13} = (1.174 - 1)\frac{13}{12} + 1 = 1.189$$

From Figure 2–32 for the NACA 64 mean line at $x = .25$

$$\frac{\Delta v}{v_0} = .258$$

and $c_{l_i} = .76$

The desired values of $\Delta v/v_0$ and c_{l_i} corresponding to the airfoil geometry are obtained by multiplying by the camber ratio, 2/6,

$$\frac{\Delta v}{v_0} = \frac{2}{6} \times .258 = .086$$

$$c_{l_i} = \tfrac{2}{6} \times .76 = .253$$

From Figure 2–31

$$\frac{\Delta v_a}{v_0} = .273 \text{ for the NACA 0012 section}$$

$$\frac{\Delta v_a}{v_0} \cong \begin{array}{l} .273 \text{ for the NACA 0013 section for} \\ \text{an additional } c_l \text{ equal to unity} \end{array}$$

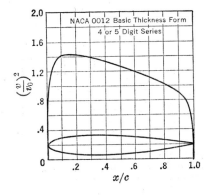

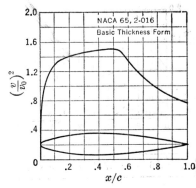

x/c	y/c	$(v/v_0)^2$	v/v_0	$\Delta v_a/v_0$
0	0	0	0	1.988
.5	—	.640	.800	1.475
1.25	1.894	1.010	1.005	1.199
2.5	2.615	1.241	1.114	.934
5.0	3.555	1.378	1.174	.685
7.5	4.200	1.402	1.184	.558
10	4.683	1.411	1.188	.479
15	5.345	1.411	1.188	.381
20	5.737	1.399	1.183	.319
25	5.941	1.378	1.174	.273
30	6.002	1.350	1.162	.239
40	5.803	1.288	1.135	.187
50	5.294	1.228	1.108	.149
60	4.563	1.166	1.080	.118
70	3.664	1.109	1.053	.092
80	2.623	1.044	1.022	.068
90	1.448	.956	.978	.044
95	.807	.906	.952	.029
100	.126	0	0	0
L.E. radius 1.58 per cent c				

x/c	y/c	$(v/v_0)^2$	v/v_0	$\Delta v_a/v_0$
0	0	0	0	1.950
.5	1.202	.560	.748	1.650
.75	1.423	.690	.831	1.500
1.25	1.796	.842	.918	1.275
2.5	2.507	1.068	1.033	.920
5.0	3.543	1.217	1.103	.680
7.5	4.316	1.287	1.134	.545
10	4.954	1.328	1.152	.480
15	5.958	1.379	1.174	.390
20	6.701	1.409	1.187	.325
25	7.252	1.433	1.197	.285
30	7.645	1.453	1.205	.255
35	7.892	1.469	1.212	.225
40	7.995	1.484	1.218	.200
45	7.938	1.497	1.224	.180
50	7.672	1.491	1.221	.160
55	7.184	1.421	1.192	.140
60	6.495	1.328	1.152	.125
65	5.647	1.235	1.111	.110
70	4.713	1.147	1.071	0.95
75	3.738	1.056	1.028	.080
80	2.759	.970	.985	.066
85	1.817	.886	.941	.050
90	.982	.816	.903	.040
95	.340	.769	.877	.025
100	0	.733	.856	0

FIGURE 2-31. NACA basic thickness forms. From NACA WR L-560, "Summary of Airfoil Data," by Ira H. Abbott, Albert E. von Doenhoff, and Louis S. Stivers, Jr.

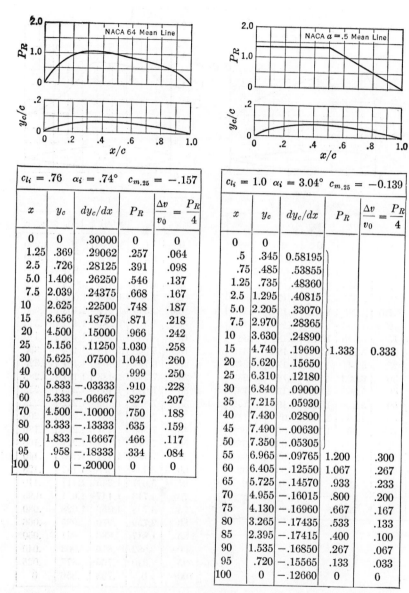

$c_{l_i} = .76$	$\alpha_i = .74°$	$c_{m.25} = -.157$

x	y_c	dy_c/dx	P_R	$\dfrac{\Delta v}{v_0} = \dfrac{P_R}{4}$
0	0	.30000	0	0
1.25	.369	.29062	.257	.064
2.5	.726	.28125	.391	.098
5.0	1.406	.26250	.546	.137
7.5	2.039	.24375	.668	.167
10	2.625	.22500	.748	.187
15	3.656	.18750	.871	.218
20	4.500	.15000	.966	.242
25	5.156	.11250	1.030	.258
30	5.625	.07500	1.040	.260
40	6.000	0	.999	.250
50	5.833	−.03333	.910	.228
60	5.333	−.06667	.827	.207
70	4.500	−.10000	.750	.188
80	3.333	−.13333	.635	.159
90	1.833	−.16667	.466	.117
95	.958	−.18333	.334	.084
100	0	−.20000	0	0

$c_{l_i} = 1.0$	$\alpha_i = 3.04°$	$c_{m.25} = -0.139$

x	y_c	dy_c/dx	P_R	$\dfrac{\Delta v}{v_0} = \dfrac{P_R}{4}$
0	0			
.5	.345	0.58195		
.75	.485	.53855		
1.25	.735	.48360		
2.5	1.295	.40815		
5.0	2.205	.33070		
7.5	2.970	.28365		
10	3.630	.24890	1.333	0.333
15	4.740	.19690		
20	5.620	.15650		
25	6.310	.12180		
30	6.840	.09000		
35	7.215	.05930		
40	7.430	.02800		
45	7.490	−.00630		
50	7.350	−.05305		
55	6.965	−.09765	1.200	.300
60	6.405	−.12550	1.067	.267
65	5.725	−.14570	.933	.233
70	4.955	−.16015	.800	.200
75	4.130	−.16960	.667	.167
80	3.265	−.17435	.533	.133
85	2.395	−.17415	.400	.100
90	1.535	−.16850	.267	.067
95	.720	−.15565	.133	.033
100	0	−.12660	0	0

FIGURE 2–32. NACA mean line pressure distribution. From NACA WR L–560, "Summary of Airfoil Data," by Ira H. Abbott, Albert E. von Doenhoff, and Louis S. Stivers, Jr.

For the desired c_l, $\Delta v_a / v_0$ is obtained by applying equation (2-51):

$$\frac{\Delta v_a}{v_0} = .273 \ (c_l - c_{l_i}) = .273(.5 - .253) = .067$$

Substituting the proper values in equation (2-49) gives for the upper surface

$$S = (1.189 + .086 + .067)^2 = 1.80$$

and for the lower surface

$$S = (1.189 - .086 - .067)^2 = 1.074$$

By the use of the pressure distribution data presented in Figures 2-31 and 2-32 important compressibility characteristics for different airfoil sections can be theoretically derived. For example, the critical Mach number of an airfoil, M_{cr} (M_0 for which $M = 1$ at a local point on the airfoil surface), is related to the low-speed peak pressure coefficient or maximum v/v_0 for a given value of lift coefficient. Using the Karman-Tsien relationship to predict the M_{cr} and the Glauert relationship to predict c_l, it is possible to obtain from the low-speed pressure distributions a plot of M_{cr} versus normal force coefficient, c_n, or c_l. By application of these data to the airplane it is possible then to predict the limitations on wing loadings and speeds to prevent serious drag rises and stability troubles. For the airplane in Figure 2-18 the M_{cr} occurs near the beginning of the transonic region.

From compressible flow theory the critical pressure ratio, P_{cr} (value of P when $M_0 = M_{cr}$), is related to M_{cr} by the following relationship:

$$P_{cr} = \frac{2}{\gamma M_{cr}^2} \left\{ \left[\frac{2}{\gamma + 1} \left(1 + \frac{\gamma - 1}{2} M_{cr}^2 \right) \right]^{\gamma/(\gamma - 1)} - 1 \right\} \quad (2\text{-}52)$$

In Figure 2-33 is presented a plot of equation (2-52). Superposed on this plot are high-speed pressure distributions for the NACA 23015 airfoil taken at various angles of attack. In this case where high-speed tunnel data are available it is possible to determine directly the intersection with the P_{cr} line for each angle of attack. From a high-speed plot of c_n or c_l versus α the relationship in Figure 2-34 can be determined. Now if high-speed wind-tunnel data for this airfoil are not available, it is still possible to estimate the M_{cr} from NACA low-speed basic thickness forms and mean line pressure distributions. For example, at a particular α or c_l the peak incompressible P, P_{inc}, is obtained from NACA data at a given $c_{l_{inc}}$. Both these values increase

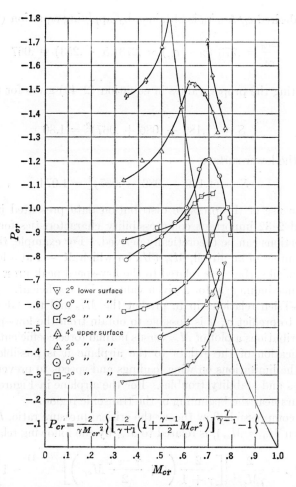

$$P_{cr} = \frac{2}{\gamma M_{cr}^2}\left\{\left[\frac{2}{\gamma+1}\left(1+\frac{\gamma-1}{2}M_{cr}^2\right)\right]^{\frac{\gamma}{\gamma-1}}-1\right\}$$

FIGURE 2–33.   Critical pressure coefficient. From NACA AR 990, "Pressure
Distribution for a 23015 Airfoil at High Speeds."

with Mach number by the Glauert relationships:

$$c_l = \frac{c_{l_{inc}}}{\sqrt{1-M^2}} \tag{2-53}$$

and $$P = \frac{P_{inc}}{\sqrt{1-M^2}} \tag{2-54}$$

Pressure coefficient variation is expressed more exactly by the Karman-

Tsien relationship:

$$P = \cfrac{1}{\cfrac{\sqrt{1 - M^2}}{P_{inc}} + \cfrac{1 - \sqrt{1 - M^2}}{2}} \qquad (2\text{--}55)$$

To determine the intersection of P with P_{cr} for the particular set of data, let $P = P_{cr}$ and $M = M_{cr}$ in equation (2–55). By substituting this value of P_{cr} in equation (2–52), a solution for M_{cr} is then obtained.

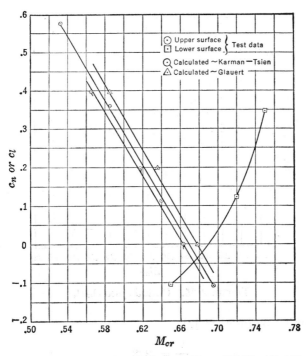

FIGURE 2–34.   23015 critical Mach number. From NACA AR 990, "Pressure Distribution for a 23015 Airfoil at High Speeds."

A general graphical solution for any value of P_{inc} or S_{inc} versus M_{cr} is presented in Figure 2–35. The correlation between calculated values of M_{cr} and high-speed test data is presented in Figure 2–34. It is interesting to note that the values computed from the Glauert relationship [equation (2–54)] and those from the Karman-Tsien relationship [equation (2–55)] bracket the experimental results.

As an example of the use of Figure 2–35 and the NACA basic thickness forms and mean line pressure distributions to determine M_{cr}, consider the NACA 65_1-212, $a = .5$, airfoil section and calculate the M_{cr} at $c_l = 0$.   From Figure 2–31

$$\text{Max.} \left(\frac{v}{v_0}\right)_{t/c\,=.16} = 1.224 \text{ at } \frac{x}{c} = .45$$

Then from equation (2–50)

$$\left(\frac{v}{v_0}\right)_{t/c\,=.12} = \left(\frac{12}{16} \times .224\right) + 1 = 1.168$$

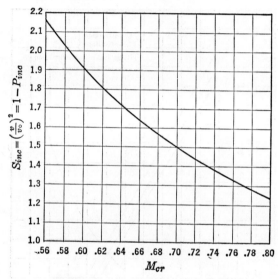

FIGURE 2–35.   Critical speed chart.

From Figure 2–32 for $c_{l_i} = 1.0$

$$\frac{\Delta v}{v_0} = .333$$

For the above section, $c_{l_i} = .2$ and $\dfrac{\Delta v}{v_0} = .2 \times .333 = .066$.   From Figure 2–31 the additional velocity ratio for $c_l = 1$ is determined.

$$\frac{\Delta v_a}{v_0} = .18$$

For $c_l = 0$ equation (2–51) is applied.

$$\frac{\Delta v_a}{v_0} = .18(0 - .2) = -.036$$

Now, applying equation (2–49),

$$S = \left(\frac{v}{v_0}\right)^2 = (1.168 + .066 - .036)^2 = \mathbf{1.435}$$

and from Figure 2–35

$$M_{cr} \cong .724$$

This value agrees closely with that listed in the Appendix for the 65_1-212 section inasmuch as the subscript 1, or third digit in the above designation, does not agree exactly with the corresponding digit in Figure 2–31. The critical Mach numbers for the airfoils listed in the Appendix are calculated in a similar manner.

In the application to the airplane, critical Mach number and lift coefficient are not as important to the designer as speed and wing loading. Speed is obtained directly from Mach number for a given altitude. Effective wing loading is related directly to Mach number and lift coefficient. By the definition of uniform level flight,

$$L = W = C_L q S = C_L \frac{\sigma V^2}{391} S \qquad (2\text{–}56)$$

where V = Speed in miles per hour. The product of dynamic pressure times lift coefficient is a constant for a given wing loading. At a given altitude the speed of an airplane in uniform level flight is fixed by the lift coefficient or angle of attack at which it is flying. Equation (2–56) may be written in terms of effective wing loading and $M^2 C_L$ which, when plotted with high-speed lift characteristics of a given wing, indicates definite limitations to safe flight. Since

$$\sigma = \frac{\rho}{\rho_0}, \quad \delta = \frac{p}{p_0}, \quad a = \text{Speed of sound} = \sqrt{\frac{\gamma p}{\rho}},$$

$$a_0 = 761 \text{ mph} \quad \text{and} \quad \gamma = 1.4$$

Equation (2–56) is rewritten as

$$\frac{W}{\delta S} = 1481 \, M^2 C_L \qquad (2\text{–}57)$$

This limitation is due to the inability of wings using high-speed airfoil sections to support more than a given value of load in pounds per square foot per absolute pressure. The maximum value of this parameter is essentially unaffected by airfoil section, thickness, and planform characteristics, and will be discussed subsequently as it affects airplane performance.

In Figure 2–36 are presented typical two-dimensional airfoil data

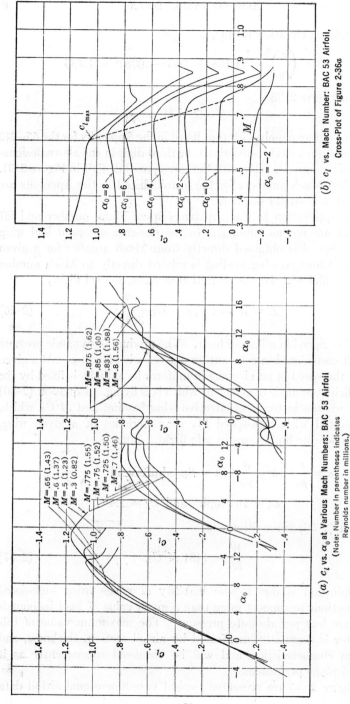

(a) c_l vs. α_0 at Various Mach Numbers: BAC 53 Airfoil

(Note: Number in parentheses indicates Reynolds number in millions.)

(b) c_l vs. Mach Number: BAC 53 Airfoil, Cross-Plot of Figure 2-36a

FIGURE 2-36. Two-dimensional high-speed lift data.

obtained at high Mach numbers in the Boeing Airplane Company wind tunnel.* In Figure 2-37 is shown high-speed drag data. In Figure 2-36b the dotted line indicates an approximate buffet boundary as defined by the lift or stability critical Mach number at each angle of attack. In Figure 2-37b the drag critical Mach number is defined approximately at the point on each curve where the drag increases rapidly. An increase in c_d of .002 may be used as a criterion for definitely establishing this point. The critical speeds established from airfoil pressure distributions are not necessarily related to the drag or lift critical speeds. Actually the drag begins to increase as soon as supersonic flow exists on any part of the wing surface. However, if separation occurs before shock, the drag increases before the M_{cr} is reached. If separation does not occur until a shock wave is formed, then no increase in drag is apparent until the M_{cr} is reached. Figure 2-36b is replotted in Figure 2-38, indicating the envelope of C_L versus M data as established by a stall limit and high-speed buffet limit. Note that upper case nomenclature are used to designate three-dimensional data. This boundary line may be determined at any speed by limiting values of lift or drag M_{cr}. Outside this boundary, flight is considered unsafe. Lines of constant $M^2 C_L$ or $W/\delta S$ are also cross-plotted on the $C_L M$ grid. Since the airplane in level flight at a given weight flies along a constant $W/\delta S$ line, the range of possible speeds between stall and buffet may be small for a highly loaded airplane at high altitude. Also, an inadvertent gust or pilot maneuver will increase the allowable $W/\delta S$ by a factor, g, equal to the applied load factor which may change the flight condition $(W/\delta S)$ outside the safe $C_L M$ boundary for the particular configuration. The limiting value of $W/\delta S$ as it affects range performance will be discussed in Chapter 4.

The discussion so far has dealt with airfoil shapes designed for sub-sonic flow. The following conclusions regarding airfoil parameters and airfoil selection will apply only to these types. In a subsequent section airfoils designed for supersonic flight will be discussed. A theoretical study of the pressure distributions as affected by changes in camber and thickness leads to some broad conclusions concerning changes in airfoil coefficients which can be verified from the experimental data presented in the NACA reports previously mentioned in this chapter. α_{OL} and c_{mac} are primarily a function of camber, both generally increasing nega-tively as the camber increases. c_{l_i} increases positively with an increase in camber. Moving the maximum camber forward decreases the negative value of c_{mac} but usually produces an abrupt separation and

* *Jour. Aero. Sciences*, "Systematic Wing Section Development," by George S. Schairer, January 1947.

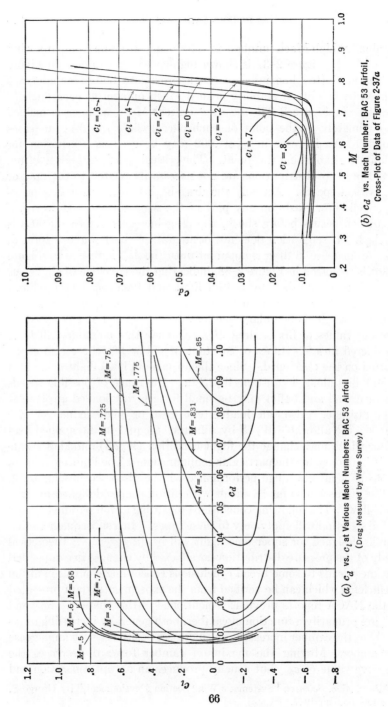

(a) c_d vs. c_l at Various Mach Numbers: BAC 53 Airfoil
(Drag Measured by Wake Survey)

(b) c_d vs. Mach Number: BAC 53 Airfoil, Cross-Plot of Data of Figure 2-37a

FIGURE 2-37. Two-dimensional high-speed drag data.

stall of the surface at high angles of attack.   Aerodynamic center and
slope of lift curve, a_0, are little affected by camber or thickness. $c_{l_{max}}$
is practically independent of camber but increases gradually for thickness ratios up to 15 to 20 per cent, then decreases.   For smooth airfoils

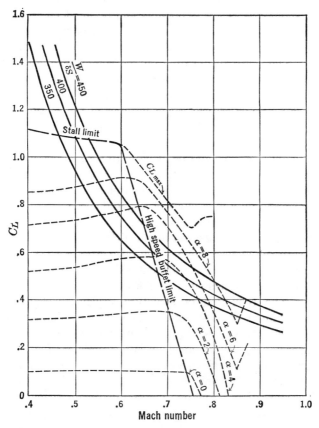

FIGURE 2–38.   Flight boundary.

$c_{d_{min}}$ decreases with a decrease in thickness ratio and is generally a
minimum when the position of maximum negative pressure is rearward
of the 50 per cent chord.   This latter effect is characterized by the
NACA 1 to 7 series sections.   However, the existence of normal roughness or waviness on the wing surface caused by practical construction
methods or even adverse operational climatic conditions makes the
selection of airfoil type sometimes immaterial.   Under these conditions
the effect of camber and thickness is very small, and use of a so-called

low-drag section often has undesirable drag increases at higher angles of attack with a $c_{d\min}$ no lower than many conventional sections.

Although the final choice of wing root and tip sections is somewhat influenced by the planform characteristics of the wing, some general rules may be stated as criteria for two-dimensional airfoil selection.

a. $c_{d\min}$ should be as low as possible and occur at c_{l_i}. c_d should be small over a large range of lift coefficients above and below c_{l_i}.

b. The maximum S or v/v_0 should be as low as possible for high-speed airplanes. The sharp rise in drag on the wing due to the formation of a compressibility shock wave near the point of maximum negative peak pressure on the wing surface can be delayed by a low value of maximum v/v_0. The critical Mach number of the section is proportional to v/v_0 which should be as low as practicable for high-speed airplanes. (See Figure 2–35.)

c. The value of $c_{m_{ac}}$ should be as close to zero as possible. This is particularly important for high-speed airplanes where an excessive tail load would be necessary to balance the large moment about the center of gravity caused by a high value of $c_{m_{ac}}$ and speed.

d. $c_{l\max}$ should be as large as possible.

e. As will be shown later, optimum airframe cruising efficiency is obtained with the highest value of the ratio c_l/c_d or L/D.

f. The section must have sufficient thickness to house structural members and other items such as gas tanks and landing gear when necessary. The reduction of structural weight, permitting larger spans for optimum cruising performance, usually dictates as large a root-section thickness as possible without unduly compromising high-speed performance. If the airplane is not designed for high speeds, the maximum possible thickness ratio may be as high as 25 per cent without serious aerodynamic limitations. Torsional rigidity is related to airfoil cross-section. Aileron reversal and flutter at high speed may be avoided by judicious chordwise location of structure even in thin sections.

g. In order to promote desirable stalling characteristics, the tip section should have a high maximum lift coefficient, a large range of angle of attack between zero and maximum lift, and a gradual stall pattern. Figure 2–39*b* illustrates a desirable stall picture. This airfoil has a flat-top lift curve. The selection of a section with a sudden break in the lift curve (Figure 2–39*a*) should be avoided if possible. Contrary to popular belief, tip stalling does not always define bad stalling characteristics. On multiengine airplanes the propeller slipstreams usually prevent root stalling. The problem then is to choose an airfoil at the tip such that a gradual stall pattern is obtained. The application of flaps often changes stall characteristics, so that test data with flaps are mandatory to insure a desirable result. Although the stalling problem is of primary importance, there is no evidence as yet that a general solution is assured by designing from available stall test data. These data are sometimes inadequate and often contradictory.

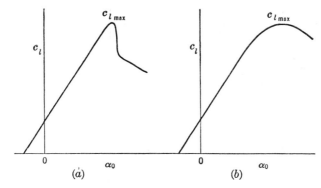

FIGURE 2–39. Typical stall patterns.

h. The choice of an airfoil with flat surfaces near the trailing edge is desirable for that section of the wing where ailerons are used. Sections with large included angles at the trailing edge will usually promote separation and undesirable aileron effectiveness and hinge moments.

i. Other airplane design considerations include such items as the selection of a sharp leading edge to reduce the wing area that is to be de-iced and the determination, if possible, of a desirable contour to reduce ice accumulation on the wing.

It is important to realize that the criteria for airfoil selection are sometimes conflicting when applied to the choice of an over-all best section. In the design of a particular airplane, the best compromise among the aerodynamic, structural, and general design criteria is desirable but often difficult to attain.

Planform Characteristics

The variation in airfoil section characteristics along the span of a wing is largely influenced by planform wing characteristics. Since the flying characteristics of the airplane are also dependent on the three-dimensional or planform wing performance the careful design of planform characteristics is usually considered very important to the designer. In a manner similar to the selection of airfoil sections, the optimum selection of planform characteristics is a compromise among

aerodynamic, structural, and design considerations. This selection is
even more difficult than that for the airfoil sections. Some of the
following characteristics defined below and illustrated in Figure 2–40
are not strictly associated with the planform of the wing but nevertheless are usually classed as planform characteristics.

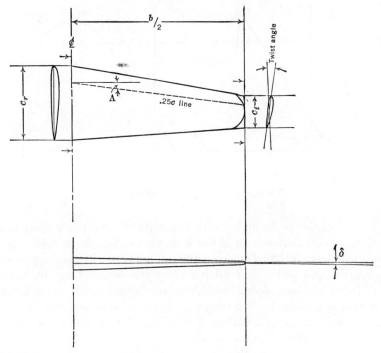

FIGURE 2–40.   Planform notation.

a. Aspect ratio, A, is the ratio of span to average chord:

$$A = \frac{b}{c_{\text{average}}} = \frac{b^2}{S}$$

The aspect ratio of modern aircraft varies from as low a value as 2 or
3 for supersonic fighters or missiles to 12 or 15 for long-range transports
and bombers. As was pointed out in a previous section, the induced
drag of the wing decreases with a corresponding increase in aspect
ratio. However, for high-speed flight the induced drag becomes very
small as compared with the profile drag, and the choice of a low-aspect-
ratio wing with corresponding high structural efficiency is usually made
by the designer.

b. Taper ratio, λ, is usually defined in terms of the ratio of tip to root chords:

$$\lambda = \frac{c_t}{c_r}$$

For a tapered planform (straight leading and trailing edges) (Figure 2–40) λ varies from values of .2 to 1. Again the choice of a desirable wing planform is a compromise between the high structural efficiency associated with a low value of λ and corresponding unfavorable tip stall characteristics. Taper in thickness is also illustrated in Figure 2–40. In this case the thickness ratio, t/c, at the root is different from that at the tip and usually higher in value for structural efficiency.

c. Sweep angle, Λ, is usually defined as the angle between the .25 chord line and the perpendicular to the airplane centerline. High values of sweep either forward or back are desirable to forestall sharp increases in wing drag due to compressibility. This will be further discussed in a later section. In general, spanwise flow associated with sweep is undesirable, particularly for sweepback where serious wing-tip stall is promoted. Sweep is sometimes utilized to relocate the horizontal position of the center of gravity with respect to the mean aerodynamic chord to permit good longitudinal stability characteristics.

d. Twist is also illustrated in Figure 2–40. It is often used as a dodge to prevent tip stall but is avoided if possible because of undesirable drag increases and structural and construction complications.

e. The tip-shape design of a wing is usually influenced by eye appeal. From a construction standpoint a rectangular tip shape may have a slight advantage over a faired tip. The difference in drag between the two most common shapes illustrated in Figure 2–40 is negligible.

f. Dihedral angle, δ, is also illustrated in Figure 2–40. It is measured in a front view between the horizontal and a line midway between the upper and lower surfaces of the wing. The dihedral affects the lateral stability characteristics of the airplane.

Of these planform characteristics, taper ratio, sweepback, and twist have an important bearing on the stall properties of the wing. In order that this phenomenon of wing stall be clearly understood, it is timely to introduce first a discussion of spanwise lift distribution. An accurate determination of this distribution is important for stress analysis as well as stall prediction.

Consider first a two-dimensional analysis of the lift per unit span along the wing. From two-dimensional theory

$$\alpha = \alpha_0$$

and

$$dL' = c_l qc$$

Since

$$c_l = a_0 \alpha_a = a_0(\alpha_0 - \alpha_{0L}) = (\alpha - \alpha_{0L})a_0$$

then

$$dL' = a_0(\alpha_0 - \alpha_{0L})qc$$

and for a given $a_0 q$

$$dL' \sim c(\alpha_0 - \alpha_{0L}) \sim c(\alpha - \alpha_{0L}) \qquad (2\text{–}58)$$

If $(\alpha - \alpha_{0L})$ is constant along the span, then the wing has zero aerodynamic twist, which means that the zero lift lines of all sections are

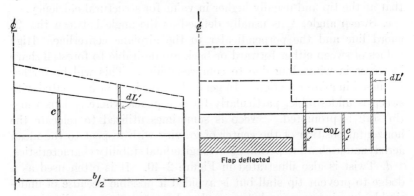

FIGURE 2–41. Two-dimensional   FIGURE 2–42. Two-dimensional spanwise
spanwise lift distribution.   lift distribution with flaps deflected.

parallel as viewed from the wing tip. A typical lift or load distribution for this case is illustrated in Figure 2–41. The lift at each section is directly proportional to the chord, and the load diagram has the same shape as the wing planform.

Now, if the chord in equation (2–58) is held constant as in a rectangular planform with zero taper, then the load distribution is proportional to $(\alpha - \alpha_{0L})$. $\alpha - \alpha_{0L}$ may vary because of change in α or α_{0L}. α_{0L} may change by a change in airfoil section or a change in effective camber due to flaps or ailerons. This latter effect is illustrated in Figure 2–42.

Now for an actual wing, instead of the foregoing idealized two-dimensional conception, the downwash must be considered. α is no longer equal to α_0, and equation (2–58) may be written

$$dL' \sim c\left(\alpha - \frac{w}{v} - \alpha_{0L}\right) \qquad (2\text{–}59)$$

It can be shown from airfoil theory that an untwisted wing with elliptical planform shape has a constant downwash. For this case the load distribution is elliptical in shape. For a rectangular untwisted wing the downwash increases from the midspan towards the tips so that the lift distribution according to equation (2–59) falls off at the wing tip. The downwash then has the effect of making the load distribution profile approach an elliptical shape, which effect smooths out any abrupt changes in load distribution caused by a change in either c or $\alpha - \alpha_{0L}$. Both theory and experiment indicate that the actual

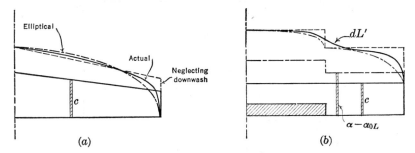

FIGURE 2–43. Three-dimensional spanwise lift distributions.

lift distribution curve lies between the curve of equation (2–58) and a semiellipse. Figure 2–43 illustrates the effect of downwash on the lift distributions shown in Figures 2–41 and 2–42.

It can also be shown by airfoil theory that only for a wing with elliptical load distribution and constant downwash is the induced drag a minimum. For most wings this condition is not true, and equations (2–16) and (2–17) for induced angle of attack and induced drag coefficient are no longer exact. Glauert originally introduced corrections to these equations in the following forms:

$$\alpha \text{ in radians} = \alpha_0 + \frac{C_L}{\pi A}(1 + \tau) \qquad (2\text{--}60)$$

$$C_D = C_{D_0} + \frac{C_L{}^2}{\pi A}(1 + \delta) \qquad (2\text{--}61)$$

where τ and δ are functions of the spanwise loading. For an untwisted wing with elliptical planform shape these correction factors are zero. The determination of spanwise loading and corresponding correction factors for all conditions of twist and planform shape is very lengthy and beyond the scope of this book. To illustrate the magnitude of these factors, consider the case of a straight-tapered, untwisted wing with various taper ratios (Figure 2–44a) and a rectangular wing with various values of aspect ratio (Figure 2–44b).

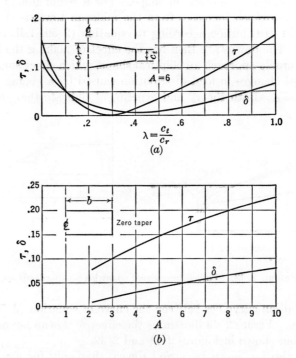

FIGURE 2–44.　Correction factors for non-elliptic lift distribution.

Now, to investigate the stall characteristics of a wing, consider the spanwise plot of load distributions and lift coefficients for the wings shown in Figure 2–45. First consider wings with zero aerodynamic twist. Since

$$dL' = c_l qc$$

$$c_l = \frac{dL'}{qc} = \frac{dL'}{c} \text{ at a given}$$

For convenience, assume the local maximum lift coefficient, $c_{l_{\max}}$, equal to one for this section. In other words, wherever this value is exceeded along the span, the wing is considered to be stalled. The wing stalls at any section where the local c_l exceeds the $c_{l_{\max}}$ for that section. The spanwise station where the stall first occurs on a rectangular wing with no aerodynamic twist is approximately expressed by the following equation:

$$\frac{y}{b/2} = 1 - \frac{c_t}{c_r} \qquad (2\text{-}62)$$

One of the principal wing planform characteristics affecting tip stall is taper ratio. It can be seen from Figure 2–45 that a high taper ratio promotes tip stall. In this case, the spanwise variation in chord decreases faster than the lift falls off, causing a local c_l near the tip

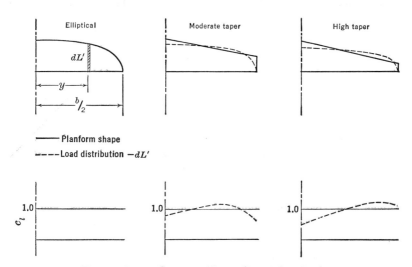

FIGURE 2–45.   Spanwise lift coefficient distribution.

much higher than one near the root of the wing. For this condition the sections near the tip stall first.

Another important factor affecting spanwise lift distribution and stall is wing twist. Figure 2–46 illustrates the basic distribution of lift coefficient, c_{l_b}, because of geometric twist (change in α) and the superposition of the additional lift coefficient distribution, c_{l_a}, which is independent of twist. The resultant spanwise lift distribution is similar to that for an untwisted wing with no taper and promotes desirable stall characteristics. This dodge which is often used by designers to prevent tip stall is commonly referred to as washout. Washout or twist is most effective for wings with moderate taper and is relatively ineffective for high values of taper ratio.

Besides the effects of taper ratio and twist on the stall characteristics of the wing the judicious choice of airfoil sections along the wing is an important criterion for good wing design. As previously discussed, the choice of a tip airfoil section with a high value of $c_{l_{max}}$ is desirable. Also, the shape of the lift curve near the stall should be similar to that shown in Figure 2-39b if an abrupt stall is to be avoided. Since the Reynolds number of the flow over the wing at any section has an influence on the magnitude of $c_{l_{max}}$, wings with little or no taper in planform are desirable. Thin sections have a tendency for more of an abrupt stall than those with a moderate thickness ratio. Wings with taper in thickness usually aggravate tip stall tendencies. A tip section with the position of maximum thickness nearer the leading edge than

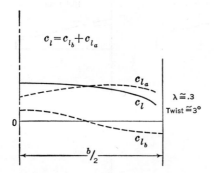

FIGURE 2-46. Basic and additional lift coefficient distribution.   FIGURE 2-47. Crossflow and stall pattern for swept-back wing.

that for the root section is desirable. Also a normal leading-edge radius should be used for tip sections. The use of wing-tip slots to increase both $c_{l_{max}}$ and the angle of attack for $c_{l_{max}}$ is most effective as a dodge to improve tip stall characteristics.

Difficulties in the control of airplanes near the stall are particularly serious for wings with high sweepback. The pressure gradients along wing surface lines perpendicular to the airplane centerline cause an outward crossflow of boundary-layer air along a swept-back wing, causing a thicker boundary layer near the trailing edge of the tip sections and hence earlier separation and stall. This phenomenon is illustrated in Figure 2-47.

The comparison of a theoretical spanwise load or lift distribution for a straight and swept-back wing (Figure 2-48) indicates the tip stall characteristic of the latter. Because of this increase in load on the tip, a rigid swept-back wing is heavier than a straight wing of the same structural aspect ratio. However, the deflection of an actual swept-back wing in flight has a relieving effect on tip load as the tip angle of attack is decreased more than sections inboard.

Tests have shown that it is almost hopeless to attempt to prevent tip stalling for highly swept-back wings only by correct selection of wing profiles along the span. In addition to all the previously described dodges usually employed on straight wings, end plates placed between the aileron and landing flap can be utilized as a last resort to prevent a building up of a strong boundary layer on the wing tips. The installation of a boundary-layer bleed-off either at this plate or along the top surface of the wing is questionable from a practical design standpoint because of mechanical complications, increase in weight, and limitations in engine power available. The installation of

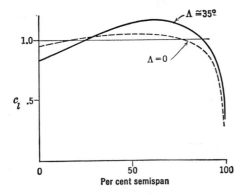

FIGURE 2–48. Spanwise lift distribution for swept-back wing.

leading-edge slots or slats is recommended as a practical and efficient method of designing a swept-back wing with good stall characteristics. The length of slotted wing inboard from the tip may be as high as 60 per cent of the semispan.

For wings with high angles of sweep forward, the inherent crossflow toward the wing root promotes a stall pattern from root to tip. The handling characteristics near the stall for an airplane with a swept-forward wing may also be undesirable if an excessive amount of lift is rapidly lost on the wing inner panels. In general the proper design of a wing, such that a breakdown of lift progresses slowly from root to tip and good aileron control is maintained until the entire wing stalls, is very difficult to attain. Even wind-tunnel tests must sometimes be interpreted with caution, as correlation with flight tests is not always good. The characteristics of swept wings as they affect performance and stability will be discussed later.

Before completing the discussion of spanwise lift distribution it remains only to mention the effect of wing mutilations and cut-outs.

Any distortion near the leading edge of a wing by mutilation or cut-outs at some spanwise location effectively reduces the lift at that point. Such an abrupt change in lift to a value approaching zero is accompanied by additional trailing vortices which in turn create additional drag. Spanwise slots or poor wing-fuselage junctures are factors in design which may cause large increases in drag. It is possible to alter the trailing edge of a wing very considerably without serious changes in lift and drag characteristics. Large cut-outs have little effect provided they are near the trailing edge. In any case the criterion by which any modification can be judged is the shape of the spanwise lift distribution. As long as the lift does not approach zero at any point along the span, the distribution of lift across the span holds nearly constant, insuring no deleterious aerodynamic effects.

High Lift and Control Devices

In order to increase the safety of airplanes during the take-off and landing, the aeronautical engineer is continually striving to reduce airplane speeds for these conditions without compromising high-speed characteristics. For this purpose the designer has some control over three airplane variables as expressed in the following equation for minimum dynamic pressure in level flight:

$$q_{min} = \frac{W}{C_{L_{max}}S} \qquad (2\text{--}63)$$

Usually the airplane weight is fixed by the load-carrying requirements of the design. It is impractical to increase the wing area beyond a certain size because of adverse effects on high speed and riding comfort. Large variable changes in wing areas have met with little success because of mechanical difficulties and weight penalties. By improving the maximum lift coefficient of the wing, the designer has accomplished a great deal. A permanent increase in the camber of an average wing section will produce a higher $C_{L_{max}}$ but will also be accompanied by a higher $C_{D_{0min}}$ and $C_{m_{ac}}$, both of which are undesirable, particularly for high-speed airplanes. Variable camber wings have been built but are impractical because of the vibration difficulties associated with such a flexible structure. Methods of increasing $C_{L_{max}}$ using boundary-layer bleed-offs or blow-offs to forestall separation to higher angles of attack have so far proved impractical. By improvements in mechanical installation such boundary-layer control arrangements may prove satisfactory. The leading-edge slot, based on the same aerodynamic principle, has had considerable practical success. The air flowing through the slot in Figure 2–49 is accelerated and moves farther toward the rear of the airfoil section before slowing down and separating from the surface. A typical plot of lift coefficient versus angle of attack for this type of device is illustrated in Figure 2–49. In this case the slat is movable. The slat is held in a closed position at low angles of

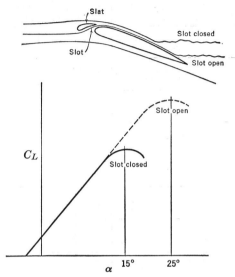

FIGURE 2–49. Typical aerodynamic characteristics with and without slot.

attack by pressure on the leading edge and automatically opens at high angles of attack by an increase in negative pressure over the slat

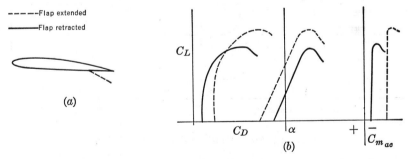

FIGURE 2–50. Typical aerodynamic characteristics with and without flap.

upper surface, causing a forward force component on the slat. The use of a leading-edge slat and slot to prevent tip stall is more important than their use as a device for lowering landing speeds. The principal disadvantage of the slot as a landing device is the high angle of attack required of the wing for stall. A conventional two-wheel landing gear raises the forward part of the airplane to an excessive height in order to permit the airplane to stall before the tail wheel contacts the ground.

Designation	Diagram	$C_{L_{max}}$	α at $C_{L_{max}}$ (degrees)	L/D at $C_{L_{max}}$	$C_{m_{ac}}$	Reference NACA
Basic airfoil Clark Y		1.29	15	7.5	−.085	TN 459
.30c Plain flap deflected 45°	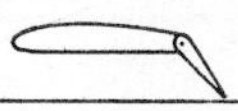	1.95	12	4.0	—	TR 427
.30c Slotted flap deflected 45°	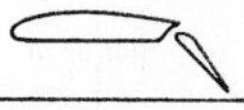	1.98	12	4.0	—	TR 427
.30c Split flap deflected 45°	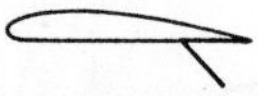	2.16	14	4.3	−.250	TN 422
.30c hinged at .80c Split flap (Zap) deflected 45°	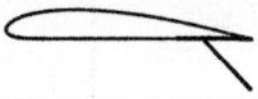	2.26	13	4.43	−.300	TN 422
.30c hinged at .90c Split flap (Zap) deflected 45°	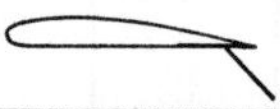	2.32	12.5	4.45	−.385	TN 422
.30c Fowler flap deflected 40°	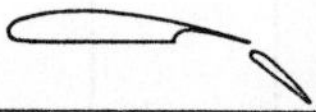	2.82	13	4.55	−.660	TR 534
.40c Fowler flap deflected 40°	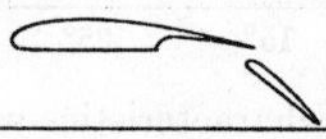	3.09	14	4.1	−.860	TR 534
Fixed slot		1.77	24	5.35	—	TR 427
Handley Page automatic slot		1.84	28	4.1	—	TN 459
Fixed slot and .30c plain flap deflected 45°		2.18	19	3.7	—	TR 427
Fixed slot and .30c slotted flap deflected 45°	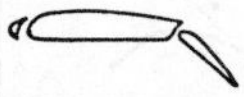	2.26	18	3.77	—	TR 427
Handley Page slot and .40c Fowler flap deflected 40°	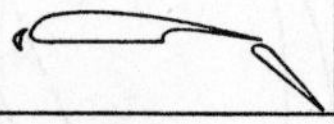	3.36	16	3.7	−.740	TN 459

NACA 7 by 10 ft tunnel data $A = 6$ $R = 609,000$

FIGURE 2–51a. Flap and slot characteristics.

To alleviate this condition slots are often used in connection with trailing-edge flaps.

The trailing-edge flap is the best and most universally used method for the reduction of landing speeds. The deflected flap in Figure 2–50a acts as a blind to change effectively the camber of the wing. In contrast to the leading-edge slot this device increases the lift coefficient for any angle of attack, maintaining approximately the same angle of attack for maximum lift coefficient as an unflapped wing. Typical aerodynamic data for an airfoil section or wing with combinations of

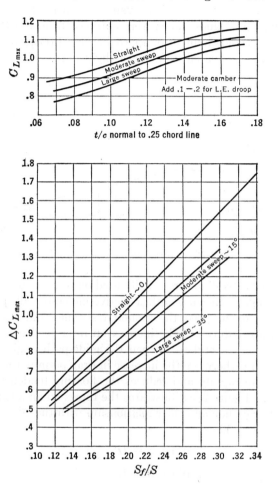

FIGURE 2–51b. C_{Lmax} and ΔC_{Lmax}.

various full-span flaps and slots are illustrated in Figure 2–51a. Flap design varies considerably in complexity and arrangement. Flaps that increase wing area when deflected produce the highest increase in maximum lift coefficient but also become more complex in mechanical design. Figure 2–51b presents generalized experimental results as obtained by the Boeing Airplane Company. C_{Lmax} for unflapped wings and $\Delta C_{L\max}$ for Fowler type flaps are given for straight and swept-back wings.   The increment in C_L due to flaps, $\Delta C_{L\max}$, is basically a function of the ratio of flap area to wing area, S_f/S. $\Delta C_{L\max}$ for a plain flap is approximately one-half the value shown for the Fowler type. The experimental values shown were obtained from the Boeing wind tunnel at a Reynolds number of approximately 1,000,000 and may be conservative, as compared to full-scale flight data.

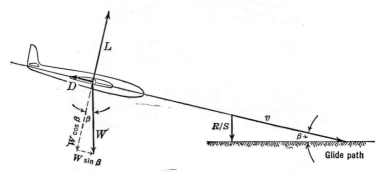

FIGURE 2-52.   Forces on airplane in steady glide.

Partially deflected flaps are usually used on heavily loaded airplanes for take-off flight.   Because of the effective increase in lift produced by the flap at a given speed the ground resistance and hence take-off ground roll are reduced.   The increase in drag coefficient associated with flaps is very useful to steepen the glide-path angle for landing approaches.   Furthermore the increase in drag permits the airplane to decelerate more rapidly and reduces the floating tendency of a very clean design.   Figure 2-52 illustrates the forces on an airplane, power off, in a uniform glide and their relationship with the glide-path angle.

From Figure 2-52,

$$W \sin \beta = D$$

$$W \cos \beta = L$$

and
$$\tan \beta = \frac{D}{L} = \frac{C_D}{C_L} \qquad (2\text{-}64)$$

From Figure 2-50b and equation (2-64) it can be seen that the glide angle, β, for maximum glide distance is increased with flaps down. From the geometry of Figure 2-52 it may also be deduced that

$$\text{Rate of sink } (R/S) \text{ in feet per minute} = \frac{60Dv}{W} \qquad (2\text{-}65)$$

For part span flaps the curves for C_L versus α and C_{mac} versus C_L are intermediate between the curves shown in Figure 2–50b. The C_D, however, is greater than the full span flap conditions at all values of C_L because of the additional increment of profile drag associated with the basic lift distribution which simulates a twist in the wing (see Figure 2–46).

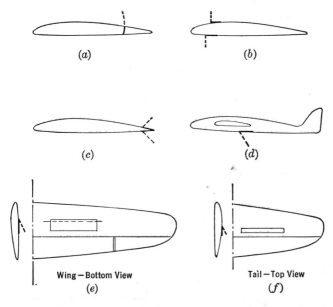

FIGURE 2–53. Typical spoiler, dive brake, and dive pull-out flap installations.

Sometimes it is advantageous to employ gadgets on wings which decrease rather than increase lift. Such devices are referred to as spoilers and are used either separately or in conjunction with ailerons to produce an airplane rolling moment. This is accomplished by spoiling, or reducing, the lift on the outboard area of one wing. Spoilers as a means for lateral control will be discussed in more detail in a later chapter. Figure 2–53a indicates a typical spoiler installation. Refinements to the surface and tip shape are usually accomplished to reduce the hinge moments necessary to extend the spoiler to the extreme position.

On very clean airplanes it is sometimes necessary to provide dive or airbrakes to limit the speed of airplanes in a dive, increase glide angle, or suddenly decrease the level-flight speed in combat maneuvers. Typical examples are illustrated in Figure 2–53b, c, and d. Dive brakes should be designed so that very little change in airplane pitching moment occurs when the brakes are extended.

As will be discussed in more detail later, it is also necessary on high-speed airplanes to provide dive pull-out flaps that produce a positive pitching moment to pull the airplane out of a steep dive. Such devices are necessary only on airplanes which reach such speeds that the effects of compressibility tend to keep the airplane in the dive. Typical wing and tail installations are illustrated in Figure 2–53e and f.

Determination of Three-dimensional Wing Data

The determination of the basic aerodynamic characteristics of the three-dimensional wing is one of the most important calculations required for stability and control analysis. The three-dimensional wing is made of airfoil sections for which two-dimensional data are immediately available in the literature. The aerodynamic data required for complete analysis of any three-dimensional wing are as follows for either flaps up or flaps down:

1. Slope of lift curve.
2. Angle of zero lift.
3. Aerodynamic center.
4. Pitching moment coefficient about a.c.

In order to calculate the above characteristics of the three-dimensional wing, it is assumed that the two-dimensional characteristics of the wing-root chord and wing-tip chord are known from airfoil section data. It will again be noted that all two-dimensional or section data are given in lower case, while three-dimensional data are given in upper case. The sketch in Figure 2–54 gives the basic terminology used. The following formulae were developed by R. F. Anderson of the NACA in 1936.*

The infinite aspect ratio or section slope of the lift curve, a_0, per degree is obtainable from section data. The three-dimensional slope of the lift curve, a, for the wing with no flap can be obtained from the following formula, where the factor f is a constant allowing for variation from elliptical lift distribution and is a function of wing aspect ratio and taper ratio as given in Figure 2–55. $\bar{a}_0$ is the mean section slope of the lift curve between root and tip section.

$$a = f \frac{\bar{a}_0}{1 + (57.3\bar{a}_0/\pi A)} \tag{2–66}$$

* NACA TR 572.

For the case of the three-dimensional wing with flaps the same formula applies, but $\bar{a}_0$ must be a weighted average of the section slope, a_0, for

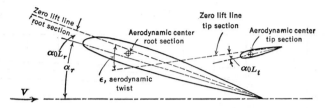

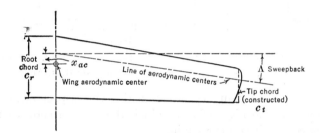

FIGURE 2-54. Basic terminology for three-dimensional wing.

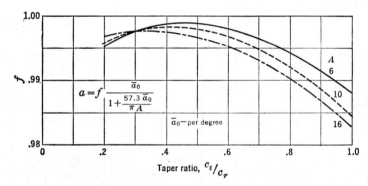

$$a = f \frac{\bar{a}_0}{1 + \frac{57.3\,\bar{a}_0}{\pi A}}$$

$\bar{a}_0$—per degree

Taper ratio, c_t / c_r

FIGURE 2-55. Slope of lift curve factor, f. From NACA TR 665, "Calculation of the Aerodynamic Characteristics of Tapered Wings with Partial Span Flaps," by H. A. Pearson and R. F. Anderson.

the section of the wing not flapped and a_{0f} for the section of the wing covered by the flap:

$$\bar{a}_0 = \frac{S_f}{S}\,a_{0f} + \left(1 - \frac{S_f}{S}\right)a_0 \qquad (2\text{-}67)$$

where S_f is the flapped wing area and S the total wing area.

The wing angle of zero lift, measured from the root chord, is as follows for the flaps-up condition:

$$\alpha_{0L} = \alpha_{0L_r} + J\epsilon \qquad (2\text{--}68)$$

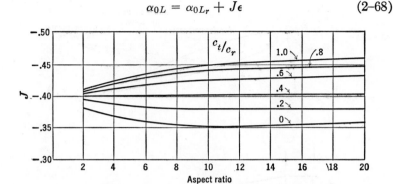

FIGURE 2–56. Angle of attack factor, J. From NACA TR 572, "Determination of the Characteristics of Tapered Wings," by R. F. Anderson.

where J is a constant depending on the wing aspect ratio and taper ratio as given in Figure 2–56, and ϵ is the angle of aerodynamic twist in degrees, positive for wash in.

The change in angle of zero lift of the three-dimensional wing with flap deflection is as follows:

$$\Delta\alpha_{0L} = -K\Delta c_l \qquad (2\text{--}69)$$

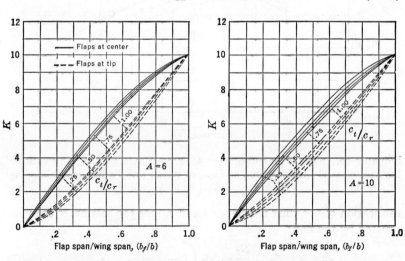

FIGURE 2–57. Angle of zero lift factor, flaps down. From NACA TR 665, "Calculation of the Aerodynamic Characteristics of Tapered Wings with Partial Span Flaps," by H. A. Pearson and R. F. Anderson.

where Δc_l is the change in section maximum lift due to the flap deflection, and K is a factor which is a function of the wing aspect and taper ratios and the flap span to wing span ratio (b_f/b). This relationship is

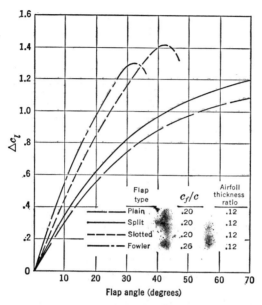

FIGURE 2-58. Increment to section lift due to flap deflection.

plotted in Figure 2-57. If the flap extends from, say .3b to .6b, then the value of K to use will be the difference between K at .6b and K at .3b. The increment in angle of zero lift should be added to the angle of zero lift with flaps retracted. The value of Δc_l for several types of flaps at various deflections is given in Figure 2-58.

The wing aerodynamic center location with respect to the aerodynamic center of the root chord can be obtained as follows:

$$x_{ac} = Hb \tan \Lambda \qquad (2\text{-}70)$$

where Λ is the angle of sweep (Figure 2-54), and H is a function of wing aspect and taper ratios as shown in Figure 2-59. With flaps down it will be assumed that the position of the section aerodynamic center will be unchanged.

The pitching moment coefficient about the wing aerodynamic center arises from two principal sources. One of these is the basic load distribution or zero lift load distribution, C_{mlb}, and the other is the airfoil section moments, C_{ms}. For the flaps-up case:

$$C_{mac} = C_{mlb} + C_{ms} \qquad (2\text{-}71)$$

where

$$C_{ms} = Ec_{mac} \text{ for } c_{mac} \text{ constant across span} \qquad (2\text{--}72)$$

or

$$C_{ms} = \frac{2b}{S^2} \int_0^{b/2} c_{mac}c^2 \, dy \text{ if } c_{mac} \text{ varies across span}$$

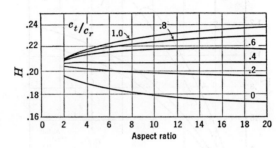

FIGURE 2–59. Aerodynamic center factor. From NACA TR 572, "Determination of the Characteristics of Tapered Wings," by R. F. Anderson.

and

$$C_{mlb} = -G\epsilon a_0 A \tan \Lambda \qquad (2\text{--}73)$$

where E and G are constants that are functions of aspect and taper ratio, as given in Figures 2–60 and 2–61.

For the flaps-down condition, again

$$C_{mac} = C_{ms} + C_{mlb}$$

where

$$C_{mlb} = N\Delta c_l A \tan \Lambda \qquad (2\text{--}74)$$

and

$$C_{ms} = Ec_{mac} + E' \Delta c_{mac} \qquad (2\text{--}75)$$

where N is a function of aspect ratio and taper ratio and flap span to wing span ratio (Figure 2–62). E and E' are given in Figure 2–63 and also are functions of the same ratios. If c_{mac} or Δc_{mac} are not constant across the span, C_{ms} must be obtained by integration:

$$C_{ms} = \frac{2b}{S^2} \int_0^{b/2} c_{mac}c^2 \, dy \qquad (2\text{--}76)$$

The value of Δc_{mac} must be obtained from section data for deflected flaps. Some typical data for Δc_{mac} are given in Figure 2–64.

The moment coefficient, C_{mac}, as developed herein, is defined as follows:

$$C_{mac} = \frac{M_{ac}}{qS^2/b} \qquad (2\text{-}77)$$

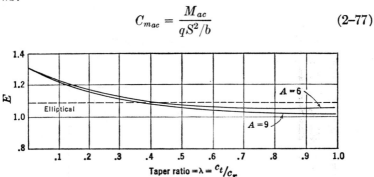

FIGURE 2-60. Section pitching moment factor. From NACA TR 572, "Determination of the Characteristics of Tapered Wings," by R. F. Anderson.

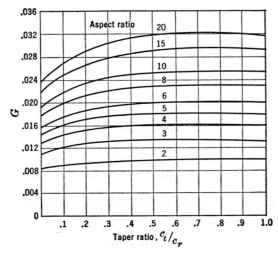

FIGURE 2-61. Pitching moment due to basic lift factor. From NACA TR 572, "Determination of the Characteristics of Tapered Wings," by R. F. Anderson.

To convert this to the coefficient based on the mean aerodynamic chord, c., or m.a.c., multiply by S/cb. The mean aerodynamic chord is the chord of an imaginary airfoil which throughout the normal flight range has the same force vectors as the three-dimensional wing. The mean aerodynamic chord can be obtained from wind-tunnel tests or, lacking these, from either the graphical development in Figure 2-65 or from equations (2-78) and (2-79).

Length of the mean aerodynamic chord is:

$$\text{m.a.c.} = \frac{2}{3}\left(a + b - \frac{ab}{a + b}\right) \qquad (2\text{-}78)$$

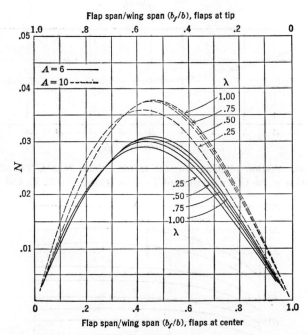

FIGURE 2–62. Effect of flap on basic pitching moment (factor). From NACA TR 665, "Calculation of the Aerodynamic Characteristics of Tapered Wings with Partial Span Flaps," by H. A. Pearson and R. F. Anderson.

Distance from root chord leading edge to leading edge of m.a.c. is:

$$m = \frac{s(a + 2b)}{3(a + b)} \qquad (2\text{-}79)$$

2-3 Estimation of Airplane Drag

Low-speed Drag Estimation

The drag estimation of experimental aircraft is of importance to the designer for the calculation of performance items such as high speed, range, and rate of climb. This drag estimation may be accomplished by several different methods, depending on the stage of development of the aircraft.

A preliminary stage of drag estimation of an airplane in an incompressible flow may be accomplished by adding up the individual drags

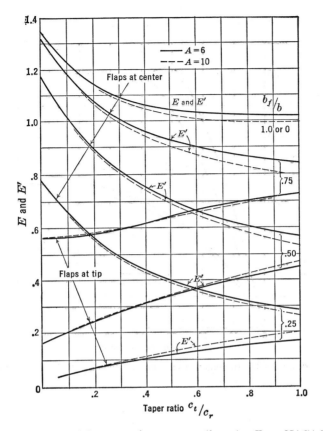

FIGURE 2–63. Effect of flap on section moments (factor). From NACA TR 665, "Calculation of the Aerodynamic Characteristics of Tapered Wings with Partial Span Flaps," by H. A. Pearson and R. F. Anderson.

of several components of the airplane. This method is commonly called drag breakdown. Consider first the fundamental character of the drag equation for the airplane. An analysis of the drag polar of a typical airplane shown in Figure 2–66 indicates that the curve of drag coefficient plotted against lift coefficient is approximately parabolic. In other words the drag coefficient can be expressed as

$$C_D = K + K'C_L{}^2 \tag{2–80}$$

A breakdown of this equation results in the sum of the parasite and induced drags:

$$C_D = C_{D_p} + C_{D_i} \tag{2–81}$$

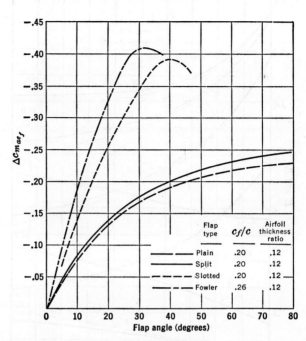

FIGURE 2-64.   Increment to section moment due to flap deflection.

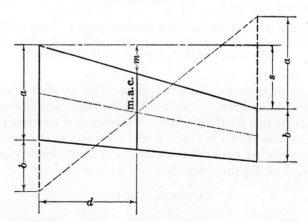

FIGURE 2-65.   Graphical solution for mean aerodynamic chord.

Since parasite drag varies with angle of attack, C_{D_p} can be expanded so that

$$C_D = C_{D_p \text{ min}} + K''C_L{}^2 + \frac{C_L{}^2}{\pi A}(1 + \delta) \qquad (2\text{-}82)$$

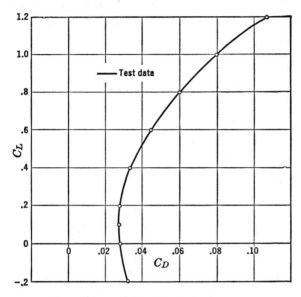

FIGURE 2–66.   Typical polar.

where $K''C_L{}^2$ represents the parabolic change of parasite drag. Since K'' and $(1 + \delta)$ both occur with $C_L{}^2$, they are usually combined so that equation (2–82) can be written as

$$C_D = C_{D_p \text{ min}} + \frac{C_L{}^2}{\pi A e} \qquad (2\text{-}83)$$

where e is sometimes defined as Oswald's efficiency factor or airplane efficiency factor and accounts for the variation of parasite drag with angle of attack and $(1 + \delta)$ in the induced drag term.

By replotting the test drag data from Figure 2–66 versus $C_L{}^2$ as in Figure 2–67, the deviation from a parabolic form is represented by the deviation from the dotted straight line. C_{D_f} is measured at the intersection of the straight line with the ordinate and is slightly lower in value than $C_{D_p \text{ min}}$, as the latter occurs (from the test data) at a value of C_L between .1 and .2. The majority of airplane polar curves can be approximated by a true polar which is mathematically expressed as

$$C_D = C_{Df} + \frac{C_L{}^2}{\pi A e} \qquad (2\text{–}84)$$

where C_{Df} and $1/\pi A e$ define the character of the drag coefficient curve for any angle of attack or lift coefficient. Only at very low or high values of lift coefficient does this approximation usually deviate from a typical polar.

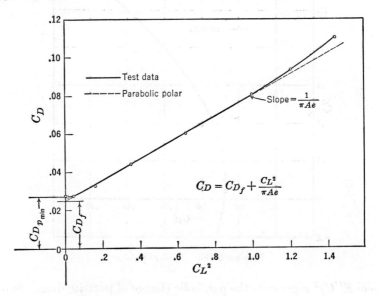

FIGURE 2–67. Parabolic polar approximation.

To obtain the value of C_{Df} for a particular airplane, the component drags are usually estimated. Since the drag of each component part may be based on a proper area, A_π, the equivalent parasite area, f, may be expressed as

$$f = C_{Df}S = C_{D\pi}A_\pi \qquad (2\text{–}85)$$

depending on the proper area used. The following table gives some average values of $C_{D\pi}$ for several components of the airplane. A_c is defined as maximum cross-sectional area. These data are for low values of Mach number and values of Reynolds number from two to six million.

The total f for the airplane is the sum of the component parts plus approximately 5 per cent of this sum to allow for mutual interference between the components. An additional allowance of 5 to 10 per cent

DRAG OF AIRPLANE COMPONENTS

Part	Description	C_{D_π}	Area on which C_{D_π} is based
Wing	Service roughness—t/c from 10 to 20 per cent	.005 to .009	S
Empennage	Service roughness—t/c from 8 to 12 per cent	.006 to .008	S_t
Fuselage	Streamline body—no excrescences	.05	A_c
Fuselage	Small plane with nose engine	.09 to .13	A_c
Fuselage	Large transport	.07 to .10	A_c
Fuselage	Bomber	.08 to .12	A_c
Nacelle (conventional)	Above wing on small airplane	.25	A_c
Nacelle (conventional)	Relatively small leading-edge nacelle on large airplane	.05 to .09	A_c
Nacelle (turbojet)	Mounted on wing	.05 to .07	A_c
Wing tanks	Suspended below wing tip	.10	A_c
Wing tanks	Centrally located at wing tip	.06	A_c
Wing tanks	Below wing, inboard (incl. support)	.19 to .21	A_c
Bomb	Suspended below wing (incl. support)	.22 to .25	A_c
Flaps	60 per cent span flaps deflected 30°	.02 to .03	S

should be made for such protuberances as antennas, radomes, and turrets. In Figure 2–68 is presented additional component data for landing gear, cowl flap, and wing flap drag. The effect of fineness ratio on C_{D_π} based on frontal or cross-sectional area is illustrated in Figure 2–69 for typical fuselage and nacelle shapes. For other shapes of a fuselage or nacelle the drag estimation should be based on wetted area as well as cross-sectional area. The value of drag coefficient, C_F, based on wetted area can be approximated from Figure 2–14 for smooth, flat surfaces. Wing profile drag coefficients for many airfoils are listed in the Appendix. The effect of variations in C_{D_0} with thickness ratio for a particular type of airfoil is also tabulated.

Airplane efficiency factor, e, may be estimated from equations (2–82) and (2–83). $C_{D_{p\min}} \cong C_{Df}$.

$$C_D = C_{Df} + K''C_L{}^2 + \frac{C_L{}^2}{\pi A}(1 + \delta) = C_{Df} + \frac{C_L{}^2}{\pi A e}$$

$$e = \frac{1}{K''\pi A + 1 + \delta} \qquad (2\text{–}86)$$

$$K'' = \frac{dC_{Df}}{dC_L{}^2} \cong .009 \text{ to } .012$$

For most airplanes the value of e ranges from .7 to .85.

A more thorough check on the value of f is obtained from flight test values of other airplanes on a wetted area chart. This chart (Figure 2–70) is simply a logarithmic plot of the equation

$$f = C_F A_W$$

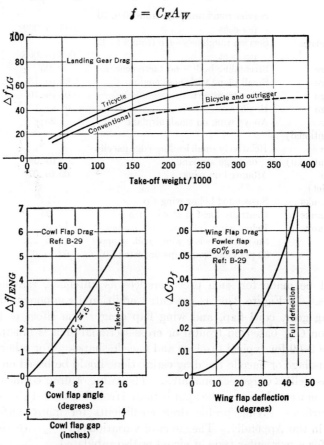

FIGURE 2–68. Drag of landing gear and flaps.

where A_W is the total surface or wetted area. The log scale is used only for convenience to cut down the size of the chart. In Figure 2–70 are plotted some approximate values for several military airplanes. No attempt is made to differentiate among different models of a particular type. By judicious comparison with airplanes of similar configuration, the designer can estimate the aerodynamic cleanness of a new type. By computing the wetted area of the new design he can then estimate from the chart the equivalent parasite area, f. It is recommended that the two methods of drag breakdown and wetted

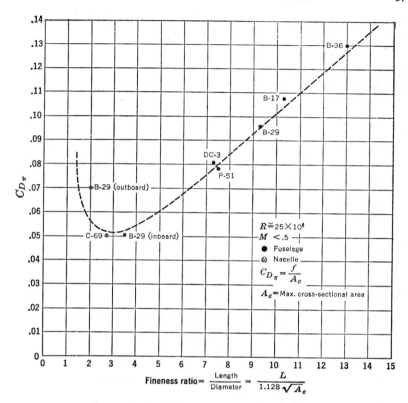

FIGURE 2–69.   Effect of fineness ratio on drag.

area comparison be used concurrently to give a good estimation of the equivalent parasite drag of the airplane.

High-speed Drag Estimation

Up to this point, only the performance of airplanes at relatively low Mach numbers has been discussed.   Since aircraft of today are traveling at supersonic speeds, some knowledge of the airplane drag variation associated with compressibility phenomena at Mach numbers approaching one and above one is required for performance estimation. In lieu of wind-tunnel data the following method* of estimating the drag coefficients of airplanes at Mach numbers approaching one is suggested.   At or near the critical Mach number of the airplane wing an abrupt change in drag coefficient occurs; this variation is presented in Figure 2–71 and is the average based on wind-tunnel data for several

* Army Air Forces Engineering MR TSEAL2-4589-12-1, dated Oct. 8, 1945.

representative military airplanes. ΔM in Figure 2–71 is equal to $M - M_{cr}{}'$, where $M_{cr}{}'$ is the average critical Mach number as defined by Figure 2–72. In this figure is represented a typical plot of M_{cr} versus C_L for an airplane wing section. The $M_{cr}{}'$ line is a straight line through the highest value of M_{cr}, point A, with half the slope of

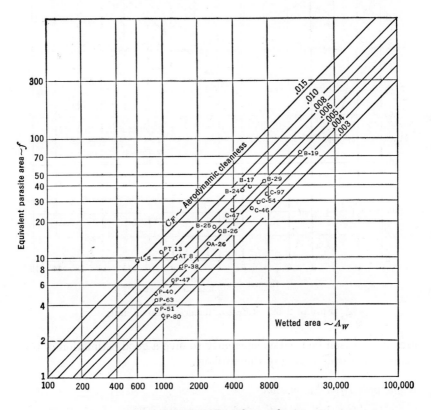

FIGURE 2–70. Wetted area chart.

the straight line variation of M_{cr} with C_L. The values of M_{cr} on which the correction is based are obtained from theoretical low-speed pressure distributions corrected to high Mach numbers by the Karman-Tsien formula and are arbitrarily taken as those of the airfoil section at the root of the airplane wing. ΔC_D in Figure 2–71 is equal to $C_D - C_{D_{inc}}$, where $C_{D_{inc}}$ is the drag coefficient of the airplane, neglecting compressibility effects. Since

$$D = C_D q S$$

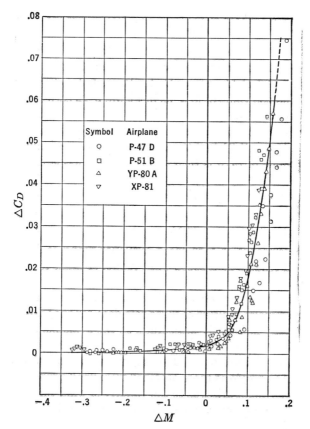

FIGURE 2–71.   Compressibility correction curve.

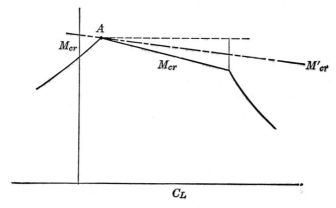

FIGURE 2–72.   Typical M_{cr} versus C_L curve.

the drag at high speeds is obtained by determining the drag coefficient at high speed in the following manner:

1. Solve for C_L and M corresponding to the chosen speed. Determine $C_{D_{inc}}$ from airplane polar or equation (2-84).

2. From a plot of M_{cr} versus C_L for the airfoil section at the root of the wing, find the value of M_{cr}' as in Figure 2-72 corresponding to the C_L calculated in (1).

3. Solve for ΔM from the relation $\Delta M = M - M_{cr}'$.

4. From Figure 2-71 find the value of ΔC_D corresponding to ΔM.

5. This value of ΔC_D is then added to the low-speed or incompressible drag coefficient, $C_{D_{inc}}$, to give the drag coefficient corrected for compressibility at the chosen speed.

This method is applicable over the range of lift coefficients normally encountered in high-speed flight and is recommended for use in estimating performance of high-speed airplanes flying at Mach numbers below one if reliable high-speed wind-tunnel or flight test data are not available.

In Figure 2-73 is presented the drag rise due to compressibility from a preliminary analysis of flight test data for the Gloster Meteor.* The curve may be slightly inaccurate, as the available thrust in flight is very difficult to calibrate with respect to pressures and temperatures in the diffuser and tail pipe. It is significant to note that these drag rises are not as severe as those obtained from wind-tunnel data (Figure 2-71). Some doubt arises as to the validity of wind-tunnel corrections at high Mach numbers where blocking effects in the tunnel are appreciable.

Estimating performance data through the transonic region are at present no more than a good guess because of the lack of reliable test data and theoretical calculations of flow phenomena in this region. Wind-tunnel data are usually restricted to a Mach number less than .95 or greater than 1.1 because of the large corrections associated with blocking effects in the test section. The National Advisory Committee for Aeronautics has obtained some aerodynamic data through a Mach number of one by dropping scale models from high altitudes, thus simulating the flight of an actual body through the transonic region under free-air conditions. Drag data of the wing or tail are obtained by connecting component parts to the body through spring balances and telemetering the deflection of these to the ground. As a further check, a time history of the speed of the body is obtained by radar optical tracking equipment. In addition to the free-fall

* *Jour. Royal Aero. Society*, May 1946.

technique rocket-powered models are sometimes used. By another method data are obtained by recording the forces on a wing model mounted vertically on the surface of an actual airplane wing over which the local flow exceeds a Mach number of one.

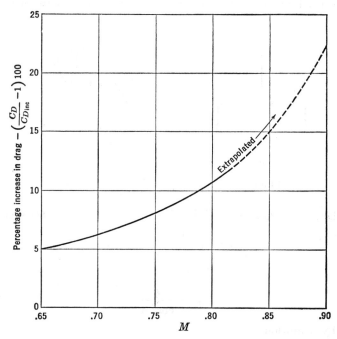

FIGURE 2–73.   Drag rise with Mach number.

Although the following theoretical conclusions for the determination of drag data through the transonic-speed region are only approximations, their validity is substantiated in most cases by the available test data. From Figure 2–71 the drag rise for the complete airplane with straight wings can be approximated. It is assumed in this case that the drag rise for the fuselage is negligible for the values of Mach number considered. For an airplane with swept wings the drag rises of the fuselage and wing usually occur nearer the same Mach number. In the following discussion the drag rise on the wing and on the fuselage will be discussed separately.

Consider first the variation in aerodynamic coefficients between a two-dimensional swept and straight wing in a potential flow field as illustrated in Figure 2–74. The component of speed, V, perpendicular to the leading edge is the same for both cases. For the swept wing the speed component parallel to the leading edge has no effect on the lift and drag pressure forces as the air flows along parallel lines of constant elevation. The pressure distribution is affected only by the flow perpendicular to the leading edge. It is apparent then that the free-stream speed for the swept wing, V_s, is increased by a factor of $1/\cos \Lambda$

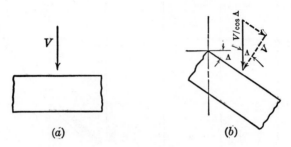

FIGURE 2-74.   Two-dimensional swept wing.

over the free-stream speed of the straight wing for equal aerodynamic forces.

$$V_s = \frac{V}{\cos \Lambda}, \quad M_{cr_s} = \frac{M_{cr}}{\cos \Lambda} \qquad (2\text{-}87)$$

$$q_s = \frac{q}{\cos^2 \Lambda} \qquad (2\text{-}88)$$

Since by definition

$$C_L = \frac{L}{qS} \quad \text{and} \quad C_D = \frac{D}{qS}$$

$$C_{L_s} = C_L \cos^2 \Lambda \qquad (2\text{-}89)$$

and

$$C_{D_s} = C_D \cos^2 \Lambda \qquad (2\text{-}90)$$

Also, by the geometry of the airfoil sections parallel to the free stream and perpendicular to the leading edge in Figure 2-74,

$$\alpha_{0_s} \cong \alpha_0 \cos \Lambda$$

and thus

$$\left(\frac{dC_L}{d\alpha}\right)_{0_s} = a_{0_s} = a_0 \cos \Lambda \qquad (2\text{-}91)$$

Three-dimensionally

$$a_s \cong a \cos \Lambda \qquad (2\text{-}92)$$

for high values of aspect ratio and

$$a_s \cong a\sqrt{\cos \Lambda} \tag{2-93}$$

for low values of aspect ratio.

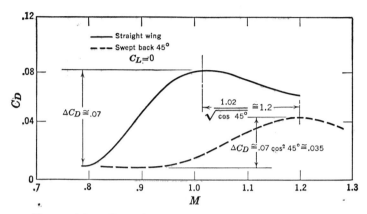

FIGURE 2-75.   Typical drag data for swept and straight wing.

Three-dimensional test data on swept-back wings indicate that the critical Mach number is increased over a straight wing by a factor of approximately $1/\sqrt{\cos \Lambda}$.

$$M_{cr_s} = \frac{M_{cr}}{\sqrt{\cos \Lambda}} \tag{2-94}$$

The full cosine effect as indicated by equation (2-87) is not realized because of boundary-layer and tip effects associated with the actual test conditions.   To build up the drag increase for a swept-back wing from straight wing data, equation (2-90) is valid for ΔC_D increases.

$$\Delta C_{D_s} = \Delta C_D \cos^2 \Lambda \tag{2-95}$$

taken at values of Mach number shifted to the right as in Figure 2-75 by

$$M_s = \frac{M}{\sqrt{\cos \Lambda}} \tag{2-96}$$

In this figure it is noticed that equations (2-95) and (2-96) apply up to the C_D peak for a typical 10 per cent thick wing at zero lift coefficient and 45° sweepback.

In Figure 2-76 is plotted the approximate peak C_D for wings of various per cent thickness for a lift coefficient equal to zero.   Very

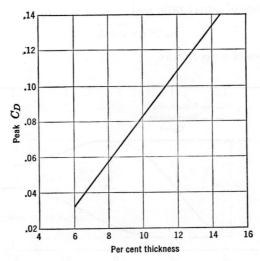

FIGURE 2–76. Peak C_D for unswept wings at $C_L = 0$.

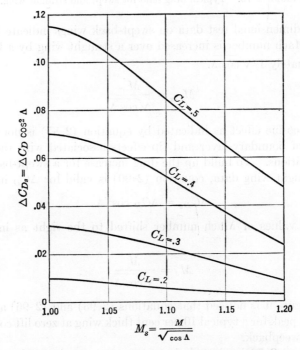

FIGURE 2–77. Peak drag rise due to angle of attack.

little information is available from tests to indicate the peak C_D of wings for values of C_L above zero.

Figure 2-77 represents an approximate variation in ΔC_D (above the peak C_D for C_L equal to zero) for various angles of sweepback.

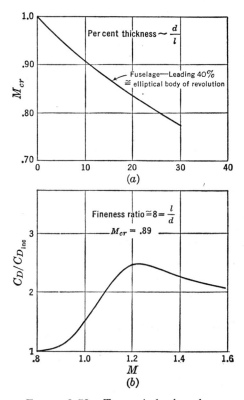

FIGURE 2-78.   Transonic fuselage drag.

Beyond the Mach number at which the peak C_D occurs it is assumed that C_D falls off approximately as the $\sqrt{M^2 - 1}$, as indicated by theoretical calculations at supersonic speeds.

To determine the complete drag of the airplane in the transonic speed region it is necessary to add the fuselage drag to that determined for the wing. Figure 2-78b represents an approximate variation of $C_D/C_{D_{inc}}$ versus M for a typical fuselage shape of fineness ratio 8 and $M_{cr} \cong .89$. For any other value of fineness ratio or thickness ratio a different M_{cr} is estimated from Figure 2-78a. The curve in Figure 2-78b is shifted parallel to the abscissa by the difference between the new M_{cr} and $M_{cr} \cong .89$. The effect of C_L on C_D is considered negligible for fuselage bodies.

In general the estimation of the peak C_D for airplanes in the transonic region is only approximate. In the supersonic speed region theory and test data agree closely. The following presentation of

$$C_D = \frac{K_1(t/c)^2}{\sqrt{M^2-1}} + C_{D_f} + \frac{4\alpha^2}{\sqrt{M^2-1}}\left[1 - \frac{1}{2A\sqrt{M^2-1}}(1-C_3A')\right]$$

$$\text{Drag} = \begin{array}{c}\text{Drag due to} \\ \text{airfoil section} \\ \text{and thickness}\end{array} + \begin{array}{c}\text{Skin} \\ \text{friction}\end{array} + \begin{array}{c}\text{Drag due to} \\ \text{angle of attack}\end{array} + \begin{array}{c}\text{Drag due to} \\ \text{tip effect}\end{array}$$

$$C_D = C_{D_t} + C_{D_f} + C_{D_i}$$

$$C_L = \frac{4\alpha}{\sqrt{M^2-1}}\left[1 - \frac{1}{2A\sqrt{M^2-1}}(1-C_3A')\right]$$

$$C_m \text{ (about (L.E.))} = \frac{2\alpha}{A(M^2-1)}\left[A\sqrt{M^2-1} - \tfrac{2}{3} - C_3A'(A\sqrt{M^2-1}-1)\right]$$

$$\text{c. p.} = \left[\frac{A\sqrt{M^2-1} - \tfrac{2}{3} - C_3A'(A\sqrt{M^2-1}-1)}{2A\sqrt{M^2-1} - 1 + C_3A'}\right]c$$

where $A' = \dfrac{\text{Airfoil cross-sectional area}}{\text{chord squared}}$ $C_3 = \dfrac{\gamma M^4 + (M^2-2)^2}{2(M^2-1)^{3/2}}$

     $A = \text{Aspect ratio}$

Type of airfoil	A'	K_1	Section
Double wedge	$\tfrac{1}{2}\,t/c$	4	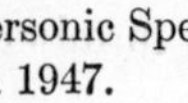
Modified double wedge	$\tfrac{2}{3}\,t/c$	6	
Biconvex	$\tfrac{2}{3}\,t/c$	5.33	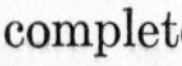

FIGURE 2–79. Wing aerodynamic characteristics at high speed (rectangular planform, sweep = 0°). Reproduced by permission from Bumblebee Series, Report 55, "Aerodynamic Characteristics of Wings at Supersonic Speeds," by R. M. Snow and E. A. Bonney, March 1947.

supersonic aerodynamic characteristics will complete the drag estimation picture through a wide range of airplane speeds.

Similarly to subsonic flow, the drag coefficient of wings at supersonic speeds may be divided into two components, one independent of the lift coefficient and one proportional to the square of the lift coefficient or angle of attack.

$$C_D = C_{D_0} + C_{D_i} \tag{2-97}$$

In this case, however, the so-called induced drag coefficient, C_{D_i}, has no simple relation to aspect ratio, as is indicated in Figure 2–79, but is a complex function of wing planform and load distribution. C_{D_i} retains the subscript i to indicate a similarity in variation with the

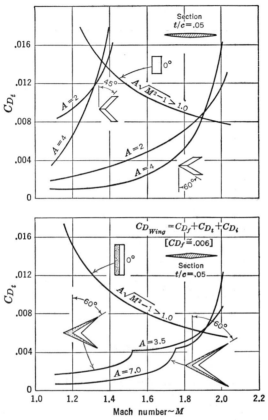

FIGURE 2–80. Thickness drag. From NACA TN 1350, "Estimated Lift-drag Ratios at Supersonic Speed," by Robert T. Jones.

$C_L{}^2$ as in the subsonic case. C_{D_0} in equation (2–97) is defined as

$$C_{D_0} = C_{D_t} + C_{D_f} \qquad (2\text{–}98)$$

where C_{D_t} is the wave drag of the section at zero lift and is greatly affected by thickness ratio. C_{D_f} is the friction drag and is usually estimated for a turbulent boundary-layer flow condition. In Figure 2–79 are summarized the important wing aerodynamic characteristics for three typical supersonic wing sections.

In Figure 2–80 are presented theoretical data for the thickness drag of various wing sections and planform shapes. The curves for the rectangular wings may be computed from the equation for C_{Dt} given in Figure 2–79. The thickness drag coefficient in this case is constant if

$$A\sqrt{M^2 - 1} > 1$$

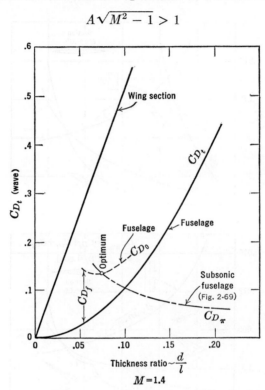

FIGURE 2–81. Comparison of wave drags of two- and three-dimensional bodies. From NACA TN 1081, "Flow over a Slender Body of Revolution at Supersonic Velocities," by Robert T. Jones and Kenneth Margolis.

and the Mach cone for each tip does not intersect the opposite tip. When the wing is swept back well beyond the Mach cone angle, subsonic flow exists over most of the wing and the drag coefficient varies inversely with aspect ratio. As the Mach cone approaches the leading edge at higher Mach numbers, the drag coefficient increases rapidly near the wing tips and the drag of the swept wing becomes greater than the straight wing.

According to Jones in NACA TN 1350, a value of C_{Df} for the wing may be assumed equal to .006. For fuselages this coefficient based on wetted area is approximately .0021 for a fully turbulent boundary layer at $R = 10^8$.   The friction drag is of the same order of magnitude as the thickness drag for 7 to 10 per cent sections.

In Figure 2–81 are plotted values of thickness or wave drag coefficient, based on cross-section area, for a typical wing section and a streamline body of revolution constructed from parabolic arcs.   For values of thickness ratio, d/l, between .05 and .1, the C_{D_f} is approximated and added to C_{D_t} of the fuselage, giving the total C_{D_0} of the fuselage.   It follows that C_{D_0} is a minimum when C_{D_f} is equal to twice C_{D_t}, as the former varies inversely with d/l and the latter directly with $(d/l)^2$ for a given cross-section.   For a supersonic design, then, a value of d/l slightly above .06 is optimum.   Cross-plotted on Figure 2–81 are values of C_{D_π} for subsonic flow of a typical fuselage shape taken from Figure 2–69.   It follows that an optimum d/l for both subsonic flow and supersonic flow is obtained near the intersection of the curves of C_{D_0} and C_{D_π}.   This value of d/l is slightly higher than the optimum for supersonic flow alone.

The drag coefficient due to angle of attack or lift, C_{D_i}, can be calculated from Figure 2–79 for a rectangular planform and various airfoil sections.   The effect of airfoil thickness, A', on C_{D_i} is very small, and for a flat-plate rectangular wing C_{D_i} reduces to

$$C_{D_i} = \frac{4\alpha^2}{\sqrt{M^2 - 1}} \left(1 - \frac{1}{2A\sqrt{M^2 - 1}} \right) \qquad (2\text{–}99)$$

Since lift acts perpendicular to the chord

$$C_{D_i} = C_L\alpha = \frac{C_{D_i}}{C_L^2} C_L^2 \qquad (2\text{–}100)$$

and

$$\frac{C_{D_i}}{C_L^2} \frac{1}{\sqrt{M^2 - 1}} = \frac{1}{4\left(1 - \dfrac{1}{2A\sqrt{M^2 - 1}} \right)} \qquad (2\text{–}101)$$

In Figure 2–82 the left side of equation (2–101) is plotted against $A\sqrt{M^2 - 1}$ for various planform shapes.   At a Mach number of 1.4 and aspect ratio of 6, C_{D_i} is about five times the subsonic value at the same value of C_L.

To determine the maximum values of L/D for supersonic aircraft, equation (2–97) is rewritten as

$$C_D = C_{D_0} + \frac{C_{D_i}}{C_L^2} C_L^2 \qquad (2\text{–}102)$$

Dividing by C_L and differentiating,

$$d\left(\frac{C_D}{C_L}\right) = -\frac{C_{D_0}}{{C_L}^2} + \frac{C_{D_i}}{{C_L}^2} \qquad (2\text{–}103)$$

Setting the derivative equal to zero, the optimum value of lift coef-

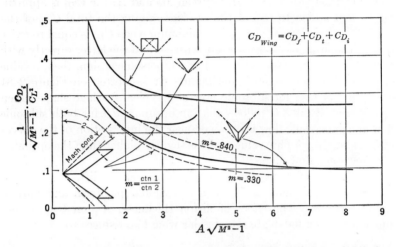

FIGURE 2–82. Drag due to lift. From NACA TN 1350, "Estimated Lift-drag. Ratios at Supersonic Speed," by Robert T. Jones.

ficient, $C_{L_{opt}}$, for maximum L/D and the value of maximum L/D result:

$$C_{L_{opt}} = \sqrt{\frac{C_{D_0}}{C_{D_i}/{C_L}^2}} \qquad (2\text{–}104)$$

and

$$\left(\frac{L}{D}\right)_{max} = \frac{1}{2}\sqrt{\frac{1}{C_{D_0}(C_{D_i}/{C_L}^2)}} \qquad (2\text{–}105)$$

In Figure 2–83 are presented values of $(L/D)_{max}$ at supersonic speeds for various airplane configurations with wings of finite thickness. Ten per cent of the wing drag is allowed for the drag of the tail surfaces. It is significant to note that reasonably high values of L/D can be obtained for long-range supersonic aircraft if careful attention is given to the design parameters at high speed. From equation (2–57),

$$\frac{W}{\delta S} = 1481 M^2 C_L$$

If reasonable values of W/S are to be obtained for flight at high Mach numbers and at the $C_{L_{opt}}$ corresponding to $(L/D)_{max}$, it is seen that low values of pressure ratio or extreme flight altitudes are necessary for maximum range performance.

Methods for estimating drag for subsonic, transonic, and supersonic flight have now been discussed. When a design has reached the wind-

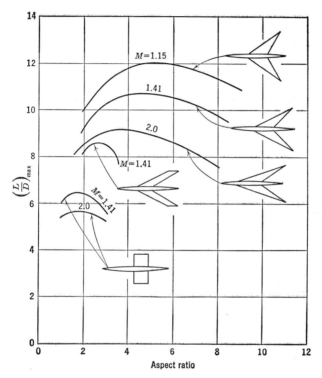

FIGURE 2–83. $(L/D)_{max}$ at supersonic speeds. From NACA TN 1350, "Estimated Lift-drag Ratios at Supersonic Speed," by Robert T. Jones.

tunnel stage, actual drag measurements are made to investigate the potentialities of the configuration. Because of difficulties in simulating free flight conditions on a model in a tunnel, and in correcting tunnel data for blocking and wall effects, the absolute magnitude of the drag in the tunnel may be erroneous. However, increment changes in drag due to the addition or replacement of component parts of the model are very accurate and of invaluable aid to the engineer in refining a design. The final phase of drag evaluation on a design is obtained by actual flight test measurements. The accuracy of these measurements is dependent on the flight test equipment used, the pilot technique, and accuracy in reducing the flight data to standard atmospheric conditions.

PROBLEMS

2–1. Find the density ratio of dry air at 12.9 in. Hg and −27.6°C by the laws of Charles and Boyle. If this is a standard atmosphere condition, determine the altitude, using the standard lapse rate. Check the density ratio from the Appendix.

2–2 *a.* For standard atmosphere conditions above the tropopause derive a general expression for:

$$(1)\ \text{Pressure ratio,}\ \frac{p}{p_0},\ \text{in terms of}\ p_t,\ p_0,\ R,\ T_i,\ h,\ h_t$$

where p_t = pressure at the tropopause.

h_t = height of tropopause in feet.

T_i = isothermal temperature, degrees absolute.

$$(2)\ \text{Density ratio,}\ \sigma,\ \text{in terms of}\ \frac{p}{p_0},\ T_0,\ T_i$$

b. Calculate (1) and (2) for 50,000 ft and check with the Appendix.

2–3. A 3000-lb airplane is flying in level flight at a constant speed of 125 mph. The altimeter reads 5000 ft, and the temperature is 90°F. If the thrust required to maintain this altitude and speed is 300 lb, find:

(1) Lift and drag.
(2) Density altitude.
(3) Density ratio.

2–4. Assume that forces on an object moving through a very viscous fluid are of the form

$$F = KV^b\mu^c l^d$$

where K, b, c, and d are constants. By dimensional analysis determine the values of the exponents b, c, and d and write the force equation.

2–5. Sound is propagated through the air by means of compression waves. The velocity of sound, a, varies with the modulus of elasticity and mass density of the air; e.g., $a = E^c\rho^d$, where c and d are constants and E and ρ have consistent units. By dimensional analysis determine the constants c and d and solve for the velocity of sound in miles per hour at sea level under standard conditions. E for air under standard conditions at sea level = 2970 lb/ft².

2–6. *a.* From the theory of dimensional analysis show how the airforce on a body moving relative to the air varies with the following parameters:

$$F = KV^b\rho^c l^d\mu^e a$$

where a = velocity of sound in air, and K, b, c, d, e, and f are constants. Assume F varies principally with V, l, and ρ.

b. Repeat part *a*, assuming F varies principally with V, l, and μ.

2–7. Find Reynolds number for an airplane wing tested under the following conditions:

Chord = 11 in.

V = 250 mph

Temperature = 50°C

Pressure = 7 atmospheres (7 times standard pressure at sea level)

Standard ρ/μ = 6380 sec/ft^2

Assume μ is independent of pressure but varies directly with the ¾ power of the absolute temperature for gases.

2–8. An airplane having characteristic three-component data as indicated in Figure 2–4 weighs 3000 lb. The minimum landing speed at sea level under standard conditions is 55 mph. If the propeller efficiency is 80 per cent, find the brake horsepower developed by the engine to keep the plane flying level at a true air speed of 120 mph at 10,000 ft standard altitude.

2–9. An airplane having a rectangular wing (NACA 4412 airfoil section) of aspect ratio 6 weighs 6000 lb. If the wing area is 300 sq ft, what thrust horsepower is required for the wing when flying 160 mph at 10,000 ft standard altitude?

2–10. *a.* Given airfoil data:

$$A = 6, \quad \alpha = 10°, \quad C_L = 1.0, \quad C_D = .065$$

If this airfoil section is used on a glider at the same lift coefficient, and $A = 12$ W = 400 lb, S = 100 ft^2, find the D, L/D, α, normal force, and chord force of the glider wing.

b. If this airfoil section is used on a monoplane at the same lift coefficient, and $A = 8$, $W = 3600$ lb, $S = 400$ ft^2, find the D and L/D of the monoplane wing.

2–11. Given an NACA 2212 airfoil to be used on a tailplane with $A = 3$, $\alpha_{0L} = -1.8°$. Find the tail lift when $\alpha_t = +5°$, V at tail = 100 mph, $S_t = 50$ ft^2, $dC_L/d\alpha$ for $A = 6 = a_6 = .0753$/degree.

2–12. *a.* Determine the maximum boundary-layer thickness and skin-friction drag per foot of span for a smooth and very thin airfoil with chord equal to 10 ft at the following speeds at standard sea level conditions:

(1) 10 mph (assume laminar flow).

(2) 1000 mph (assume turbulent flow for entire chord length).

b. Plot thickness of boundary layer versus chord length for (1) and (2).

2–13. $C_{m0.25} = C_{mac} = -.08$ for a particular airfoil: $C_{L\max} = 1.6$. Plot the C_p versus C_L for $C_{L\max}$ and ±⅛, ±¼, and ±½$C_{L\max}$.

2–14. The following data are taken from a wind-tunnel test:

C_L	$C_{m0.25}$
.2	−.114
.6	−.102

a. Find the C_p at $C_L = .8$.

b. Determine the C_L at which the wing just balances in the airstream. Assume axis of drag force on wing acts through the hinge point.

c. Is this condition stable?

2–15. A 180-lb wing is glided from a cliff against a 10-mph headwind and lands 700 ft from the base of the cliff in 18.3 sec.

Wing data: b = 18 ft, c = 3 ft, $A = 6$, NACA 2412 airfoil

Find: (1) The air speed of the wing.

                     (2) The height of the cliff.

2-16. An airplane model is tested in a wind tunnel under the following conditions:

$$V = 75 \text{ mph}, \quad T = 80°\text{F}, \quad p = 27.5 \text{ in. Hg}$$

Test results: $L = 6.8$ lb, $D = .51$ lb, $M_0 = -8.35$ in.-lb

Wing characteristics: $b = 24$ in., $c = 4$ in., $\dfrac{dC_L}{d\alpha_0} = a_0 = .10/\text{degree}$,

$$\alpha_{0L} = -4°, C_{mac} = -.068$$

Find: (1) C_L.
          (2) C_{D_p}.
          (3) $C_{m.25}$
          (4) Aerodynamic center.
          (5) α.
          (6) α_0.

2-17. Find the pressure coefficient, S, at the station $x = .5$ on the upper and lower surfaces of the NACA 65,2-215 $a = .5$ airfoil at a lift coefficient of .2. Plot a curve of c_l versus calculated M_{cr} for both upper and lower surfaces.

2-18. An airplane is in a steady glide at a C_L of .7 and has a sinking speed of 300 ft/min. $A = 8$. A previous wind-tunnel test of the same airplane, with the one exception that the aspect ratio was equal to 6, had given a $C_D = .04$ at $C_L = .7$.

Find: The wing loading of the plane. (Assume standard sea level conditions and $L \cong W$.)

2-19. An airplane weighing 3000 lbs requires 75 hp to take off at its stalling speed of 50 mph at Denver (5000 ft standard altitude). If 500 lb fuel are burned in flying to Cheyenne (6000 ft standard altitude), what is the minimum landing speed of the plane at Cheyenne? What is the horsepower required to fly in level flight at this landing speed at Cheyenne?

2-20. Given the following wing characteristics:

> NACA 2412 section (Figure 2-20)
> $S = 300$ ft^2
> $A = 6$
> Taper ratio $= c_t/c_r = .5$
> Aerodynamic twist $= -2°$ (washout)
> 30% chord split flap from .3 semispan to .6 semispan
> Sweep of leading edge only equals 20 degrees

a. Determine the following three-dimensional wing characteristics with flaps deflected 30 degrees:

> (1) Slope of lift curve.
> (2) Angle of zero lift.
> (3) Aerodynamic center.
> (4) Pitching moment about a.c.

b. Determine length and position of mean aerodynamic chord.

2-21. Given that $C_D = C_{D_f} + \dfrac{C_L{}^2}{\pi A e}$, where C_{D_f} and e are constants.

a. Find the ratio D/L as a function of C_L.
b. Find C_L for minimum D/L which corresponds to maximum L/D ratio.

c. Find maximum L/D and ratio of (effective induced/parasite) drag at maximum L/D.

2-22. Given the following airplane characteristics:

$$f = 3.3 \text{ (entire airplane)}$$
$$S = 250$$
$$A = 6$$
$$e = .8$$
$$\Lambda = 0$$

Airfoil as in Problem 2-17
Frontal area of fuselage $= 30 \text{ ft}^2$
Fineness ratio $= 7$

a. Plot curves of airplane C_D versus M from $M = .25$ to $M = 1.2$ for values of C_L equal to .1, .2, and .3.

b. If weight of airplane equals 10,000 lb, plot curves of C_L, C_D, L/D, and D versus M from $M = .25$ to $M = 1.2$. Assume airplane is in level flight at 35,000 ft altitude.

2-23. Given the following missile characteristics: $S = 100 \text{ ft}^2$, biconvex airfoil, $t/c = .05$, $A = 4$, zero taper, $\Lambda = 60$ degrees, parabolic fuselage of revolution, maximum diameter $= 2$ ft, fineness ratio $= 15$, power plant contained in fuselage.

Plot C_D, C_L, and L/D from:

$$M = .5 \text{ to } M = 2 \text{ for uniform level flight at 60,000 ft altitude}$$

Assume $W/S = 30$.

CHAPTER 3

PROPULSION

3-1 Power-plant Efficiency

As was shown in Chapter 2, the performance of an airplane depends on the magnitude of the forces acting on the airplane in some attitude of flight. In uniform level flight the drag is equal to the thrust and the lift is equal to the weight. In order that the airplane accelerate or climb, the thrust must be greater than the drag. Some knowledge of the nature of the lift, weight, and drag forces has been obtained by previous discussions. Before a performance analysis of the airplane can be attempted, it is necessary that the magnitude and direction of the thrust force be known. The nature of this force, of course, depends on the type of power plant selected for propulsion. It is the purpose of this chapter to discuss some of the fundamental concepts of propulsion and performance characteristics for several different types of power plants.

In recent years the practical methods of propulsion have increased to include in addition to the reciprocating engine as a power plant four other basic types: turbine propeller (turboprop), turbojet, ramjet, and rocket, characterized by different features of construction, operation, and performance. In Figure 3-1 are presented typical performance data for these different types, including schematic diagrams and performance curves. All these power plants produce thrust by Newton's fundamental principle that the net force on a body is proportional to the mass per unit time and the amount the velocity of the mass is increased with respect to the body. All these propulsive devices can be considered as heat engines that convert fuel energy into useful work done on the airplane. The over-all efficiency of each power plant can be expressed as the per cent of total heat energy of fuel and oxidizer that is converted into useful work.

$$\eta_o = \text{Over-all efficiency} = \frac{\text{Useful work done on plane}}{\text{Heat energy of fuel and oxidizer}} \quad (3\text{-}1)$$

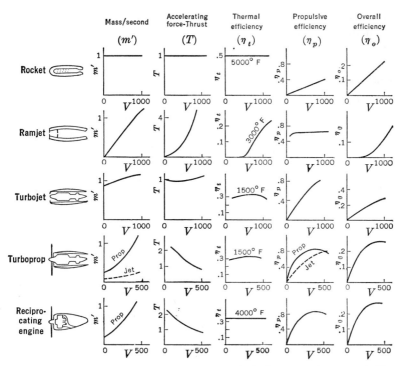

FIGURE 3–1.   Power-plant characteristics.   Method of presentation suggested by Westinghouse "Jet Propulsion" brochure, B3834, 5M-2-47.

This efficiency can be broken down further as a multiple of thermal efficiency, η_t, and propulsive efficiency, η_p, where

$$\eta_t = \frac{\text{Mechanical energy produced in system}}{\text{Heat energy of fuel and oxidizer}} \qquad (3\text{-}2)$$

and

$$\eta_p = \frac{\text{Useful work done on plane}}{\text{Mechanical energy produced in system}} \qquad (3\text{-}3)$$

Again, the over-all efficiency is the net measure of the power-plant utility and is expressed as

$$\eta_o = \eta_t \eta_p \qquad (3\text{-}4)$$

In short, all propulsion devices consist of a heat engine that converts fuel energy into mechanical energy in combination with a system for converting the mechanical energy into useful thrust.   The parameters presented in Figures 3–2 and 3–3 depend on the over-all efficiency of each power plant and permit a basis by which each type can be fairly compared. These data are representative of average power plant characteristics.

3–2   Power-plant Data

Rocket

Perhaps the most fundamental type of reaction power plant is the rocket (Figure 3–1). It is not dependent upon atmospheric oxygen because both the fuel and oxidizer are carried within the system. In its basic form the rocket consists of a combustion chamber in which is contained either separate supplies of liquid fuel and oxidizer or both

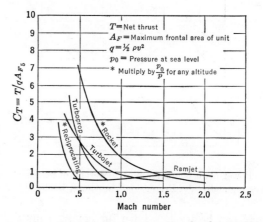

FIGURE 3–2.   Propulsion comparison (sea level to 35,000 ft).

combined as a solid substance, a means for ignition, and a nozzle through which the heat energy of combustion is converted into mechanical energy. Solid-fuel rockets are sometimes used to propel missiles and assist airplanes during take-off. This type has the advantages over the liquid-fuel type of simplicity, safety in handling, and ease of storage. It has, however, the disadvantages of lower energy and jet velocity, loss of control once combustion begins, and maximum efficiency for only one condition of the variable-size combustion chamber, which consists of a cavity in the fuel. Rockets as aircraft prime movers utilize liquid fuels and oxidizers which are pumped into the combustion chamber from tanks. A typical example of this type is the motor that propels the German V-2 missile. The fuel is usually any of the common petroleum products such as gasoline,

kerosene, or acetylene. Alcohol and liquid hydrogen are other possibilities. The oxidizer is usually liquid oxygen. Aniline and nitric
acid form another usable combination. Some other disadvantages of
the liquid-fuel type are danger in handling and extremely high combustion temperatures which impose rigorous metallurgical problems.

In a rocket engine the rate of fuel burning and jet velocity are approximately constant for all speeds. Hence the rocket thrust, fuel
consumption, and thermal efficiency are essentially constant for all
speeds. Thrust and fuel consumption improve slightly with altitude,

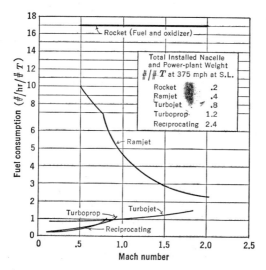

FIGURE 3–3. Fuel-consumption comparison (sea level to 35,000 ft).

since reduced atmospheric pressure permits a greater expansion of the
exhaust gases in the nozzle and hence a greater utilization of heat
energy. Because the amount of fuel and oxidizer that can be carried
is limited by weight, maximum thrust utility can be developed only
by the use of fuels of maximum energy release (high temperature per
unit weight), which produce a high jet velocity. At values of Mach
number below one the thrust capabilities of the rocket are considerably
greater than any of the other power-plant types. See Figure 3–2.
At high altitudes the thrust capabilities are superior for all speeds.

From equation (3–2) the thermal efficiency of the rocket can be
expressed as

$$\eta_t = \frac{\dfrac{w}{2g} v_w^2}{J H_f w_f} \tag{3–5}$$

where w = weight of fuel and oxidizer per second.

w_f = weight of fuel consumed per second.

v_w = velocity of wake or jet in feet per second.

J = mechanical equivalent of heat = 778 (1 Btu = 778 ft-lb).

H_f = heat value of fuel = 18,500 Btu per lb.

The rocket, having no highly stressed moving parts, permits operation at high temperatures and pressures. A high thermal efficiency is obtained because high pressure energy is built into the oxidizer before it is placed in the rocket. In other engines the oxidizer (air) is compressed during the cycle.

From equation (3–3) the propulsive efficiency of the rocket can be expressed as the percentage of mechanical energy produced in the system that is converted to useful work. Thrust is equal to $\dfrac{w}{g} v_w$ as both the fuel and oxidizer experience a change in speed equal to v_w. The kinetic energy input is equal to $\dfrac{1}{2} \dfrac{w}{g} (v_0^2 + v_w^2)$. In this case $\dfrac{1}{2} \dfrac{w}{g} v_0^2$ represents the energy that has been required to bring the fuel and oxidizer up to the flight speed of the rocket. Although equation (3–6) is not a rigorous expression for propulsive efficiency of a complete rocket flight, it is representative of an instantaneous condition and may be used to compare with the expression later developed for the ideal propulsive efficiency of a propeller, turbojet, or ramjet.

$$\eta_p = \frac{Tv_0}{KE \text{ of system}} = \frac{\dfrac{w}{g} v_w v_0}{\dfrac{w}{2g}(v_w^2 + v_0^2)}$$

$$= \frac{2\dfrac{v_0}{v_w}}{1 + \left(\dfrac{v_0}{v_w}\right)^2} = \frac{2\dfrac{v_w}{v_0}}{1 + \left(\dfrac{v_w}{v_0}\right)^2} \tag{3–6}$$

where v_0 = velocity of airplane in feet per second.

v_w = velocity of wake or jet in feet per second.

T = thrust in pounds.

w = weight of fuel and oxidizer per second.

A plot of equation (3–6) is presented in Figure 3–4. At zero airplane speed all the energy of the rocket is in the jet, $(v_w/v_0) = \infty$, and the η_p is zero. As the speed of the airplane increases, the energy is divided between the rocket and the wake or jet. When $v_0 = v_w$, or $v_w/v_0 = 1$, all the energy is utilized for propelling the airplane and the ideal

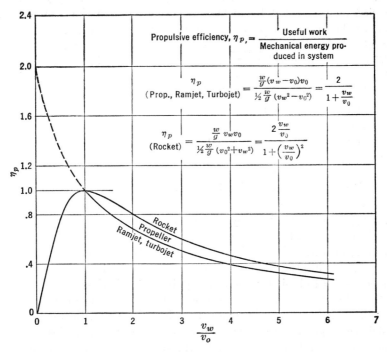

FIGURE 3–4. Propulsive efficiency versus v_w/v_0.

η_p is 100 per cent. For this condition no energy is consumed for propelling the particles in the wake, as their velocity is zero with respect to the earth. If $v_0 > v_w$, the η_p begins to decrease. Only for very high-speed flight conditions does the ratio v_w/v_0 become less than one. Since thrust may be produced by various combinations of jet velocity and mass, the η_p is a maximum for a given thrust if the jet velocity is a minimum, since the mechanical energy produced in the system varies as the first power of the mass and the second power of the jet velocity. The rocket η_p is necessarily low as a result of the high jet velocity required. Rocket performance is characterized by a low power-plant weight and form drag and high fuel consumption per pound of thrust. See Figures 3–2 and 3–3. The consumption of fuel alone would be much lower than that indicated in Figure 3-3 if the consumption of the oxidizer were ignored.

Ramjet

Since the largest single component by weight of the rocket is the oxidizer, attempts have been made to lighten the weight penalty carried by the rocket by utilizing the oxygen from the atmosphere. One of the most basic types of engines utilizing the air is the ramjet or athodyd. This thermal-jet engine is essentially an expanding diffuser entry in which the air compresses, a tube-shaped combustion chamber into which gasoline is forced, and a nozzle exit. Only at high speeds is sufficient compression obtained by ram to permit operation. At transonic speeds and above, the specific fuel consumption approaches that for a turbojet. Power-plant weight per pound of thrust is low. Since negligible ram pressure is available at take-off speeds and consequently little thrust, assisted take-off and acceleration to operating speeds must be obtained by an auxiliary engine.

A variation of the ramjet is the pulsejet or intermittent ramjet, which was used successfully in the German V-1 or "buzz" bomb. By valving the air in the inlet by shutters which alternately open and shut by the pressure of the intake air and back pressure from the combustion process, operation is possible at lower speeds than the ramjet. Objectionable vibration characteristics from the intermittent combustion process seem to limit the use of the pulsejet to pilotless aircraft.

The thrust of the ramjet varies considerably with airplane or flight speed. The mass of air handled varies directly with v_0, thermal efficiency depends on compression ratio which in turn varies with v_0, and the thrust is proportional to the difference between v_w and v_0. In Figure 3-2 the thrust coefficient, C_T, increases with M and falls off at high M (high compression ratio) because of temperature limitations on the strength of materials, and some losses in the intake diffuser at supersonic speeds.

For a ramjet, turbojet, or propeller, the thrust is equal to the mass of air and fuel passing through the system per unit time multiplied by the change in speed. The mechanical energy produced in the system is the difference between the kinetic energy of the wake and the free stream. The thermal efficiency of the ramjet may be expressed as:

$$\eta_t = \frac{\dfrac{w}{2g}(v_w{}^2 - v_0{}^2)}{JH_f w_f} \tag{3-7}$$

where w = weight of air per second, and the propulsive efficiency as:

$$\eta_p = \frac{\dfrac{w}{g}(v_w - v_0)v_0}{\dfrac{w}{2g}(v_w^2 - v_0^2)} = \frac{2\dfrac{v_0}{v_w}}{1 + \dfrac{v_0}{v_w}} = \frac{2}{1 + \dfrac{v_w}{v_0}} \qquad (3\text{–}8)$$

For simplicity the fuel flow is considered negligible.

The thermal efficiency of the ramjet becomes attractive only at high speeds where high compression ratio is obtained by ram. Also, as the mass flow increases with speed, it is possible to burn more fuel for a given top temperature. At very high speeds ($M > 1$) the mass flow is limited by compression shocks at the duct inlet. Because of the relatively low ram and jet velocity the propulsive efficiency of the ramjet increases rapidly at low speeds (Figure 3–1). At higher speeds the η_p remains essentially constant because the peak cycle pressure is the result of ram, and therefore the wake velocity is nearly proportional to flight velocity.

Turbojet

Like the ramjet, the turbojet makes use of the air as a working gas. Kerosene or gasoline may be used as the fuel. The air is compressed by a diffuser and an exhaust-driven turbine-compressor combination. After combustion and re-expansion the energy is extracted by a turbine which runs the compressor. The remaining energy is exhausted as a high-velocity jet through a nozzle, producing thrust. The thrust of a turbojet is nearly constant with speed. Even though the mass flow increases with speed, the acceleration ($v_w - v_0$) decreases with speed, thus producing a nearly constant thrust.

The thermal efficiency of the turbojet is much better than that of the ramjet at low speeds because the compressor maintains an appreciable compression ratio (4 to 5) even at zero airplane speed. Because of a higher wake or jet velocity the turbojet has a lower propulsive efficiency than the ramjet. Equations (3–7) and (3–8) apply to the turbojet as well as the ramjet. Power-plant weight is higher than the ramjet and rocket, but fuel consumption is considerably lower. Only at very high speeds ($M = 2$ to 3) does the fuel consumption for the ramjet approach that which could be obtained by the turbojet. The thrust coefficients for a typical turbojet and ramjet are about the same at an M slightly above 1. (See Figure 3–2.) In Figure 3–5 are plotted estimated turbojet engine characteristics for a Rolls-Royce "Nene." These data are representative of a 1949 power plant.

In Figure 3–6 is presented a schematic sketch of the component parts of a typical turbojet installation. The performance characteristics of this type of power plant can be obtained from non-di-

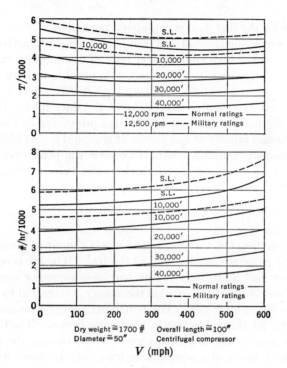

FIGURE 3–5. Estimated "Nene" performance. Estimated from Rolls-Royce Turbojet power-plant brochure published by Taylor Turbine Corporation, August 1946.

mensional plots of engine data. The basic performance curves of gross thrust, F, airflow, w_a, and fuel flow, w_f, are derived in terms of:

1. Compressor inlet total pressure, P_2.
2. Ambient or free-stream static pressure, p_{am}.
3. Turbine RPM, N.
4. Compressor inlet total temperature, T_2, absolute.
5. Linear dimension, D.

As an example, setting up F as a function of these variables, three non-dimensional parameters are determined by the use of the π

theorem (dimensional analysis theory) such that

$$\pi_1 = \frac{P_2}{p_{am}}$$

$$\pi_2 = \frac{ND}{\sqrt{T_2}}$$

$$\pi_3 = \frac{F}{D^2 P_2}$$

or

$$\frac{F}{D^2 P_2} = f\left(\frac{ND}{\sqrt{T_2}}, \frac{P_2}{p_{am}}\right)$$

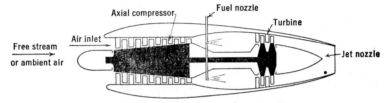

FIGURE 3–6.   Schematic sketch of turbojet.

For a particular power plant the dimension, D, may be eliminated for presentation of the performance data. Also, by the use of suitable constants and by defining

$$\delta_2 = \frac{P_2}{p_0}$$

and

$$\theta_2 = \frac{T_2}{T_0}$$

where the subscript 0 refers to standard sea level conditions, then

$$\frac{F}{\delta_2} = f\left(\frac{N}{\sqrt{\theta_2}}, \frac{P_2}{p_{am}}\right) \tag{3–9}$$

which is in such a form as to take care of non-standard conditions of pressure and temperature.

In a similar way w_a and w_f can be set up as

$$\frac{w_a \sqrt{\theta_2}}{\delta_2} = f\left(\frac{N}{\sqrt{\theta_2}}, \frac{P_2}{p_{am}}\right) \tag{3–10}$$

and

$$\frac{w_f}{\delta_2 \sqrt{\theta_2}} = f\left(\frac{N}{\sqrt{\theta_2}}, \frac{P_2}{p_{am}}\right) \tag{3–11}$$

Typical non-dimensional engine characteristics are presented in Figure 3–7. To determine the net thrust, T, of the power plant, the ram drag must be subtracted from the gross thrust, F:

$$T = F - \text{Ram drag}$$

The ram drag is the free-stream momentum of the air entering the compressor inlet per second, so that the preceding equation may be expanded in the following form:

$$T = F - \frac{w_a v_o}{g} \qquad (3\text{--}12)$$

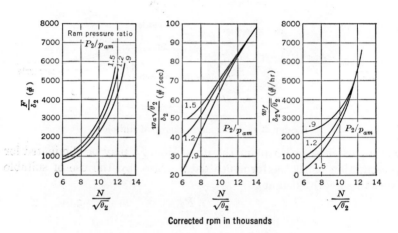

Corrected rpm in thousands

FIGURE 3–7. Typical non-dimensional engine characteristics.

A typical example illustrating the use of Figure 3–7 and equation (3–12) to calculate the net T of the power plant follows.

Performance data:

$$h = 35,000 \text{ ft} \qquad P_2 = 8.44 \text{ in. Hg}$$

$$N = 11,000 \qquad T_2 = 475°\text{F abs.}$$

$$V = 400 \text{ mph} \qquad v_0 = 586 \text{ ft/sec}$$

$$T_{am} = -25°\text{F} \qquad q_c{}^* = 127 \text{ lb/ft}^2 = 1.8 \text{ in. Hg}$$

$^* q_c = q \left(1 + \dfrac{M^2}{4} + \dfrac{M^4}{40} + \dfrac{M^6}{1600} \cdots \right)$: *Introduction to Aerodynamics of a Compressible Fluid*, by Liepmann and Puckett, p. 27, John Wiley and Sons, 1947.

p_{am} @ 35,000 ft = 7.036 in. Hg   Ram η = .78 [see equation (3–13)]

$$\delta_2 = \frac{P_2}{p_0} = \frac{8.44}{29.92} = .282$$

$$\theta_2 = \frac{T_2}{T_0} = \frac{475}{518.4} = .916$$

$$\frac{N}{\sqrt{\theta_2}} = \frac{11,000}{\sqrt{.916}} = 11,490$$

$$\frac{P_2}{p_{am}} = \frac{8.44}{7.036} = 1.20$$

From Figure 3–7:

$$\frac{F}{\delta_2} = 4300, \quad \frac{w_a\sqrt{\theta_2}}{\delta_2} = 79, \quad \frac{w_f}{\delta_2\sqrt{\theta_2}} = 4800$$

$$F = 4300 \times .282 = 1210 \text{ lb}$$

$$w_a = \frac{79 \times .282}{\sqrt{.916}} = 23.3 \text{ lb/sec}$$

$$w_f = 4800 \times .282 \times \sqrt{.916} = 1298 \text{ lb/hr}$$

$$T = F - \frac{w_a v_0}{g} = 1298 - \frac{23.3 \times 586}{32.2} = 1298 - 423 = 875 \text{ lb}$$

In deriving equation (3–8) it is assumed that the weight of fuel is negligible compared to the weight of air. By this assumption the energy required to bring the fuel up to the flight speed of the airplane is neglected. In the preceding example the air to fuel ratio for the turbojet is approximately 55 to 1. For a ramjet this ratio varies from 15 to 30 but is still sufficiently high to justify the assumption used in deriving equation (3–8).

As can be seen from Figure 3–7, a high ram pressure ratio in a turbojet intake or diffuser is conducive to high F and w_a and low w_f. For a given flight speed the ram pressure ratio is directly proportional to ram efficiency. Since ram efficiency, as defined below, is a measure of the percentage recovery of total pressure in the diffuser, the energy losses in the diffuser should be kept to a minimum.

$$\text{Ram efficiency, } \eta_R = \frac{P_2 - p_{am}}{q_c} \tag{3–13}$$

where P_2 = the average total pressure at the compressor inlet.

p_{am} = static ambient pressure.

q_c = free-stream dynamic pressure, including compressibility effects.

Figure 3–8 indicates the increase in net thrust, T, and decrease in specific fuel consumption, c', that can be gained by a high ram efficiency for a given power plant.

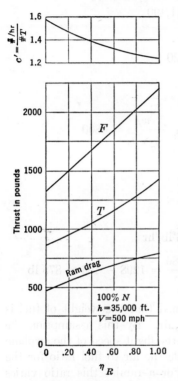

FIGURE 3–8. Variation of engine characteristics with ram efficiency.

A high ram efficiency can be maintained in an intake duct if the friction loss and turbulence due to separation are kept to a minimum. In order to reduce internal losses a low inlet velocity ratio, v_1/v_0, is desirable. In Figure 3–9 is presented the variations of ram η with v_1/v_0 for typical intake ducts. The diffuser should be designed to give good values of efficiency for all flight speeds. For a given engine N, the quantity flow of air, Q, into the compressor must be a constant; therefore the v_1/v_0 varies as speed changes. The merits of the scoop and flush duct are about the same, but both are inferior to the nose intake if the boundary layers at the scoop or flush intake are sufficient to cause internal separation. At low values of v_1/v_0 the high adverse pressure gradient ahead of the duct intake usually causes separation and reduces the ram η, as indicated by the dotted curve in Figure 3–9. To alleviate this condition boundary-layer bleeds are often used on scoop and flush ducts.

At high values of v_1/v_0 the ram η falls off because of the high internal friction drag associated with high velocity. A low value of v_1/v_0 is desirable for low internal friction losses but often causes separation on the external duct lip because of rapid expansion of the air around the outside of the lip. The intake duct installation for a given configuration is limited to a minimum v_1/v_0 by external separation and a maximum v_1/v_0 by internal friction drag. Of course some internal separation may occur for low values of v_1/v_0 for the scoop or flush intake if boundary-layer bleeds are not installed. The optimum design of diffuser is thus dictated by engine requirements, airplane speeds, and compromises between internal and external energy losses.

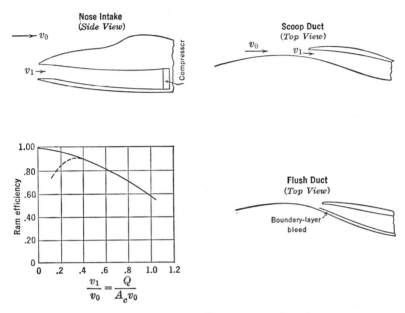

FIGURE 3–9.   Variation in ram efficiency with inlet velocity ratio.

Turboprop

As in the turbojet, the gas turbine with propeller drive, or turboprop, consists of a diffuser, combustion chamber, turbine-compressor combination, and exhaust nozzle.   In the turboprop, however, the turbine takes more heat energy than is required to run the compressor, the excess being used to run a propeller.   Of the net useful energy going into propulsion about 80–95 per cent is derived from the propeller and the remainder from the jet reaction thrust in the exhaust nozzle.   As is typical of all propeller-driven aircraft, the power is nearly constant with speed and consequently the thrust decreases as the speed increases.   The thermal efficiency of the turbine propeller is fairly high owing to the favorable compression ratio obtained by the compressor. Like the turbojet, the thermal efficiency is limited by the maximum temperature that the stressed rotating parts can withstand.   The propulsive efficiency at low speeds is very high because of the large mass of air handled and low wake velocity. Because of the large mass of air, supercharging is impractical, and, as in the turbojet, the thrust decreases with altitude. The propulsive efficiency of the turbine propeller may be calculated by equation (3-8). The thermal efficiency is expressed in terms of the sum of the shaft horsepower, SHP, and differential jet kinetic energy, all divided by the heat energy added to the system:

$$\eta_t = \frac{\dfrac{w}{2g}\,(v_w{}^2 - v_0{}^2) + 550SHP}{JH_f w_f} \qquad (3\text{-}14)$$

In Figure 3-10a are plotted typical non-dimensional turbine-propeller characteristics. The parameters used are similar to those used for the turbojet in Figure 3-7. However, in this case a speed parameter, $V/\sqrt{\theta}$, is used in conjunction with an assumed ram efficiency of .80. In the previous case (Figure 3-7) the characteristic curves can be used for various ram efficiencies as P and T are measured at the compressor inlet. The pressure ratio for Figure 3-10a, δ, is defined as

$$\delta = \frac{p_{am}}{p_0}$$

where p_{am} = static ambient pressure.

p_0 = sea level standard static pressure.

The temperature ratio, θ, is defined as

$$\theta = \frac{T_{am}}{T_0}$$

where T_{am} = absolute ambient temperature.

T_0 = absolute sea level temperature.

Standard values of δ and θ are noted in Figure 3-10a. In Figure 3-10b are plotted typical curves of equivalent shaft horsepower and specific fuel consumption as calculated for specific conditions from the general curves in Figure 3-10a. A propeller efficiency, η_p, is assumed at 85 per cent.

Ducted Fan

From Figure 3-1 it is noticed that the propulsive efficiency for propeller-driven engines approaches a value of .9 at or near speeds of 450 mph. Beyond this speed region the propulsive efficiency decreases rapidly because of compressibility losses near the propeller blade tips.

The propulsive efficiency of the turbojet is much lower than that of propellers up to this point owing to the high energy losses associated with the small-diameter high-speed wake of the jet. Since the diameter of a jet is approximately one-third that of an equal-thrust-

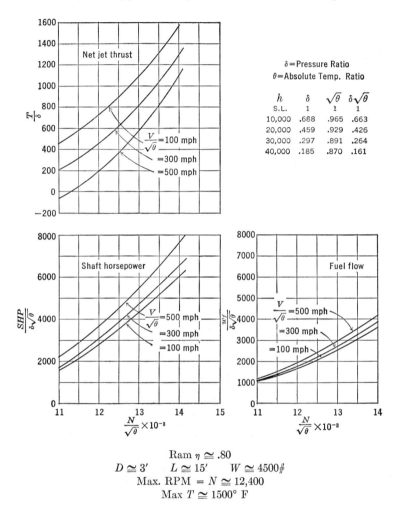

δ = Pressure Ratio
θ = Absolute Temp. Ratio

h	δ	$\sqrt{\theta}$	$\delta\sqrt{\theta}$
S.L.	1	1	1
10,000	.688	.965	.663
20,000	.459	.929	.426
30,000	.297	.891	.264
40,000	.185	.870	.161

Ram $\eta \cong .80$
$D \cong 3'$ $L \cong 15'$ $W \cong 4500\#$
Max. RPM $= N \cong 12,400$
Max $T \cong 1500°$ F

FIGURE 3–10a.   Typical non-dimensional turboprop characteristics.

producing propeller, the wake speed is about three times as much. From Figure 3–4 it can be seen that a v_w/v_0 of 1.25 is required for an η_p of approximately .90. For a propeller this is possible, but a jet airplane would require a supersonic flight speed before a comparable

v_w/v_0 could be attained. For subsonic flight, then, a turbojet never realizes its propulsive efficiency potential.

In Figure 3-11 are presented typical efficiency curves for a turboprop, turbojet, and ducted fan. In the speed region between 500 and 700 mph the ducted fan bridges the gap where neither the propeller nor turbojet is as efficient. For subsonic airframe design 700 mph represents the upper limit. Even at this speed the turbojet efficiency is not

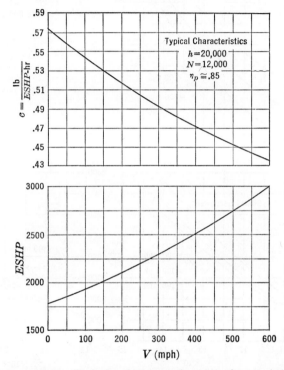

FIGURE 3-10b. Typical turboprop performance characteristics.

as high as that for the ducted fan. For conventional propeller design 500 mph represents the upper limit. This region represents a speed requirement that is important for the development of both commercial and military airplanes for many years to come.

A possible configuration of this proposed engine type is indicated in Figure 3-12. Energy is taken from the wake of a typical turbojet to run a free turbine and fan combination that is separately mounted directly in the rear of the first turbine which is integral with the compressor. The fan operates in a concentric duct constructed around the basic unit.　The wake speed of the combination is lower than that of the straight jet but higher than that of a turboprop.　In this manner a propulsive efficiency is attained which is a compromise between the two

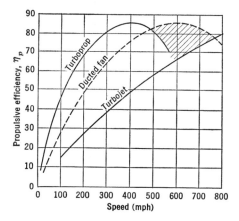

FIGURE 3-11.　Propulsive efficiency of ducted fan.

basic units.　At take-off flight speeds, augmentation theoretically produces thrust values 150 per cent as much as the basic turbojet installation.

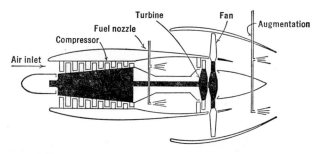

FIGURE 3-12.　Schematic sketch of ducted fan.

Although the ducted fan is not mechanically perfected, it presents a favorable possibility for the attainment of high propulsive efficiencies at high subsonic flight speeds.　Since the flight of long-range supersonic airplanes is still a conjecture, the use of this type of engine in airplanes designed to meet military and commercial requirements of high subsonic speeds and long flight ranges is very promising.

With the preceding example of turbojet engine performance as calculated from Figure 3–7 as a basis, the following analysis indicates the gains that may be expected from a ducted fan of the same fuel flow.

Turbojet performance:

$$v_0 = 586 \text{ ft/sec} = 400 \text{ mph}$$

$$w_a = 23.3 \text{ lb/sec}$$

$$w_f = 1298 \text{ lb/hr} = .36 \text{ lb/sec}$$

$$T = 875 \text{ lb}$$

$$c' = \frac{1298}{875} = 1.485 \text{ lb/hr/lb } T$$

Computing wake speed for above turbojet conditions:

$$v_w = \frac{g}{w}\left(T + \frac{w}{g}v_0\right)$$

$$v_w = \frac{32.2}{23.66}\left(875 + \frac{23.66}{32.2} \times 586\right) = 1780 \text{ ft/sec}$$

Assuming that the η_p of an installation similar to that shown in Figure 3–12 is approximately 75 per cent (see Figure 3–11) and the wake speed from the fan and turbine, v_w', is the same:

$$v_w' = \frac{2v_0}{\eta_p} - v_0 = \frac{2 \times 586}{.75} - 586 = 977 \text{ ft/sec}$$

If the efficiency of the free turbine is assumed equal to 80 per cent, the energy that it absorbs can be calculated as the difference between the kinetic energy corresponding to the original wake speed, v_w, and the new wake speed, v_w':

$$KE = \frac{\eta w}{2g}\left(v_w{}^2 - v_w'{}^2\right)$$

$$KE = \frac{.80 \times 23.66}{64.4}(1780^2 - 977^2) = 648{,}000 \text{ ft-lb/sec}$$

To calculate the mass flow of air through the fan, w', an efficiency for the fan of 95 per cent is assumed. The kinetic energy absorbed by the fan is the same as that for the free turbine.

$$648{,}000 = \frac{.95 \times w'}{64.4}\left(v_w'{}^2 - v_0{}^2\right)$$

$$w' = \frac{64.4 \times 648{,}000}{.95(977^2 - 586^2)} = 71.8 \text{ lb/sec}$$

The thrust for the ducted fan, T', can now be calculated as a function of the total mass flow of the entire unit and the difference between the free-stream and wake speeds.

$$T' = \frac{w + w'}{g} \, (v_w' - v_0)$$

$$T' = \frac{23.66 + 71.8}{32.2} \, (977 - 586) = 1160 \text{ lb}$$

For the same fuel flow the percentage gain in thrust over the turbojet becomes

$$\frac{1160 - 875}{875} \cong 32.6$$

The specific fuel consumption decreases by the same percentage and becomes

$$\frac{1298}{1160} \cong 1.12 \text{ lb/hr/lb } T$$

By augmentation both the fuel flow and thrust can be considerably increased. The efficiency curve for the ducted fan presented in Figure 3–11 represents one given design. By varying the fan diameter it is possible to compute a family of curves for any compromise between the efficiency curve for the turboprop and that for the turbojet.

Although the preceding analysis does not account for the transfer of heat energy in the system, it does give a first approximation of the performance gains possible with a ducted fan turbojet. It is believed that the added drag, weight, and mechanical complexity associated with this installation, as compared to the basic turbojet, do not outweigh the advantages of the increase in thrust and decrease in specific fuel consumption for certain applications.

Reciprocating Engine

The reciprocating or piston engine is the most highly developed aircraft power plant. Combustion of a gasoline-air mixture takes place in the cylinders essentially as a constant volume process. The power developed is converted into thrust by means of a propeller. A portion of the heat energy remaining in the exhaust gases is sometimes recovered as jet thrust or alternately is used to drive a turbosupercharger. The latter increases the mass flow of air through the engine so that an essentially constant power output is maintained to altitudes as high as 40,000 feet. This engine is characterized by a relatively high power-plant weight and form drag. Since the piston engine must be cooled, its drag is somewhat higher than that for the turbojet and turbine propeller. Fuel consumption is very low and essentially unaffected by altitude. The thermal efficiency of the reciprocating engine is high and constant with flight speed. Each piston is designed for a high compression ratio (high temperatures and pressures) and is not dependent on flight speed for increased ram as in the case of the thermal jet engines. The maximum power output of this type of engine is limited by the piston capacity to swallow air, whereas the turbojet and turbine propeller gulp air in proportion to the inlet diameter and air velocity. A typical engine specification is presented in Figure 3-13.

The propulsive efficiency of the reciprocating engine may again be shown by the general expression of equation (3-8). A general expression for thermal efficiency may be developed for the reciprocating engine which is also applicable to the other power plants previously discussed. The numerator of equation (3-2) or (3-7), differential jet kinetic energy, may be expressed also as SHP times a constant. Thermal efficiency may then be written as

$$\eta_t = \frac{SHP}{w_f} \frac{550 \times 3600}{H_f J}$$

where w_f is fuel flow in pounds per hour.

Since w_f/SHP is specific fuel consumption, c, in pounds per shaft horsepower per hour, and H_f and J are constants,

$$\eta_t \cong \frac{.138}{c} \tag{3-15}$$

assuming an average value for $H_f = 18,500$ Btu/lb.

Another type of power plant that offers considerable merit is the so-called compound engine. This engine utilizes components of both the reciprocating engine and gas turbine. A feedback turbine utilizes the waste energy in the exhaust of a reciprocating engine by converting this energy into shaft horsepower by a direct connection between the turbine and the engine drive shaft. Such a combination has the highest of all power-plant installed weights but higher powers and lower specific fuel consumptions than the reciprocating engine alone.

Power-plant Selection

The selection of a suitable power plant to meet airplane speed and range requirements is not only dependent on the thrust and fuel-consumption characteristics of the various power plants but also on the weight of each type. For example, in Figure 3-14, assuming a constant

PRATT & WHITNEY AIRCRAFT
Division United Aircraft Corporation
East Hartford 8, Connecticut, U. S. A.

A.T.C. No. — See Special Notes
Fuel Grade — 115/145

Model — Wasp Major TSB3-G
Spec. No. 7052

WASP MAJOR TSB3-G ENGINE SPECIFICATION
A Single Stage, Single Speed Engine
Suitable for Use with Exhaust Driven Supercharger

GUARANTEED DYNAMOMETER PERFORMANCE
Power Curve No. T-984

Metric				English		
BHP	RPM	ALT. meters		BHP	RPM	ALT. feet
			Take-off Power			
3550	2700	150	With Water Injection	3500	2700	500
3295	2700	215	Without Water Injection	3250	2700	700
2685	2550	1675	Normal Rated Power	2650	2550	5500
2840	2550	1070	Maximum Continuous Power	2800	2550	3500

Cruising Power
Maximum cruising power shall be approved in writing after consultation with Pratt & Whitney Aircraft .

DESCRIPTION AND DIMENSIONS
Installation Drawing No. 97801

Type — 28 Cylinder Air-Cooled Radial

146 mm	Bore	5.75 inches
152.5 mm	Stroke	6.00 inches
71.5 liters	Total Displacement	4363 cu. inches
.381	Propeller Gear Ratio	.381
Clockwise	Propeller Shaft Rotation, viewed from antipropeller end	Clockwise
60-A	Propeller Shaft Spline SAE No.	60-A
1359 mm	Engine Diameter, Maximum	53.50 inches
2457 mm	Engine Length, Maximum	96.75 inches
115/145	Fuel Knock Value C.F.R. Method, F3 Lean, F4 Rich	115/145
100	Oil Grade S.U.S. at 210 F	100
11.2 grams/BHP/hr	Oil Consumption Maximum @ Normal Rated Power and Speed	.025 lb/BHP/hr
6.7 grams/BHP/hr	@ 1850 B.H.P. and 2270 R.P.M.	.015 lb/BHP/hr
See Installation Drawing	Approximate Center of Gravity	See Installation Drawing

DRY WEIGHT

1574 kg	Including Standard Accessory Equipment	3470 lb

Date: 7-19-45
Revised: 6-14-46   Additional details on page 2   Page 1

FIGURE 3-13a. Reproduced by permission of The Pratt & Whitney Aircraft Division, United Aircraft Corporation.

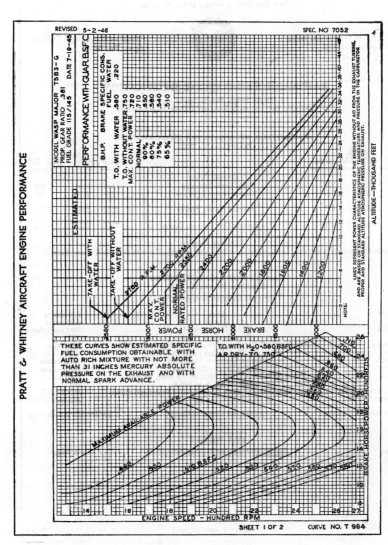

FIGURE 3-13b. Reproduced by permission of The Pratt & Whitney Aircraft Division, United Aircraft Corporation.

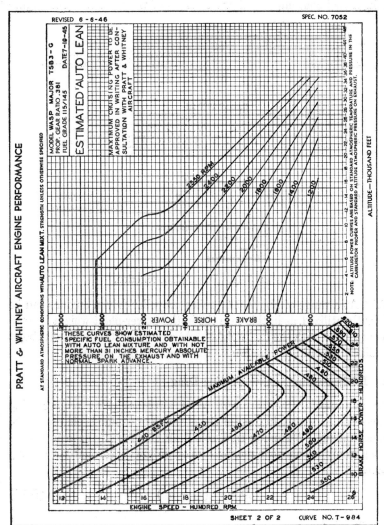

FIGURE 3-13c. Reproduced by permission of The Pratt & Whitney Aircraft Division, United Aircraft Corporation.

structure, fixed equipment, and gross weight of a series of airplanes, the useful load capabilities of a rocket airplane might be superior to any other type at a range of 50 miles because the combined weight of power plant and fuel is lower than in any other engine type in spite of the high fuel-consumption characteristics of the rocket. (See Figure

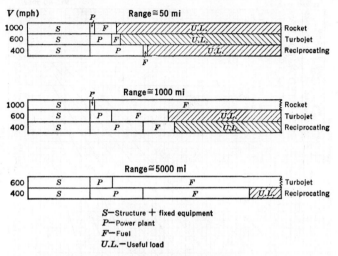

S—Structure + fixed equipment
P—Power plant
F—Fuel
U.L.—Useful load

FIGURE 3–14. Power-plant selection.

3–3.) At a range of above 5000 miles propeller-driven engines can be efficiently utilized, as in this case the initial power-plant weight is small in comparison with the fuel weight. At a range of 1000 miles a turbojet installation should be superior to the reciprocating engine.

3–3 Reciprocating Engine Cooling Drag

In determining the performance of airplanes using reciprocating engines some increase in drag of the airplane or engine is chargeable to engine cooling. Assuming that this drag is equal to the reaction resulting from the change in momentum of air flowing through the cooling system,

$$D = \frac{w}{g} (v_0 - v_w) \qquad (3\text{–}16)$$

where D = drag in pounds.

w = flow of cooling air in pounds per second.

v_0 = free-stream velocity in feet per second.

v_w = wake velocity at atmospheric pressure due to energy removed internally.

Since

$$v_0 = \sqrt{\frac{2q_0}{\rho}}$$

and

$$v_w = \sqrt{\frac{2}{\rho}(q_0 - \Delta p)}$$

where q_0 = free-stream dynamic pressure.

Δp = baffle pressure drop in system in pounds per square foot.

Then

$$\frac{v_w}{v_0} = \sqrt{1 - \frac{\Delta p}{q_0}}$$

and

$$D = \frac{w}{g} v_0 \left(1 - \sqrt{1 - \frac{\Delta p}{q_0}}\right) \tag{3-17}$$

To determine the increase in nacelle drag coefficient as a function of $\Delta p/q_0$, divide equation (3-17) by $q_0 A$ and express

$$w = \rho g A_e \sqrt{\frac{2\Delta p}{\rho}}$$

where A = nacelle frontal area in square feet.

A_e = equivalent orifice area of engine in square feet.

Then

$$\frac{D}{q_0 A} = C_D = 2\frac{A_e}{A}\sqrt{\frac{\Delta p}{q_0}}\left(1 - \sqrt{1 - \frac{\Delta p}{q_0}}\right) \tag{3-18}$$

Figure 3-15 is obtained by plotting C_D versus $\Delta p/q_0$ from equation (3-18) for representative values of A_e/A equal to .09 and .125 and adding to the Basic Nacelle C_D. These calculated values agree closely with flight test data. For cowl flaps full open, $\Delta p/q_0 = 1$, the increase in nacelle C_D is approximately four times that for the close position. Only a very small part of this drag is attributed to external cowl-flap drag. For reasonable installations the cooling drag effect on airplane performance for climb conditions may be approximated by reducing the engine brake horsepower by 6 per cent.

3-4 Propeller Charts

Both the reciprocating engine and turbine propeller employ a propeller to convert the mechanical energy produced in the power plant to useful work done on the airplane. In order to calculate the useful work or thrust horsepower for these two types of power plants some knowledge of propeller characteristics is now necessary. As defined by equation (3-3), the propulsive efficiency may be written as:

$$\eta_p = \frac{Tv_0}{BHP} \tag{3-19}$$

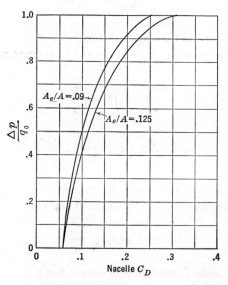

FIGURE 3-15. Effect of cooling drag on typical nacelle installation.

for the reciprocating engine, and

$$\eta_p = \frac{Tv_0}{ESHP} \tag{3-20}$$

for the turbine propeller, where $ESHP$ is the equivalent shaft horsepower developed by both the turbine and the jet. In either case, equation (3-8), that

$$\eta_p = \frac{2}{1 + \dfrac{v_w}{v_0}}$$

still applies. Equation (3-8) usually appears in the literature in a slightly different form for propeller systems and is sometimes defined, for potential flow, as ideal efficiency, η_i. This form of the equation is obtained by analyzing the efficiency for a propeller disc operating in an ideal fluid wherein a continuous longitudinal velocity exists through the disc with no slipstream rotation.   (See Figure 3–16.)   If in this system half the percentage increase in v_0 occurs at the disc and is defined as a, with the total percentage increase defined as b, then from equation (3–8)

$$n_i = \frac{2}{1 + \dfrac{v_0(1 + 2a)}{v_0}} = \frac{1}{1 + a} \tag{3-21}$$

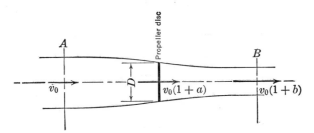

FIGURE 3–16.   Momentum flow through propeller.

Here again it is shown that, for a given thrust, highest efficiencies are obtained by accelerating a large mass with a small increase in slipstream velocity, as in the case of a propeller system.   The assumption that a is equal to one-half b can readily be verified by an analysis of the kinetic energies in the slipstream at two stations in Figure 3–16.

Applying the equation of continuity, the kinetic energy per second, KE/s, at point A is

$$KE/s \text{ at } A = \tfrac{1}{2}m'v^2 = \tfrac{1}{2}\rho Av_0(1 + a)v_0{}^2 = \tfrac{1}{2}\rho Av_0{}^3(1 + a)$$

where m' = mass per second, and $A = \pi D^2/4$.

At point B behind the disc,

$$KE/s \text{ at } B = \tfrac{1}{2}\rho Av_0{}^3(1 + a)(1 + b)^2$$

Since the power output or gain in kinetic energy per second, $Tv_0(1 + a)$, is the difference between the energies at points A and B

$$Tv_0(1 + a) = KE/s \text{ at } B - KE/s \text{ at } A$$

$$Tv_0(1 + a) = \tfrac{1}{2}\rho Av_0{}^3(1 + a)(1 + b)^2 - \tfrac{1}{2}\rho Av_0{}^3(1 + a)$$

$$T = \tfrac{1}{2}\rho Av_0{}^2[(1 + b)^2 - 1] \tag{3-22}$$

Also, since

$$T = m'\Delta v_0$$

$$T = \rho Av_0(1 + a)bv_0 = \rho Av_0{}^2(1 + a)b \tag{3-23}$$

Then, by equating equations (3-22) and (3-23)

$$\tfrac{1}{2}\rho A v_0^2(2b + b^2) = \rho A v_0^2(1 + a)b$$

$$b + \frac{b^2}{2} = (1 + a)b$$

and

$$b = 2a \qquad\qquad (3\text{-}24)$$

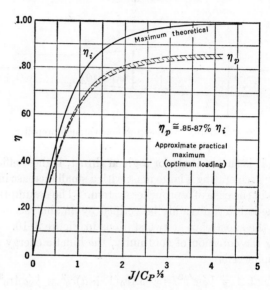

FIGURE 3-17. Propeller efficiency (incompressible).

As shown in Figure 3-17, actual propulsive efficiencies are somewhat less than that predicted by equation (3-21) for the ideal case, principally because of the parasite drag on the actual propeller blades, the interference between blades, and the energy loss in the rotation of the slipstream. Propulsive efficiency as defined in propeller charts is usually obtained from experimental tests on a propeller with spinner mounted in front of a nacelle or fuselage. (See Figure 3-18.) Because of the flow of air around the propeller and over the body (Figure 3-18b) a resultant pressure field is produced yielding an increment of thrust, ΔT, and an additional increment of drag, ΔD, from spinner and shanks plus that from the increase in v_0 over the body from the slipstream. Propulsive efficiency as obtained from experimental tests and presented in the propeller charts which follow may then be defined as

$$\eta_p = \frac{(T' + \Delta T - \Delta D)v_0}{2\pi n Q} \tag{3-25}$$

where Q = engine torque.

n = engine revolutions per second.

T' = thrust of propeller alone as illustrated in Figure 3-18a.

$$\Delta T \cong \Delta D$$

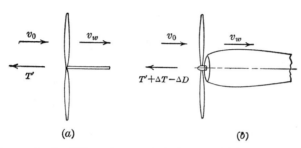

(a) (b)

FIGURE 3-18.   Effect of test installation on propulsive efficiency.

With this understanding, future references to thrust for propeller systems will be net thrust as defined in the numerator of equation (3-25). In order to develop some of the special terminology used in propeller charts (see Figure 3-17), consider the element of a propeller blade at some radius as indicated in Figure 3-19. This element is then shown on the unwrapped plane of the helical surface generated by the motion of the element.

In Figure 3-19 are represented the differential forces that act on a blade element at radius, r, in feet; v and n are respectively the forward speed in feet per second and revolutions per second of the blade element. The angle, β, is the setting or pitch of the blade and is measured between the tangential velocity and the chord line. The angle of attack of the blade is a function, then, of β and $\tan^{-1}\dfrac{v}{\pi n d}$, where d is equal to twice r. The η of the blade element may be expressed as

$$\eta = \frac{dT\ v}{2\pi n\ dF\ r}$$

where dT is a differential thrust force, and dF a differential torque force. For a given v and n, maximum η occurs when dT/dF or dL/dD

is a maximum. In other words the optimum angle of attack for each element is that which gives maximum L/D. On a fixed pitch propeller the twist (angle β) is usually adjusted along the blade so that optimum efficiency occurs at each radius for a given design condition of v and n. For example, at radius r', in Figure 3–13, the angle β is larger than for radius r, to give optimum blade element efficiency at this point.

Since the design range for both v and n is wide, low efficiencies are obtained, for example, at take-off if the fixed pitch propeller is designed for cruising conditions. In a variable or adjustable pitch propeller

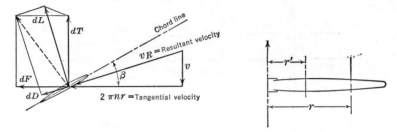

FIGURE 3–19. Differential forces on blade element.

the whole blade may be rotated, changing β all along the blade radius. This permits the attainment of high efficiencies for all operating conditions. The optimum efficiency, however, is obtained for only one combination of v and n.

From the foregoing discussion it is evident that the over-all operating conditions of any propeller can be expressed as a function of α or the two variables, β and $\tan^{-1}\dfrac{v}{\pi n d}$. When multiplied by π, the latter is usually defined as advance ratio, J, where

$$J = \frac{v}{nD} = \frac{88V}{ND}$$

where N is revolutions per minute, and V is miles per hour. β and J are two dimensionless parameters, then, which govern the geometrical conditions under which a propeller of a given type operates. They are usually measured and defined for a given propeller at three-quarters of the distance from the blade axis to the tip.

Other propeller coefficients or parameters can be derived from the forces that operate on each blade element. The total T and F forces are obtained, of course, by summing up along the blade radius all the differential forces on each blade element. From dimensional analysis the general form of these two forces may be written

$$T = T_C \rho D^2 v^2 \tag{3-26}$$

$$F = T_F \rho D^2 v^2 \tag{3-27}$$

where D is the propeller diameter, and T_C and T_F are non-dimensional coefficients of thrust and torque. In order to change these coefficients into a more usable form, equation (3-26) is rewritten in the following manner:

$$T = T_C \left(\frac{v}{nD}\right)^2 \left(\frac{nD}{v}\right)^2 \rho D^2 v^2$$

$$T = C_T \rho n^2 D^4 \tag{3-28}$$

where

$$C_T = T_C \left(\frac{v}{nD}\right)^2$$

v/nD or J is analogous to angle of attack for a wing section. In other words, T and F can be derived at a fixed value of J so that any multiplication or division by J does not introduce a variable. The equation for F now becomes

$$F = C_F \rho n^2 D^4 \tag{3-29}$$

Since engine or propeller torque is a function of F and some radius, an equation for torque, Q, may be written in terms of a torque coefficient, C_Q,

$$Q = Fr = \frac{C_F}{2} \rho n^2 D^4 D = C_Q \rho n^2 D^5 \tag{3-30}$$

Since power is a function of torque and rotation

$$P = 2\pi n Q = C_P \rho n^3 D^5 \tag{3-31}$$

Propeller efficiency is a function of J, C_T, and C_P and may be expressed, as previously indicated, in the following manner:

$$\eta = \frac{Tv}{BHP}$$

By substituting values of T and P from equations (3-28) and (3-31)

$$\eta = \frac{C_T \rho n^2 D^4 v}{C_P \rho n^3 D^5} = \frac{C_T}{C_P} \frac{v}{nD}$$

$$\eta = \frac{C_T J}{C_P} \tag{3-32}$$

For engineering units

$$J = \frac{88V}{ND}$$

$$C_T = .1518 \frac{(T/1000)}{\sigma(N/1000)^2 (D/10)^4}$$

$$C_P = .5 \frac{BHP/1000}{\sigma(N/1000)^3 (D/10)^5}$$

For determining the THP output of a propeller-engine combination it is necessary to compute the propeller efficiency for given conditions of engine RPM and BHP. Many types of propeller charts using combinations of the foregoing propeller parameters are used for this purpose. In Figure 3-20 is presented one of the most useful and fundamental types of propeller chart. The parameter $J/C_P^{1/3}$ is the same as that used in Figure 3-17 and is derived in the following paragraph.

From equations (3-23) and (3-24) the power produced by the propeller disc of Figure 3-16 may be expressed as

$$P = 2A\rho v_0{}^3(1 + a)^2 a \qquad (3\text{-}33)$$

since

$$\eta_i = \frac{1}{1 + a}$$

$$a = \frac{1}{\eta_i} - 1$$

Substituting for a in equation (3-33) and rearranging,

$$\frac{P}{\rho A v_0{}^3} = 2\left(\frac{1 - \eta_i}{\eta_i}\right)\frac{1}{\eta_i{}^2} = 2\left(\frac{1 - \eta_i}{\eta_i{}^3}\right) \qquad (3\text{-}34)$$

Also, since

$$P = C_P \rho n^3 D^5$$

and

$$A = \frac{\pi D^2}{4}$$

equation (3-34) may be rewritten as

$$\frac{C_P \rho n^3 D^5}{\rho \dfrac{\pi D^2}{4} v_0{}^3} = 2\left(\frac{1 - \eta_i}{\eta_i{}^3}\right)$$

and reduced to

$$\frac{\eta_i}{(1 - \eta_i)^{\frac{1}{3}}} = \left(\frac{\pi}{2}\right)^{\frac{1}{3}} \frac{J}{C_P^{\frac{1}{3}}} \tag{3-35}$$

In Figure 3–20*a and b, η versus $J/C_P^{\frac{1}{3}}$ is plotted for various values of effective power coefficient, C_{PX}. As in all propeller problems the

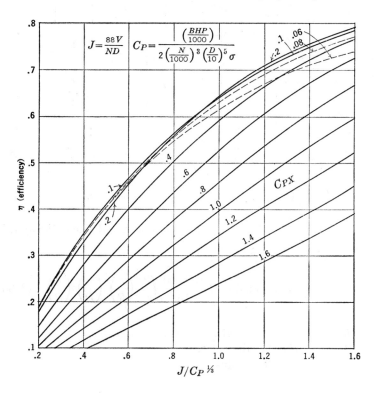

FIGURE 3–20a.   General propeller chart #1.

power absorbed by the propeller must equal the BHP of the engine for given flight conditions, and C_{PX} is the parameter that governs this condition.   Figure 3–20 represents characteristics of an average typical propeller tested by the National Advisory Committee for Aeronautics in the 20-ft propeller research tunnel and may be used for other types and numbers of blades by judicious selection of the adjustment factor, X, which is obtained in the lower half of Figure 3–20b as a function of

* Adapted from Boeing Airplane Company General Propeller Chart.

propeller activity factor. The term activity factor is a non-dimensional function of the blade planform that expresses the integrated

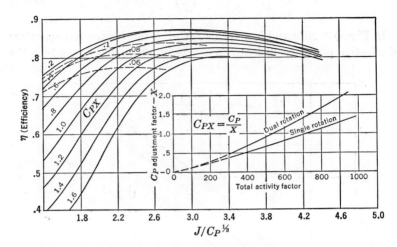

FIGURE 3–20b.   General propeller chart #2.

capacity of the blade elements to absorb power. Mathematically it is
calculated by the following expression:

$$\text{Activity factor} = \text{No. blades} \times \frac{100,000}{16} \int_{.2}^{1.0} \left(\frac{r}{R}\right)^3 \left(\frac{b}{D}\right) d\left(\frac{r}{R}\right) \quad (3\text{–}36)$$

where b = the blade width at a given r.
      D = the propeller diameter.
      R = the propeller radius.

It is based on the fact that the ability with which a blade absorbs
power is a function of the blade area multiplied by the third power of
the speed of the air passing over it.   The power-absorption ability of
a blade element is in turn proportional to the radius as measured from
the axis of rotation.   An increase in blade width near the propeller tip
is therefore relatively more effective for power absorption than an
increase in width near the shank.   The total activity factor, of course,
is also a function of the number of blades.   Activity factor for a single
blade varies approximately between 90 and 140.   To use Figure 3–20,
then, for any type of propeller for which the activity factor is known,
the effective C_P, C_{PX}, is obtained by dividing C_P by the adjustment

factor, X. In Figure 3–20 the use of X and activity factor accounts only for variations in blade width and number of blades for equivalent geometrical blade planforms. It is reasonable to expect that the same method would apply to other planforms. Variations in shank and hub

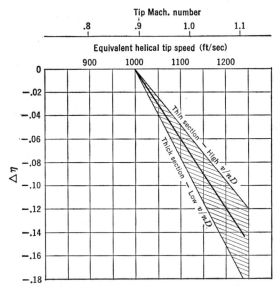

FIGURE 3–21. Propeller compressibility data.

drag change the η slightly from that obtained in Figure 3–20 and may be considered negligible for most propellers for which the activity factor is not unduly high.

Another correction to η, $\Delta\eta$ loss due to compressibility drag, may be very severe for high-speed aircraft. This correction is presented in Figure 3–21 and is based on an average of a large number of experimental tests. The data presented in Figure 3–21 may be conservative as the design of special high-speed propellers is continually improving. Little is known about the improvements in propeller efficiencies that may be possible with swept-back blades, thin blade sections, and different propeller shank or root shapes. In the design of propellers care must be taken to limit the diameter of the propeller and propeller-engine gear ratio to values that will keep the tip Mach number below a $M \cong .9$ if possible. In the design of propellers for high-speed airplanes employing high-power engines, high activity factors are required to absorb high powers. Since blade diameters are limited by permissible tip speeds, more blades must be used to absorb this power.

Mutual interference with more blades also causes a loss in efficiency. It is therefore necessary that a proper compromise between D and number of blades be made in the selection of a propeller to give optimum η for any given set of design conditions of engine power and N. Of course gear ratio and the selection of blade activity factor are also variables. The selection of a propeller blade, number of blades, blade activity factor, and gear ratio may be made by determining the optimum η from Figure 3–20 for the best combination of these variables for a given engine BHP and N. To make the selection even more difficult the engine conditions usually vary for different flight conditions, and the best all-around propeller for take-off, climb, cruise, and high-speed is sometimes required. The following procedure for obtaining η from Figure 3–20 is illustrated below for a particular set of variables, assuming D, type of blade, and gear ratio are preselected.

1. Select a V and altitude for which the THP is desired:

$$THP = \eta BHP$$

2. Determine the engine BHP and N for this speed and altitude for the desired operating engine conditions (e.g., maximum power or cruise power).

3. Compute propeller N:

$$\text{Propeller } N = \text{Engine } N \times \text{Gear ratio}$$

4. Compute J:

$$J = \frac{88V}{ND}$$

5. Compute C_P:

$$C_P = .5 \frac{BHP/1000}{\sigma(N/1000)^3(D/10)^5}$$

6. Determine X and C_{PX}:

$$C_{PX} = \frac{C_P}{X}$$

7. Compute

$$\frac{J}{C_P^{\frac{1}{3}}}$$

8. Enter chart for value of $J/C_P^{\frac{1}{3}}$ and C_{PX}, and read off η.

9. Correct for loss in η from Figure 3–21 if necessary.

From the above procedure a curve of THP available versus speed may be determined.

Now that some information of the nature of the propelling force associated with different power plants is known, the actual performance of an airplane may be determined. In the following chapter the mechanics of an airplane in flight is studied, and simple formulas and charts are developed to determine airplane performance.

PROBLEMS

3-1. *a.* From Figure 3-2 determine the maximum continuous thrust at $M = .5$ at 30,000 ft altitude for the following power plants:

(1) Reciprocating engine
(2) Turbine propeller   Nacelle diameter $= 50$ in.
(3) Turbojet

b. For a range of 400 miles and a cruising speed of 300 mph at 30,000 ft altitude, which of the above power plants is most efficient (e.g., minimum weight of power plant plus fuel)? See Figures 3-3 and 3-14.

c. Draw a plot of cruising speed versus range (200–1000 miles), indicating the zones in which each power plant is most efficient.

3-2. *a.* Determine from Figure 3-7 the net thrust of the turbojet power plant for the following operating conditions:

$N = 11,500$ Altitude $= 25,000$ ft
$V = 450$ mph   Standard ambient T and p at 25,000 ft
$P_2 = 13.5$ in. Hg $T_2 = -25°C$

b. What is the ram efficiency?

c. If the heating value of the fuel is 18,500 Btu/lb, and the wake velocity, v_w, is 2000 ft/sec, determine the thermal and propulsive efficiencies.

d. Assuming η_p of a ducted fan (Figure 3-12) installation (built from unit in Part *a*) is .80, what is the improvement in specific fuel consumption over the basic unit? Assume η of free turbine $= .80$ and η of fan $= .95$.

3-3. From Figure 3-10a, determine the necessary data to draw curves in Figure 3-10b of c and $ESHP$ versus V for sea level and 40,000 ft altitude. Assume $N = 12,000$, and propeller efficiency, $\eta_p \cong .85$.

3-4. Prove that $b = 2a$ in Figure 3-16 by solution of the following equations:

(1) Bernoulli's equation of flow from Point A to a point just in front of the propeller disc.
(2) Bernoulli's equation of flow from a point just to the rear of the disc to Point B.
(3) Equation of thrust in terms of pressure differential across the disc.
(4) Equation of thrust in terms of mass flow at the disc.

3-5. Assuming an airplane powered by two Wasp Major engines (Figure 3-13), determine from Figure 3-15 the approximate cooling THP required for climb at $V_i = 160$ mph, flaps full open. Assume

$$A = 1.1 \frac{\pi D^2}{4}, \text{ and } A_e = A$$

$$V_i = \sqrt{\sigma} V$$

What percentage of the maximum continuous *BHP* available does this power represent?

3-6. Given the following airplane characteristics:

(1) Four engines rated at:

> 3500 bhp each for take-off, $N = 2700$
> 2000 bhp each for best cruise, $N = 2500$
> 2650 bhp each maximum continuous, $N = 2550$

(2) Four-bladed propellers, diameter $= 16.5$ ft, activity factor per blade $= 120$
(3) Gear ratio $= .5$

Find η at

(1) Cruise speed of 275 mph at 10,000 ft altitude.
(2) Take-off speed of 120 mph at sea level.
(3) High speed of 375 mph at 20,000 ft altitude
   (use maximum continuous *BHP*).

3-7. Design a 3-bladed propeller for the airplane in Problem 3-5 to give optimum efficiency in climb at $V_i = 160$ mph at 5000 ft altitude.

> Choice of blade activity factor: 90 and 110
> Choice of gear ratio: .381 and .4375

CHAPTER 4

AIRPLANE PERFORMANCE

4-1 Performance Computation

The performance characteristics of an airplane in various attitudes of flight can be easily determined by analyzing the nature of the forces that act on the airplane in a steady-state condition of level flight, glide, climb, or turn. (See Figure 4-1.) For example, the sinking speed or rate of sink of an airplane, power off, is determined in Chapter 2 by analyzing the forces acting on the airplane in a steady-state glide. Most of the principal items of airplane performance can be determined by the drag or power requirements of the airplane in level flight at a uniform speed. The available thrust or power from the engine at this same speed determines, then, the climb, descent, or level flight characteristics of the airplane. Conventional methods of performance computation for airplanes using reciprocating or turbine engines driving propellers adopt either mathematical or graphical solutions of the power required by and the power available for the airplane in uniform level flight at various speeds. The use of the available power instead of available thrust is convenient in dealing with this type of an engine, as the engine data are usually presented in terms of power. Turbojet, ramjet, and rocket engine data, however, are presented in units of thrust, so that the use of available and required airplane thrust in determining airplane performance is both convenient and fundamental.

Thrust Required and Available

In Figure 4-1a, the forces on an airplane in equilibrium flight at a constant altitude and speed are represented by the vectors lift, drag, thrust, and weight, acting through the airplane center of gravity. From the summation of horizontal forces,

$$D = T = C_D q S \qquad (4-1)$$

Assuming the parabolic variation of C_D versus C_L as developed in Chapter 2,

$$D = \left(C_{D_f} + \frac{C_L{}^2}{\pi A e} \right) \frac{\sigma V^2}{391} S$$

155

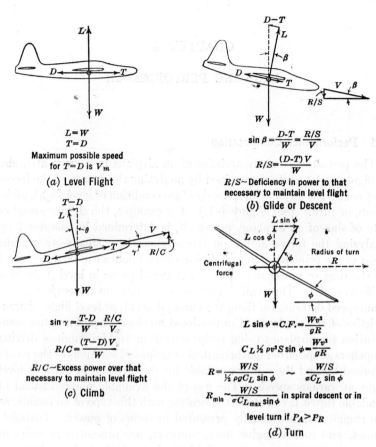

FIGURE 4–1. Airplane in equilibrium in various attitudes of flight.

and since

$$f = C_{D_f}S$$

and

$$L = W$$

then

$$D = T_R = \frac{\sigma f}{391} V^2 + \frac{124.5}{\sigma e} \left(\frac{W}{b}\right)^2 \frac{1}{V^2} \qquad (4-2)$$

Equation (4–2) is a mathematical expression for the thrust required, T_R, for the airplane in level flight. The first term is defined as parasite thrust required, and the second term as effective induced thrust required. From this expression it can be seen that four fundamental parameters affecting the drag of the airplane at a given speed are

$$\frac{W}{b}, f, e, \text{ and } \sigma$$

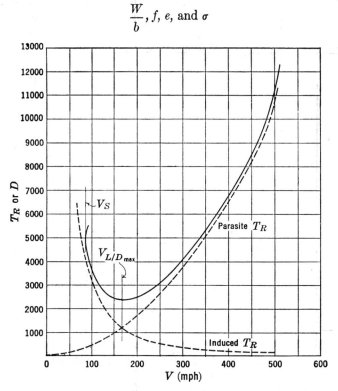

FIGURE 4–2. Thrust required versus speed.

If wind-tunnel data indicate radical departures from a parabolic variation of C_D with C_L, then a graphical plot of drag may be determined
from equation 4–1, in which C_D is determined corresponding to a given
speed or C_L.

In order to determine graphically airplane performance items for
airplanes with engines for which performance is presented in terms of
thrust, it is convenient to use a plot of fundamental drag or thrust
required, as defined by equation 4–2 or wind-tunnel data, as presented
in Figure 4–2.   Here again the final curve is the sum of the components
of minimum parasite drag and effective induced drag.   At the higher
ranges of speed common to jet-propelled airplanes the T_R increases at
a greater rate than that indicated by equation (4–2) because of compressibility effects.   Of course at transonic and supersonic speeds the
T_R curves depart radically from a parabolic variation.   It is also noted

that the T_R curve hooks up rapidly near the stalling speed, V_S, of the airplane. Near this speed, separation occurs on the wing and the drag increases at a greater rate than that predicted by the parabolic polar. The speed for maximum lift to drag ratio, $V_{L/D \max}$, is easily determined as that corresponding to minimum T_R.

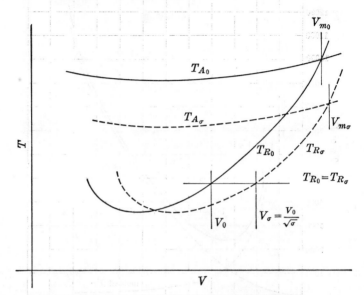

FIGURE 4-3. T_R and T_A at sea level and altitude.

Figure 4-3 shows a plot of thrust required and available at sea level and altitude. To obtain thrust required at some altitude, $T_{R\sigma}$, equation (4-2) may be used with the appropriate density ratio, σ, for the altitude selected. Since in level flight,

$$W_0 = C_{L0} \frac{V_0^2}{391} S = W_\sigma = C_{L\sigma} \frac{\sigma V_\sigma^2}{391} S \qquad (4\text{-}3)$$

this method mathematically determines points on the new T_R curve at altitude at constant values of V and changing values of C_L. Another way to determine the same curve at altitude is to keep C_L constant and to increase the values of V at sea level by the factor $1/\sqrt{\sigma}$, at the same time holding drag or thrust constant. Mathematically this can be verified by analyzing the relationship of W_0 and W_σ from equation (4-3). By definition in this case,

$$C_{L0} = C_{L\sigma}, \quad C_{D0} = C_{D\sigma}$$

therefore

$$q_0 = q_\sigma$$

or

$$V_\sigma = \frac{V_0}{\sqrt{\sigma}} \qquad (4\text{-}4)$$

also

$$T_{R0} = C_{D0}q_0 S$$

and

$$T_{R\sigma} = C_{D\sigma}q_\sigma S$$

therefore

$$T_{R0} = T_{R\sigma} \qquad (4\text{-}5)$$

Actually, of course, an airplane, in flying from one altitude to another at approximately the same weight, experiences changes in both C_L and V. However, it would be difficult to determine mathematically T_R conditions at different altitudes by letting both these parameters vary at the same time. By keeping either one constant and varying the other the mathematical solution is greatly simplified.

Thrust-available curves in Figure 4-3 are typical for a turbojet power plant. The resulting system of T_A and T_R curves is fundamental for determining level flight, climb, and descent performance of this type of airplane. Reference to Figure 4-1 again indicates the relationships consistent with maximum speed, V_m, rate of climb, R/C, and rate of sink, R/S, performance. Maximum and minimum speeds occur at the highest and lowest speeds for which available and required thrusts are equal. Stalling speed, V_S, may dictate minimum flying speed. Rate of climb in units of feet per minute is expressed thus:

$$R/C = \frac{88\Delta T V}{W} \qquad (4\text{-}6)$$

where $\Delta T = T_A - D$. It is of interest to note that speeds for maximum R/C and minimum R/S cannot be readily determined from Figure 4-3, as both R/C and R/S are proportional to power, and power contains speed as a variable. R/C and R/S versus speed plots are therefore necessary to determine optimums. As will be shown later, the maximum and minimum distances between the basic power-available and power-required curves determine these optimums.

The total time to climb to any altitude, t, may be obtained by adding increments in time between altitude regions in which the rate of climb varies linearly. Consider the time to climb between the two altitudes, h_1 and h_2, in Figure 4–4. In general

$$dt = \frac{dh}{dh/dt} = \frac{1}{R/C} dh$$

$$\int_{t_1}^{t_2} dt = \int_{h_1}^{h_2} \frac{1}{R/C} dh$$

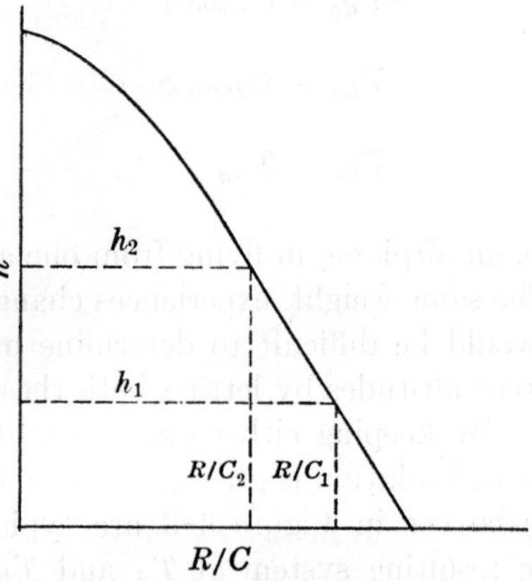

FIGURE 4–4. Time to climb.

Since R/C is assumed to be a linear function of h between altitudes h_1 and h_2

$$R/C = R/C_1 - \frac{R/C_1 - R/C_2}{h_2 - h_1}(h - h_1)$$

$$= \frac{R/C_1 h_2 - R/C_2 h_1 + h(R/C_2 - R/C_1)}{h_2 - h_1}$$

$$t_2 - t_1 = \int_{h_1}^{h_2} \frac{(h_2 - h_1)dh}{R/C_1 h_2 - R/C_2 h_1 + h(R/C_2 - R/C_1)}$$

$$= \frac{h_2 - h_1}{R/C_2 - R/C_1} \log_e \left[R/C_1 h_2 - R/C_2 h_1 + h(R/C_2 - R/C_1) \right]_{h_1}^{h_2}$$

$$= \frac{h_2 - h_1}{R/C_2 - R/C_1} \log_e \frac{R/C_2 h_2 - R/C_2 h_1}{R/C_1 h_2 - R/C_1 h_1}$$

$$= \frac{h_2 - h_1}{R/C_2 - R/C_1} \log_e \frac{R/C_2}{R/C_1}$$

$$t_2 - t_1 = 2.3 \frac{h_2 - h_1}{R/C_1 - R/C_2} \log_{10} \frac{R/C_1}{R/C_2} \qquad (4\text{-}7)$$

$\approx \Delta$ time in minutes to climb between altitudes h_1 and h_2.

Equation (4-7) may be used to determine the time to climb between sea level and any altitude provided an approximate linear variation in R/C to that altitude exists. If small increments in altitude are chosen, an approximate and simplified method may also be used. In this case

$$t_2 - t_1 = \frac{2}{R/C_1 + R/C_2} (h_2 - h_1) \qquad (4\text{-}8)$$

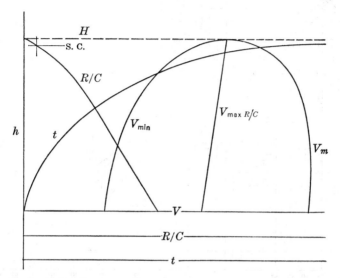

FIGURE 4-5. Summary of performance items for turbojet airplane.

Expressions similar to equations (4-7) and (4-8) with R/S substituted for R/C may be used to determine the time to descend between two altitudes. Since an airplane usually descends at more nearly a constant rate than that for climb, equation (4-8) is usually used for descent problems.

A typical summary of some of the important performance items for a turbojet airplane appears in Figure 4-5. In this figure, V_m, V_{min}, maximum R/C, V for maximum R/C, and time to climb are plotted against altitude. The absolute ceiling of the airplane, H, is the altitude at which the rate of climb is zero. Service ceiling, S.C., is defined as the altitude at which the rate of climb is 100 feet per minute.

Sometimes it is required to compute the speed performance of an airplane for a given amount of available thrust. A graphic solution such as indicated by Figure 4-3 would give the desired result, but the plotting of the basic thrust-available and thrust-required curves to obtain a particular speed intersection is time consuming and unnecessary. Of course the direct solution of equation (4–2) is possible but somewhat difficult. Consider now dividing equation (4–2) by a value of thrust available equal to the thrust required.

$$1 = \frac{\sigma f}{T}\frac{V^2}{391} + \frac{124.5}{V^2}\frac{(W/b)^2}{\sigma e T} \tag{4-9}$$

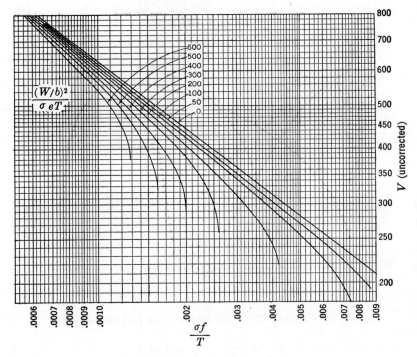

FIGURE 4–6. Speed solution chart.

Equation (4–9) is graphically solved in Figure 4–6 in terms of the parameters

$$\frac{\sigma f}{T}, \quad \frac{(W/b)^2}{\sigma e T}, \quad \text{and } V$$

This chart is particularly useful for determining any speed corresponding to a given thrust available. The speed read from Figure 4–6 is uncorrected for compressibility effects.

Power Required and Available

Assuming again a parabolic variation of C_D versus C_L, the basic power-required equation is derived directly from the thrust-required relationship.

$$P_R = \frac{DV}{375} = \frac{T_R V}{375}$$

and multiplying equation (4-2) by $V/375$

$$P_R = \frac{\sigma f}{146{,}600} V^3 + \frac{.332}{\sigma e} \left(\frac{W}{b}\right)^2 \frac{1}{V} \tag{4-10}$$

Equation (4-10) is a mathematical expression for power required, P_R, for the airplane in level flight. The first term is defined as parasite

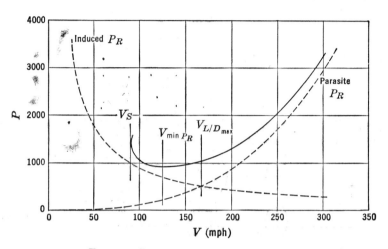

FIGURE 4-7.   Power required versus speed.

power required, and the second term as effective induced power required. Graphical plots of each term in equation (4-10) and the sum
of both terms are represented in Figure 4-7.

Several fundamental characteristics of the basic P_R curve are of
interest. At a speed corresponding to the intersection of a straight line
through the origin and tangent to the P_R curve the parasite P_R is
equal to the effective induced P_R. This speed corresponds to that for
maximum L/D and is defined as $V_{L/D_{max}}$. From the basic thrust-
required curve in Figure 4-2 it can also be seen that parasite T_R is
equal to effective induced T_R at $V_{L/D_{max}}$. As the tangent to the P_R
curve at this point dictates the minimum ratio of sinking speed, R/S,
to V, it can be again seen from Figure 4-1b that for this speed the glide
angle, power off, is a minimum, and the ratio L/D is a maximum. To
verify mathematically the relationship of parasite and induced powers,

drags, or drag coefficients at this point, consider the parabolic drag polar divided by C_L.

$$\frac{C_D}{C_L} = \frac{C_{Df}}{C_L} + \frac{C_L^2/\pi Ae}{C_L} \tag{4-11}$$

Differentiating C_D/C_L with respect to C_L and setting to zero

$$\frac{d(C_D/C_L)}{dC_L} = -\frac{C_{Df}}{C_L^2} + \frac{1}{\pi Ae} = 0$$

and for minimum C_D/C_L or maximum L/D,

$$\frac{C_{Df}}{C_L^2} = \frac{1}{\pi Ae}$$

or

$$C_{Df} = \frac{C_L^2}{\pi Ae} \tag{4-12}$$

By substituting the relationship from equation (4-12) back into equation (4-11) and inverting, an expression for $(L/D)_{max}$ results.

$$\left(\frac{L}{D}\right)_{max} = \left(\frac{C_L}{C_D}\right)_{max} = \frac{\sqrt{\pi AeC_{Df}}}{2C_{Df}}$$

$$\left(\frac{L}{D}\right)_{max} = \sqrt{\frac{\pi Ae}{4C_{Df}}} = .886b\sqrt{\frac{e}{f}} \tag{4-13}$$

The corresponding V for $(L/D)_{max}$ is obtained by substituting a value for C_L from equation (4-12) in the expression

$$V = \sqrt{\frac{391W}{\sigma C_L S}}$$

$$V_{L/D_{max}} = \sqrt{\frac{391W/S}{\sigma\sqrt{\pi AeC_{Df}}}} = \frac{14.85}{\sqrt[4]{fe}}\sqrt{\frac{W}{\sigma b}} \tag{4-14}$$

Thrust and power at $V_{L/D_{max}}$ can be obtained also from the preceding relationships.

$$D_{L/D_{max}} = T_{L/D_{max}} = \frac{W}{L/D_{max}} = 1.132\frac{W}{b}\sqrt{\frac{f}{e}} \tag{4-15}$$

$$P_{L/D_{max}} = \frac{T_{L/D_{max}}V_{L/D_{max}}}{375} \tag{4-16}$$

From Figure 4–1b it is seen that the speed for minimum P_R, V_{minP_R}, in Figure 4–7 is also the speed for minimum sinking speed, power off. It is of practical importance to a pilot to recognize that the speed for minimum R/S is not the best speed at which to make a dead-stick approach over an obstacle to a landing field. It is apparent that the V for L/D_{max} would give the flattest glide and give the airplane the best chance to make the field, even though the R/S for this condition is not a minimum.

Referring again to Figure 4–7, it is of interest to note that at the speed for minimum P_R the parasite P_R is one-third of the induced P_R, or in terms of drag coefficient

$$C_{D_f} = \frac{1}{3} \frac{C_L{}^2}{\pi A e} \tag{4-17}$$

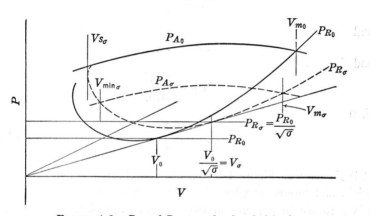

FIGURE 4–8. P_R and P_A at sea level and altitude.

This relationship can be mathematically verified by expressing P_R at a given altitude as a function of $C_D/C_L{}^{3\!/_2}$ and then differentiating the expression for $C_D/C_L{}^{3\!/_2}$ with respect to C_L to find a value of C_L for minimum P_R. To obtain an expression for V at minimum P_R a technique comparable to that used in deriving equation (4–14) yields

$$V_{minP_R} = \frac{11.28}{\sqrt[4]{fe}} \sqrt{\frac{W}{\sigma b}} \tag{4-18}$$

Now to obtain other performance items, such as maximum speed and speeds for best rates of climb and descent for various amounts of power available, consider curves of P_A and P_R for different altitudes (Figure 4–8) and plots of rate of sink and rate of climb for a given altitude (Figure 4–9). In Figure 4–8 curves of thrust horsepower available at sea level, P_{A0}, and at altitude, $P_{A\sigma}$, are obtained from engine-performance curves and propeller charts. Engine brake or shaft horsepower is converted to thrust horsepower by multiplying by propeller efficiency

$$SHP \times \eta = THP = P_A \qquad (4\text{--}19)$$

Power required for an airplane at a given weight at any altitude, $P_{R\sigma}$, is computed directly from equation 4–10 for a constant value of speed. If lift coefficient is assumed constant, values of V_0 and P_{R0} for a given point on the sea level curve in Figure 4–8 are increased by a factor of $1/\sqrt{\sigma}$ for the corresponding point on an altitude curve. It has already been shown that

$$V_\sigma = \frac{V_0}{\sqrt{\sigma}}$$

and

$$T_{R0} = T_{R\sigma}$$

and since

$$P_R = \frac{T_R V}{375}$$

then

$$P_{R\sigma} = \frac{P_{R0}}{\sqrt{\sigma}} \qquad (4\text{--}20)$$

It is also seen in Figure 4–8 that any radial straight line through the origin and intersecting the P_{R0} curve can be extended by the ratio $1/\sqrt{\sigma}$ to obtain a point on the $P_{R\sigma}$ curve. Relationships between either thrust or power-required curves for different values of weight can be derived from the fundamental weight expression indicated by equation (4–3).

The system of P_R and P_A curves in Figure 4–8 is fundamental in determining climb and descent performance of an airplane. Minimum speed and maximum speed are determined by the intersection of the two curves at any one altitude. At speeds below and above these two speeds the airplane does not have sufficient P_A to maintain level flight. It is of interest to note in Figure 4–8 that the stalling speed at altitude, $V_{S\sigma}$, cannot be reached in level continuous flight. From Figure 4–1b and c it is also noted that the rate of climb or sink is directly proportional to the difference between the available and required power. In Figure 4–9 are plotted rates of climb for maximum P_A and rates of sink for some percentage reduction in P_A, all at one altitude.

A speed solution chart in terms of shaft horsepower appears in

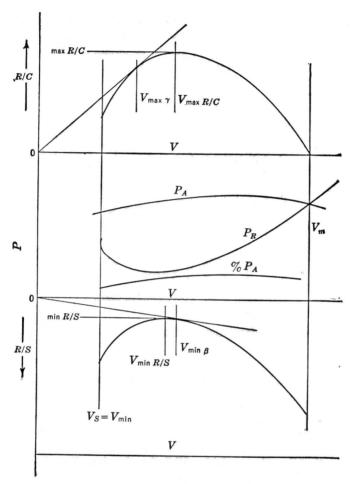

FIGURE 4-9.   Climb and descent performance.

Figure 4-10.   By dividing equation (4-10) by P_R or the equivalent of ηSHP, equation (4-21) results:

$$1 = \frac{\sigma f V^3}{\eta SHP \; 146,600} + \frac{.332}{V} \frac{(W/b)^2}{\sigma e \eta SHP} \qquad (4\text{-}21)$$

Equation (4-21) is graphically solved in Figure 4-10 in terms of parameters

$$\frac{\sigma f}{\eta SHP}, \quad \frac{(W/b)^2}{\sigma e \eta SHP}, \quad \text{and} \quad V$$

This graph is then particularly useful in determining any cruising or high speed for any amount of brake or shaft horsepower available. Plotting of the basic P_R and P_A curves to obtain a particular speed intersection as in Figure 4–8 is more laborious and unnecessary.

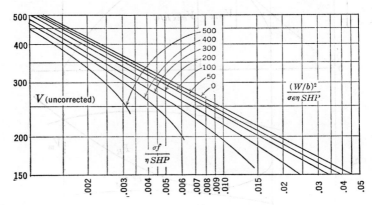

FIGURE 4–10.   Speed solution chart.

4–2   Generalized Performance Method

Speed Performance

Many attempts have been made in the literature to minimize the number of computations necessary to plot curves of T_R or P_R for various weights and altitudes. For example, in Figure 4–3, a curve of T_R is necessary for each altitude. If T_R were plotted against $V/\sqrt{\sigma}$ as the abscissa, then one curve would suffice for all altitudes, provided, of course, that the other variables, $W/b, f$, and e, that affect the airplane drag or power required remain constant. By an additional simplifying step the T_R curve may be completely generalized or non-dimensionalized, so that one curve is applicable for all altitudes and any combination of the airplane characteristics of $W/b, f$, and e. Equation (4–2) may be written as

$$T = AV^2 + \frac{B}{V^2} \qquad (4\text{--}22)$$

where A and B are constants which depend on altitude and airplane characteristics.

$$A = \frac{\sigma f}{391} \qquad B = \frac{124.5}{\sigma e}\left(\frac{W}{b}\right)^2$$

By analyzing the nature of the two terms in equation (4–22) or Figure 4–2, it is noticed that at one and only one speed are the parasite and effective induced thrusts equal. This occurs at the speed for minimum drag or thrust required. This speed is also defined as speed for maximum lift to drag ratio, $V_{L/D\max}$. To non-dimensionalize equa-

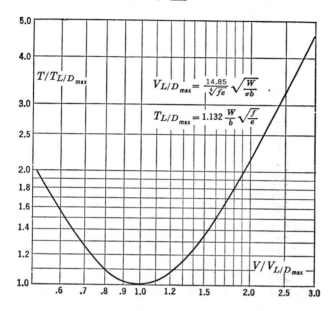

FIGURE 4–11.   Generalized thrust-required curve.

tion (4–22), it is suspected then that the divisor is one-half $T_{L/D\max}$, which is common to both the parasite and effective induced terms in equation (4–22) or Figure 4–2 only at $V_{L/D\max}$. The following form then results:

$$\frac{2T}{T_{L/D\max}} = \left(\frac{V}{V_{L/D\max}}\right)^2 + \frac{1}{\left(\dfrac{V}{V_{L/D\max}}\right)^2} \qquad (4\text{–}23)$$

which is a non-dimensional form of airplane thrust required versus speed. Mathematically this is proved by dividing equation (4–22) by $\sqrt{AB}$

$$\frac{T}{\sqrt{AB}} = \left(\frac{V}{\sqrt[4]{B/A}}\right)^2 + \frac{1}{\left(\dfrac{V}{\sqrt[4]{B/A}}\right)^2}$$

where

$$\sqrt{\overline{AB}} = \frac{1}{2} T_{L/D\max} = .565 \frac{W}{b} \sqrt{\frac{f}{e}} \qquad (4\text{-}24)$$

and

$$\sqrt[4]{B/A} = V_{L/D\max} = \frac{14.85}{\sqrt[4]{fe}} \sqrt{\frac{W}{\sigma b}} \qquad (4\text{-}25)$$

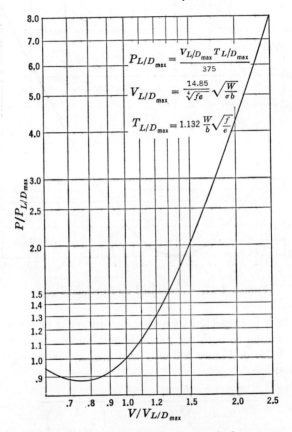

FIGURE 4-12. Generalized power-required curve.

The non-dimensional form of the generalized thrust-required curve is presented in Figure 4–11. For convenience, $T_{L/D\max}$ instead of one-half this value is used in this figure. To develop the generalized power-required curve presented in Figure 4–12,

$$P = \frac{DV}{375} = \frac{TV}{375}.$$

and

$$\frac{P}{P_{L/D\max}} = \left(\frac{T}{T_{L/D\max}}\right)\left(\frac{V}{V_{L/D\max}}\right) \qquad (4\text{--}26)$$

Both these generalized curves are applicable for all altitudes and any combination of airplane characteristics.

To use Figures 4–11 and 4–12, values of available thrust and power are divided by the same constants that are used to non-dimensionalize the thrust and power-required curves. Conventional methods can then be used to determine speed performance. For example, $V_{L/D\max}$ and $T_{L/D\max}$ are computed at a given altitude in terms of the airplane parameters, W/b, f, and e as expressed in equations (4–14) and (4–15) and indicated on Figure 4–11. Maximum $\frac{V}{V_{L/D\max}}$ is obtained from Figure 4–11 at the intersection of the $\frac{V}{V_{L/D\max}}$ curve and the maximum $\frac{T}{T_{L/D\max}}$ ratio as determined from maximum engine thrust available. Maximum V uncorrected for compressibility effects is evaluated by multiplying maximum $\frac{V}{V_{L/D\max}}$ by $V_{L/D\max}$. Considerable time is saved in routine performance calculations by reproducing large numbers of the non-dimensional T_R curve in Figure 4–11 and imposing thereon the particular values of $\frac{T_A}{T_{L/D\max}}$ peculiar to the airplane and engine.

Compressibility Speed Corrections

The determination of speeds by the preceding performance methods is based on the assumption that the drag of an airplane follows a parabolic polar relationship. At or near the critical Mach number of the airplane the drag increases at a greater rate than that predicted by the parabolic variation as evidenced by the high-speed drag data presented in Chapter 2. By use of the compressibility drag correction methods presented in Chapter 2 and analysis of wind-tunnel data it is possible to generalize compressibility corrections for typical airplanes in terms of lift coefficient, wing-thickness ratio, and flight altitude (temperature), which are the primary variables affecting the wing critical Mach number.

In Figure 4–13 are presented speed correction curves for straight wing airplanes for altitudes of sea level, 20,000 feet and 35,000 feet. A corrected speed V_{corr}, is obtained from this chart for any incompressible speed, V, as determined from the preceding speed-solution or generalized thrust and power curves. M_{cr} is defined as V_{corr}/a, where

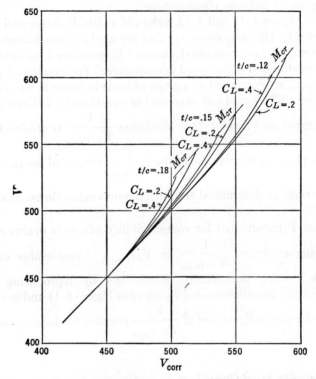

FIGURE 4–13a. Speed correction curve, constant thrust (sea level).

a is the speed of sound in the standard atmosphere. The assumption of constant thrust between V and V_{corr} applies approximately for turbojet airplanes. In this chart an incompressible speed is determined in terms of a generalized ratio of $C_{D_{corr}}/C_D$ versus V_{corr} and M_{cr} as obtained from test data.

$$T = D = C_D q S = C_{D_{corr}} q_{corr} S \qquad (4\text{--}27)$$

and for a particular airplane and altitude

$$V = V_{corr} \sqrt{\frac{C_{D_{corr}}}{C_D}} \qquad (4\text{--}28)$$

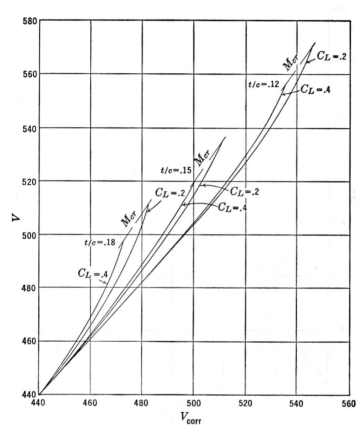

FIGURE 4–13b.   Speed correction curve, constant thrust ($h = 20,000$ ft).

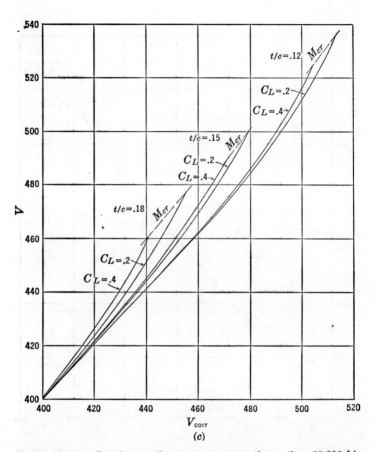

FIGURE 4–13c. Speed correction curve, constant thrust ($h = 35{,}000$ ft).

In Figure 4–14 constant power is assumed between V_{corr} and V such that

$$V = V_{\text{corr}} \sqrt[3]{\frac{C_{D_{\text{corr}}}}{C_D}} \qquad (4\text{–}29)$$

These curves give approximate corrections for propeller-driven airplanes. In Figure 4–15 is presented speed corrections at constant

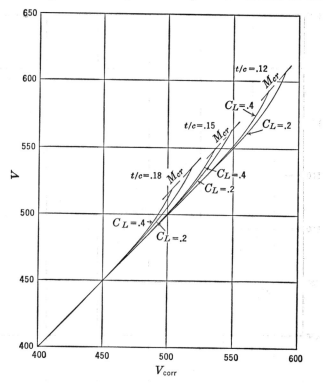

FIGURE 4–14a. Speed correction curve, constant power (sea level).

thrust for typical turbojet airplanes with wings swept back 35 degrees. Constant power corrections are not given, as it is assumed that propeller-driven aircraft will not be designed with highly swept wings. The foregoing charts should be used with caution, as generalized subsonic high-speed data are still controversial.

Climb and Descent Performance

For constant thrust-available engines, airplane speeds for maximum rate of climb and minimum rate of sink may be obtained from the

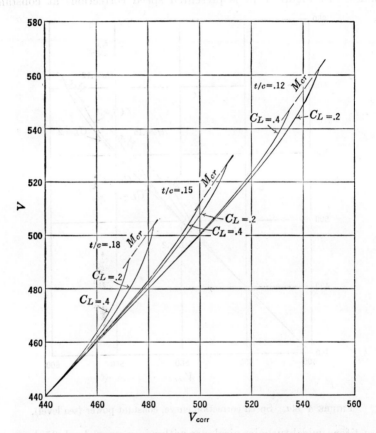

FIGURE 4–14b. Speed correction curve, constant power ($h = 20,000$ ft).

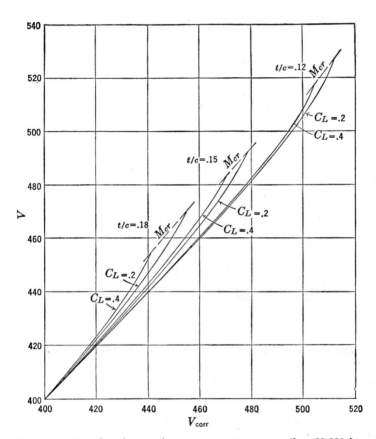

FIGURE 4–14c. Speed correction curve, constant power (h = 35,000 ft).

non-dimensional curves presented in Figure 4–16.   From Figure 4–11
and equation (4–6)

$$R/C = \frac{88 T_{L/D\max} V_{L/D\max} \dfrac{\Delta T}{T_{L/D\max}} \dfrac{V}{V_{L/D\max}}}{W} \qquad (4\text{–}30)$$

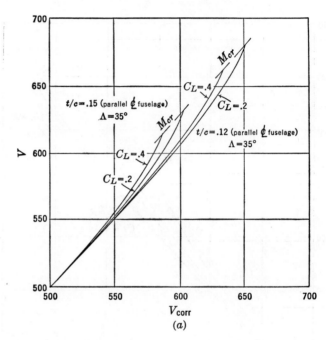

FIGURE 4–15a.   Speed correction curve, constant thrust (sea level).

where

$$\frac{\Delta T}{T_{L/D\max}} = \frac{T_A}{T_{L/D\max}} - \frac{T_R}{T_{L/D\max}}$$

and

$$\frac{2 T_R}{T_{L/D\max}} = \left(\frac{V}{V_{L/D\max}}\right)^2 + \frac{1}{\left(\dfrac{V}{V_{L/D\max}}\right)^2} \qquad (4\text{–}31)$$

For a given altitude and airplane

$$R/C \sim \frac{\Delta T}{T_{L/D\max}} \frac{V}{V_{L/D\max}} \qquad (4\text{–}32)$$

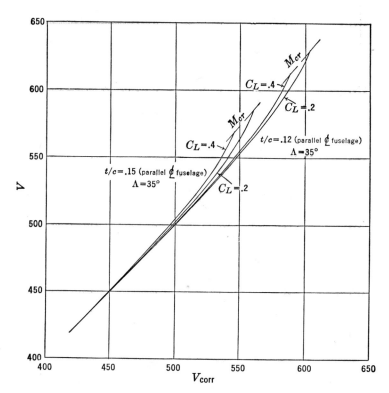

FIGURE 4-15b.   Speed correction curve, constant thrust ($h = 20,000$ ft).

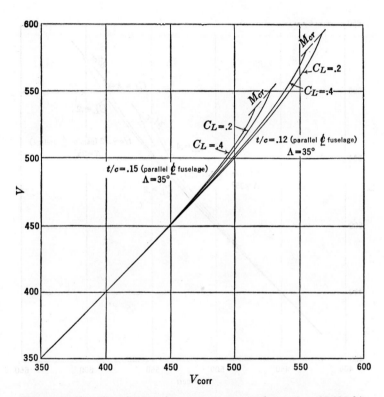

Figure 4–15*c*. Speed correction curve, constant thrust (h = 35,000 ft).

Letting

$$\frac{2T_R}{T_{L/D\max}} = y$$

$$\frac{2T_A}{T_{L/D\max}} = y_1$$

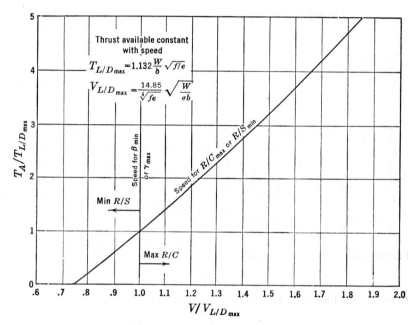

FIGURE 4–16.   Climb and descent performance.

and

$$\frac{V}{V_{L/D\max}} = x$$

Equation (4–31) can then be written

$$y = x^2 + \frac{1}{x^2}$$

From equation (4–30) and Figure 4–17 the maximum R/C or minimum R/S occurs at maximum $x\ \Delta y$.   To find the values of x for maximum

$x \, \Delta y$, the conventional method of differential calculus is applied:

$$x \, \Delta y = x(y_1 - y) = xy_1 - x^3 - \frac{1}{x}$$

$$\frac{d(x \, \Delta y)}{dx} = y_1 - 3x^2 + \frac{1}{x^2} = 0$$

$$x \text{ for maximum } R/C = \sqrt{\frac{y_1 + \sqrt{y_1{}^2 + 12}}{6}} \qquad (4\text{--}33)$$

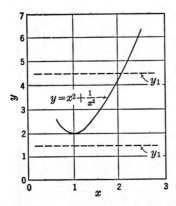

FIGURE 4–17.   Non-dimensional
Δ thrust.

From equation (4–33) values of $V/V_{L/D_{\max}}$ for $R/C_{\max}$ for several values of y_1 or $T_A/T_{L/D_{\max}}$ can be computed and plotted as in Figure 4–16. If $T_A/T_{L/D_{\max}}$ is less than one, speeds for $R/S_{\min}$ are obtained.

Speeds for minimum angle of glide, $\beta_{\min}$, and maximum angle of climb, $\gamma_{\max}$, occur at $V/V_{L/D_{\max}} = 1$ if $\beta_{\min}$ and $\gamma_{\max}$ are small. For example, from Figure 4–1c

$$\gamma = \sin^{-1} \frac{(T - D)}{W}$$

and in terms of non-dimensional parameters of Figure 4–17 and for small values of γ

$$\gamma = \text{Constant} \left(y_1 - x^2 - \frac{1}{x^2} \right) \qquad (4\text{--}34)$$

$$\frac{d\gamma}{dx} = 0$$

and

$$x = 1 = \frac{V}{V_{L/D_{\max}}} \text{ for } \gamma_{\max}$$

By methods of analysis comparable to that used to derive the curves in Figure 4–16, climb and descent performance for constant-power conditions are presented in Figure 4–18. In this chart $V/V_{L/D_{\max}}$ for $\gamma_{\max}$ or $\beta_{\min}$ can be determined for any value of $P_A/P_{L/D_{\max}}$. The speed for $R/C_{\max}$ or $R/S_{\min}$ occurs at $V/V_{L/D_{\max}} = .76$, which is the speed for minimum power required.

4-3   Special Performance Problems

Range and Endurance

The determination of range and endurance for airplanes using reciprocating engines is relatively simple if propulsive efficiency, η, and specific fuel consumption in pounds per brake horsepower-hour, c, are

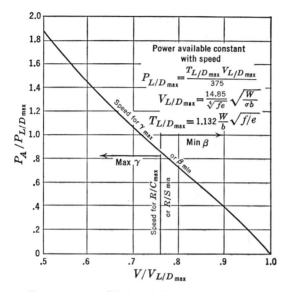

FIGURE 4–18.   Climb and descent performance.

considered constant for cruising speeds and corresponding powers delivered by the engine. This assumption is valid for a first approximation, as this type of an engine usually operates for maximum range performance at a constant cruising power with an auto lean fuel-to-air mixture ratio which results in a relatively constant variation in c between 50 and 65 per cent of the normal rated power output of the engine. For corresponding cruising speeds propulsive efficiency is constant within 1 or 2 per cent. At high altitudes the power requirements for optimum cruising conditions sometimes exceed the cruising power available. For these conditions the engine must operate at a high per cent of normal power at a high relative value of fuel-to-air mixture. Because of the high values of specific fuel consumption associated with these conditions the range of the airplane is less than at lower altitudes.

Fundamentally, for any airplane-engine combination the endurance, E, may be computed by dividing the total available fuel, F, by the average pounds per hour consumed by the engine:

$$E = \frac{F}{(\text{lb/hr})_{\text{average}}}$$ (4–35)

Similarly, range, R, may be computed by multiplying the total fuel by the average miles traveled per pound of fuel consumed.

$$R = \left(\frac{\text{Miles}}{\text{lb}}\right)_{\text{average}} (F)$$ (4–36)

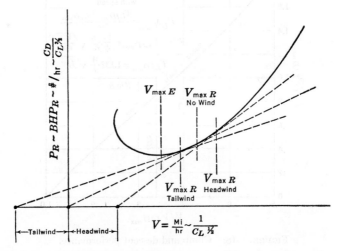

FIGURE 4–19.   Maximum range and endurance conditions for reciprocating-engine airplane.

By referring to Figure 4–19 maximum R and E may be determined for a reciprocating-engine airplane in terms of the fundamental power-required characteristics for an average weight condition of the airplane. By assuming a constant c and η, the ordinate in Figure 4–19 is proportional to lb/hr since

$$\frac{P_R}{\eta} = BHP_R$$

and

$$cBHP_R = \text{lb/hr}$$

and the curve of lb/hr would have the same shape as the P_R curve but displaced vertically by a constant amount. The V for maximum E corresponds to the point on the curve for minimum lb/hr. For maximum R a tangent to the curve through the origin locates the V for maximum R, as this point on the curve has the lowest ratio of $(\mathrm{lb/hr})/V$ or the highest ratio of miles per pound.

$$\frac{V}{\mathrm{lb/hr}} = \frac{\mathrm{Miles/hr}}{\mathrm{lb/hr}} = \frac{\mathrm{Miles}}{\mathrm{lb}}$$

As was pointed out before, this speed is also $V_{L/D_{\max}}$. In Figure 4–19 the ordinate is proportional to $C_D/C_L^{3/2}$ and the abscissa to $1/C_L^{1/2}$. Therefore V for maximum range is identical to $V_{L/D_{\max}}$, since the tangent through the origin to the P_R curve also defines the minimum ratio of $C_D/C_L^{3/2}$ divided by $1/C_L^{1/2}$ which is $L/D_{\max}$. With a tailwind or headwind the V for maximum range may be determined by locating the new origins on the abscissa corresponding to the new ground speeds and drawing tangents to the same curve. For both these conditions, of course, the airplane flies at some other airspeed than that for maximum L/D to obtain maximum range. Maximum endurance is a function only of minimum lb/hr and is therefore unchanged by the tailwind or headwind.

A consideration of the range problem from a classical standpoint as originally developed by Breguet many years ago leads to the same conclusions. The fundamental relationship for both range and endurance is

$$dt = dE = \frac{d\,\mathrm{lb}}{\mathrm{lb/hr}} = \frac{dF}{\mathrm{lb/hr}}$$

where dF is the differential weight of fuel burned. Expanding lb/hr,

$$dt = dE = \frac{dF}{BHPc} = \frac{-dW}{BHPc} \tag{4–37}$$

where $W = W_0 - F$, and $dW = -dF$, and represents the change in weight of the airplane as a result of fuel burned in time dt. W_0 is the initial weight and remains a constant. Expanding equation (4–37),

$$dt = \frac{-dW}{\dfrac{Pc}{\eta}} = \frac{-dW}{\dfrac{DV\,c}{375\,\eta}}$$

and since

$$D = \frac{W}{L/D}$$

$$dt = \frac{-375}{V}\frac{L}{D}\frac{\eta}{c}\frac{dW}{W} \tag{4–38}$$

By assuming flight at constant angle of attack which corresponds to constant C_L and C_D, V is allowed to vary while L/D remains constant. η/c is also assumed constant, and since

$$V \, dt = ds = dR$$

equation (4–38) may be integrated between the limits of the initial and final weights, W_0 and W_1, to obtain R:

$$\int_0^R dR = -375 \frac{L}{D} \frac{\eta}{c} \int_{W_0}^{W_1} \frac{dW}{W}$$

$$R = -375 \frac{L}{D} \frac{\eta}{c} \log_e \frac{W_1}{W_0}$$

and maximum R in miles

$$R = 375 \left(\frac{L}{D}\right)_{\max} \frac{\eta}{c} \log_e \frac{W_0}{W_1} \qquad (4\text{–}39)$$

or

$$R = 863.5 \left(\frac{L}{D}\right)_{\max} \frac{\eta}{c} \log_{10} \frac{W_0}{W_1} \qquad (4\text{–}40)$$

Thus, for a reciprocating engine, assuming constant η/c, maximum R occurs at the V for maximum L/D and is independent of altitude. Equation 4–40 in a modified form can be used to compute the range of an airplane with any power-plant installation. From Chapter 3

$$c \cong \frac{.138}{\eta_t}$$

and

$$R \cong 6280 \, \eta_p \eta_t \frac{L}{D} \log_{10} \frac{W_0}{W_1} \qquad (4\text{–}40a)$$

To facilitate the solution of equation (4–39) or (4–40) a plot of W_1/W_0 versus range for representative values of $L/D_{\max}$ may be computed from equations (4–13) and (4–14). In Figure 4–20 a typical value of $\eta/c = 1.89$ is assumed. For other values of η/c the range is corrected by a direct ratio.

An alternate graphical solution of range usually takes the form of a plot of miles/lb versus W as indicated in Figure 4–21 for a typical reciprocating-engine airplane configuration. The straight line variation in miles/lb versus W is usually the result of the small variation in η/c determined from detailed calculations and differs but slightly from the logarithmic variation indicated by equation (4–40). The range is determined from Figure 4–21 by the area under the curve. Both equation (4-40) and Figure 4-21 indicate the favorable effect on range of a high ratio of W_0/W_1. Near the end of a flight, approaching a value of $W = W_1$, the miles/lb is considerably greater than that at the

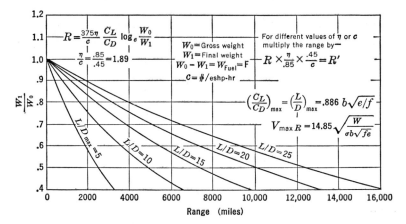

FIGURE 4-20.   Range chart for reciprocating engine airplanes.

beginning of the flight because of the decrease in power requirements at the lower weight.   It can also be seen from Figure 4-21 that weight control on airplanes with a high ratio of W_0/W_1 is very important if

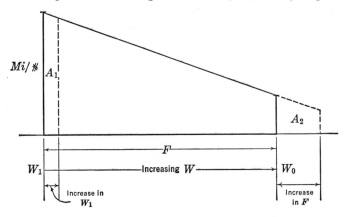

FIGURE 4-21.   Graphical solution of range.

range is to be kept a maximum.   Any increase in W_1 (operating empty weight) in the form of structure or equipment requires a proportionately greater increase in fuel weight if the total range is to be maintained constant.   For example, in this figure, A_1 must equal A_2 if R is to be maintained constant.   The low value of miles/lb at the beginning of the flight requires a proportionately greater amount of fuel than that required at the end of the flight. For some large bombers an increase of one unit in weight empty requires an increase in fuel weight of two units if range is to be maintained constant.

Endurance may also be computed by the Breguet method if the V in equation (4-38) is expressed in terms of $\sqrt{391W/\sigma C_L S}$, and the resulting form integrated between the limits of W_0 and W_1.

$$E = \frac{-375C_L/C_D}{\sqrt{\dfrac{391}{\sigma C_L S}}} \frac{\eta}{c} \int_{W_0}^{W_1} \frac{dW}{W^{3/2}}$$

$$E = 37.9 \frac{\eta}{c} \frac{C_L{}^{3/2}}{C_D} \sqrt{\frac{\sigma S}{W_0}} \left[\left(\frac{W_0}{W_1}\right)^{1/2} - 1 \right] \qquad (4\text{-}41)$$

Equation (4-41) is solved graphically in Figure 4-22 in terms of W_1/W_0 and $W_0/\sigma S$. Representative values of η/c, C_L, and C_D are

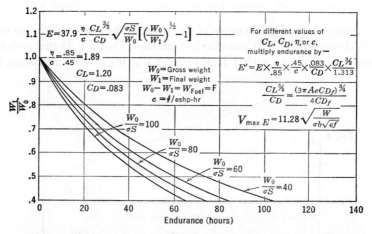

FIGURE 4-22. Endurance chart for reciprocating engine airplanes.

assumed constant. For other values of these parameters the E is corrected by a direct ratio. It is of interest to note that altitude, σ, directly affects the endurance. At sea level conditions the endurance is a maximum. Speed for maximum endurance is equal to speed for minimum power required and is expressed by equation (4-18).

For airplanes powered by other than reciprocating engines equations (4-40) and (4-41) are no longer valid for estimating range and endurance performance because specific fuel consumption in lb/eshp-hr varies with speed. Reference to Figure 3-3 indicates fuel-consumption comparisons for different engines in terms of pounds per hour per pound of thrust, $\text{lb/hr/lb}\,T = c'$. From equation (4–40) it is significant to note that $V_{L/D_{\max}}$ is the optimum speed for the airframe if maximum range is to be obtained. For the reciprocating-engine airplane this speed is also the optimum for maximum range because η/c is constant with speed,

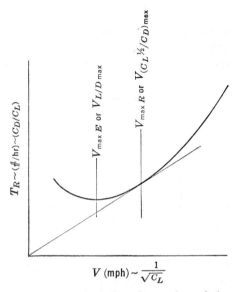

FIGURE 4–23. Graphical solution of range for turbojet airplane.

and the airplane-engine combination then operates at the speed for best airframe efficiency. For turbojet airplanes the rate of burning fuel is proportional to thrust output, and in general c' in $\text{lb/hr/lb}\ T$ may be considered approximately constant with speed in the speed region for maximum range. In other words, fuel consumption in lb/thp-hr varies inversely with speed. The speed for maximum range for a given airplane and altitude, then, is a compromise between the speed for best airframe efficiency and the maximum speed of the airplane, provided, of course, that a parabolic polar is assumed. This relationship is presented graphically in Figure 4–23. By assuming a constant c', the ordinate in Figure 4–23 is proportional to lb/hr and C_D/C_L.

$$c' T_R = \text{lb/hr}$$

and

$$D = T_R = \frac{W}{L/D} = \frac{W}{C_L/C_D}$$

For minimum ratio of lb/hr/V or maximum miles/lb a tangent to the

T_R curve through the origin locates the $V_{\max R}$. In terms of airplane characteristics, this point is also

$$V_{(C_L^{1/2}/C_D)_{\max}}$$

$$\text{Minimum } \frac{C_D/C_L}{1/C_L^{1/2}} = \text{Maximum } \frac{C_L^{1/2}}{C_D}$$

The speed for maximum endurance corresponds to the speed for minimum thrust required or $V_{L/D_{\max}}$. To evaluate $(C_L^{1/2}/C_D)_{\max}$ and the corresponding speed, consider the parabolic polar divided by $C_L^{1/2}$:

$$\frac{C_D}{C_L^{1/2}} = \frac{C_{D_f}}{C_L^{1/2}} + \frac{C_L^{3/2}}{\pi A e}$$

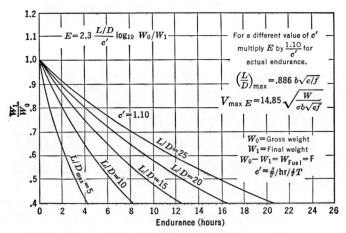

FIGURE 4-24. Endurance chart for turbojet airplanes.

Differentiating with respect to C_L and setting equal to zero

$$C_{D_f} = 3 \frac{C_L^2}{\pi A e}$$

and

$$\left(\frac{C_L^{1/2}}{C_D}\right)_{\max} = \frac{\sqrt[4]{\pi A e C_{D_f}/3}}{4/3 C_{D_f}} \tag{4-42}$$

and

$$V_{(C_L^{1/2}/C_D)_{\max}} \text{ or } V_{\max R} = \frac{19.5}{\sqrt[4]{fe}} \sqrt{\frac{W}{\sigma b}} \tag{4-43}$$

An adaptation of Breguet's methods can be used to derive mathematical forms for range and endurance for jet airplanes based on the preceding assumptions.

$$dt = dE = \frac{dF}{\text{lb/hr}} = \frac{-dW}{\text{lb/hr}}$$

Expanding lb/hr,

$$dt = \frac{-dW}{c'D} = -\frac{1}{c'}\frac{L}{D}\frac{dW}{W}$$

$$E = -\frac{1}{c'}\frac{L}{D}\int_{W_0}^{W_1}\frac{dW}{W} = -\frac{1}{c'}\frac{L}{D}\log_e\frac{W_1}{W_0}$$

$$E_{max} = \frac{2.3}{c'}\left(\frac{L}{D}\right)_{max}\log_{10}\frac{W_0}{W_1} \tag{4–44}$$

Equation (4–44) is graphically presented in Figure 4–24. It is of interest to note that maximum endurance is independent of altitude. For range

$$dR = V\,dt = -\frac{V}{c'}\frac{L}{D}\frac{dW}{W} \tag{4–45}$$

From equation (4–45) it is seen that maximum range occurs at the maximum product of $V\,L/D$. For jet airplanes this maximum product occurs at the highest altitude at which $V_{(L/D)_{max}}$ is equal to V_{max}. For an optimum range flight path, then, the jet airplane continues to climb as the weight is decreased. By an analysis of the flight boundary curve as presented in Figure 2–38 a maximum allowable altitude for any particular configuration may be determined in terms of $W/\delta S$. For example, for cruising flight, assuming a maximum angle of bank of 40 degrees, a value of approximately 1.3 $(W/\delta S)$ may be permitted within the flight boundary limit. For high-speed or ceiling limitations a 3^r-degree bank may be assumed, which permits a maximum value of 1.15 $(W/\delta S)$ within the flight boundary limit. The airplane then climbs along a flight path as limited by allowable $W/\delta S$.

For all practical purposes equation (4–45) can be expanded to a maximum range expression at a constant altitude. This expression gives values of range within 3 per cent of the absolute optimum for most airplane-engine combinations.

$$dR = - \frac{\sqrt{391W/\sigma SC_L}}{c'} \frac{C_L}{C_D} \frac{dW}{W}$$

$$R = - \frac{\sqrt{\frac{391}{\sigma S}}}{c'} \frac{C_L^{1/2}}{C_D} \int_{W_0}^{W_1} \frac{dW}{W^{1/2}}$$

and

$$R_{\max} \text{ in miles} = \frac{2}{c'} \left(\frac{C_L^{1/2}}{C_D}\right)_{\max} \sqrt{\frac{391W_0}{\sigma S}} \left[1 - \left(\frac{W_1}{W_0}\right)^{1/2}\right] \quad (4\text{-}46)$$

Equation (4–46) is graphically presented in Figure 4–25 at an altitude of 35,000 feet in terms of W_1/W_0 and initial wing loading, W_0/S.

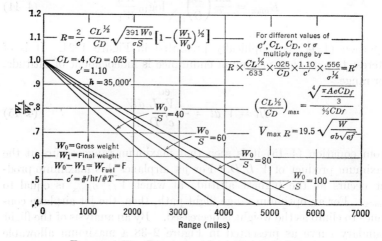

FIGURE 4–25.   Range chart for turbojet airplanes.

Typical values of C_L, C_D, and c' are assumed. For other values of these parameters the range is corrected by a direct ratio. Equation (4–46) indicates that maximum range is obtained with increasing altitude and wing loading. The airplane flies faster with an increase in these variables and thus takes advantage of the better economy associated with a fuel flow rate proportional to thrust. The maximum cruising speed is sometimes limited at some altitude at which cruising speed approaches critical speed.

For airplanes powered by turbine propellers, endurance and range can be best estimated by use of equations (4–35) and (4–36). Engine rate of fuel flow may be obtained for any given speed from typical engine characteristic curves as presented in Figure 3–10. By reference to Figure 3–10b it is seen that specific fuel consumption, c, in lb/eshp-hr is not constant as in the case of the reciprocating engine. Also, c' in lb/hr/lb$T$ is not constant as in the case of the turbojet engine. The speed for maximum range occurs between $V_{L/D_{\max}}$ and $V_{(C_L^{1/2}/C_D)_{\max}}$. As in the case of the turbojet engine, range increases with increasing altitude.

Maximum range for an airplane powered by a rocket or ramjet occurs at a speed which gives a maximum product of $\dfrac{V}{c'}\dfrac{L}{D}$ (see equation 4–45). For subsonic flight (parabolic polar), equation (4–46) will apply for maximum range of a rocket airplane at a constant altitude. As was shown in Chapter 2, the V for maximum range for supersonic flight occurs at an optimum lift coefficient, $C_{L_{\text{opt}}}$, for which the drag due to lift is equal to the sum of the parasite and thickness drags. Flight at this speed represents flight at $L/D_{\max}$. Again, by expressing the uniform level flight condition of the airplane as

$$ W = C_L q S $$

or

$$ \frac{W}{\delta S} = 1481 M^2 C_L $$

it is seen that either the W/S or flight altitude in terms of pressure ratio, δ, must be high in order for the airplane to fly at $C_{L_{\text{opt}}}$ at supersonic speeds.

From Figure 3–3 it is seen that c' for a ramjet is considerably lower than that for a rocket and decreases in magnitude with an increase in speed. The optimum product of $\dfrac{V}{c'}\dfrac{L}{D}$ for a ramjet would occur at a slightly higher speed than that for a rocket airplane.

In general, the range performance of aircraft and power-plant combinations may be compared on a basis of three parameters, assuming a constant value of propulsive efficiency η_p for propeller-driven aircraft

$$ \frac{V}{c'}, \quad \frac{L}{D}, \quad \text{and} \quad \frac{W_0}{W_1} $$

V/c' may also be expressed as a function of η_p/c, since

$$ \frac{V}{375c'} = \frac{\eta_p}{c} $$

Values of $(L/D)_{\max}$ for subsonic flight are approximately double that which can be theoretically calculated for supersonic flight. Values of weight ratio, $\dfrac{W_0}{W_1}$, are also more favorable for subsonic flight because of the great quantity of fuel required to reach optimum altitude and speed conditions for supersonic flight combined with the high structural weights required for high indicated air speeds. Values of $\dfrac{V}{c}$ or for η_p/c for a ramjet-propelled supersonic aircraft are comparable to those for subsonic aircraft powered by reciprocating, turboprop, or turbojet engines. Because of the above limitations, it appears that the maximum range of a ramjet and/or rocket-propelled supersonic aircraft is limited to a value of approximately one-fifth to one-tenth that of a subsonic aircraft, assuming comparable values of gross weight and flight within the atmosphere. With the advent of more powerful turbojet engines it also appears that a supersonic turbojet configuration may increase this range considerably but never to a value comparable with optimum subsonic conditions.

Take-off and Landing Distances

Airplane take-off distance is perhaps the most difficult performance item to predict accurately. Most analyses of this problem, although mathematically rigorous, are based on assumptions that are accurate only for special conditions of pilot technique, ground conditions, airplane attitude and drag, and average variations in effective thrust. Experimental data on a given airplane are often widely dispersed as a result of these variables, and an average of several runs is usually used as a basis for correlation with theoretical analyses. Before presenting charts based on a correlation of experimental data and some of the important take-off parameters, consider first some of the physical aspects of the take-off ground run and climb to a 50-foot altitude.

Mr. E. P. Hartman's analysis* of airplane ground run is fundamental and may be applied with good accuracy to both propeller and jet airplanes. As in any acceleration, time, and distance problem,

$$dv = a\, dt$$

and

$$ds = v\, dt = \frac{v}{a}\, dv$$

$$s = \int_0^{v_{TO}} \frac{v}{a}\, dv = \frac{1}{2} \int_0^{v_{TO}} \frac{1}{a}\, d(v)^2 \qquad (4\text{–}47)$$

*NACA TN 557, "Considerations of the Take-off Problem," February 1936.

If the acceleration, a, varies in a non-linear manner up to take-off speed, v_{TO}, equation (4–47) may be integrated by steps to obtain the total take-off run. If $1/a$ is assumed to vary linearly with v^2 up to v_{TO}, then

$$s = \frac{1}{2}\frac{v_{TO}^2}{a \text{ @ } .5v_{TO}^2} = \frac{1}{2}\frac{v_{TO}^2}{a \text{ @ } .7v_{TO}}$$

Now the accelerating force or effective thrust, T_E, is

$$T_E = T - R = \frac{W}{g}a$$

where T is engine thrust, and R is the air and ground resistance. Hence, ground run

$$s = \frac{Wv_{TO}^2}{64T_E \text{ @ } .7v_{TO}} \qquad (4\text{--}48)$$

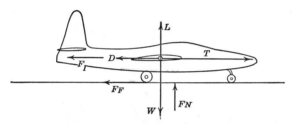

FIGURE 4–26.   Airplane ground run.

Equation (4–48) may be used with reasonable accuracy provided the T_E at $.7v_{TO}$ is greater than 10 per cent of the take-off weight and the take-off speed is not greater than 20 per cent above the stalling speed of the airplane in the take-off configuration. If these two conditions are not satisfied, it is recommended that a step-by-step integration of equation (4–47) be used to obtain the ground run.

To obtain the airplane resistance, R, consider the air and ground forces acting on the airplane in Figure 4–26.

$$F_N = W - L$$

and

$$R = D + F_F = D + \mu F_N = C_D qS + \mu(W - L)$$

where $\mu =$ coefficient of rolling friction.

The polar of the airplane during ground roll is approximated by

$$C_D = C_{D_f}' + \frac{C_L^2}{\pi A}$$

Note that C_{D_f}' is greater than C_{D_f} because of landing-gear and wing-flap drag. For reciprocating engines an additional increment is added for cowl-flap drag. Also e is $\cong 1$ to account for ground effect and change in e due to wing-flap deflection. Expanding,

$$R = C_{D_f}'qS + \frac{C_L^2}{\pi A} qS + \mu W - \mu C_L qS$$

and

$$R = \mu W + qS \left(C_{D_f}' + \frac{C_L^2}{\pi A} - \mu C_L \right) \qquad (4\text{-}49)$$

Since $\left(\dfrac{C_L^2}{\pi A} - \mu C_L \right)$ is much less than C_{D_f}', and since the airplane C_L along the ground varies so much with pilot technique, an optimum C_L for minimum ground resistance is assumed. Differentiating equation (4-49),

$$\frac{dR}{dC_L} = qS \left(\frac{2C_L}{\pi A} - \mu \right) = 0$$

$$C_L = \frac{\mu \pi A}{2}$$

and

$$R_{\min} \cong \mu W + q \left(f' - \frac{\mu^2 \pi b^2}{4} \right) \qquad (4\text{-}50)$$

In equation (4-50) $f' = C_{D_f}'/S$ and approximately equals two to four times the basic f of the airplane, depending upon the additional drag due to landing gear, wing-flap position, and cowl-flap position. Figure 2-68 presents some typical drag increases for these items.

By assuming the take-off speed to be approximately 20 per cent above the stalling speed of the airplane for the take-off configuration, equation (4-48) can be used to estimate a minimum safe ground run before take-off. The distance to climb to 50 ft is largely dependent on the available excess power during climb and the flight path of the airplane during transition from take-off to steady climb. The latter is very difficult to analyze accurately because of varying pilot techniques. It is suggested that the distance to make a steady climb from the ground to 50 ft be added directly to the ground run to obtain the total distance over a 50-ft obstacle. An additional safety factor of approximately 200 ft would probably account for long transitions.

In light of the preceding discussion it can be seen that even a detailed analysis of the take-off problem is at best only as accurate as the accuracy of the assumptions made. For the first approximation of ground run or total distance over 50 ft the use of Figure 4–27 is considered reliable. These curves are based on reliable flight test data from a large number of airplanes and are presented in terms of the most important airplane and engine parameters affecting take-off. For jet planes maximum static thrust, T, is used as a parameter. For propeller-driven airplanes a more convenient parameter is engine brake-horsepower, BHP. The ground run distance is approximately .8 of the total over 50 ft for propeller-driven airplanes and approximately .9 for jet airplanes. This difference is due to the different variation

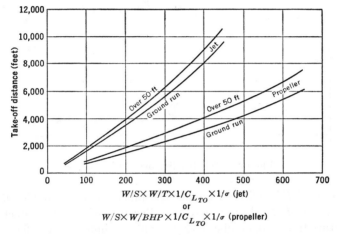

FIGURE 4–27. Take-off chart.

in available thrust with speed for the two types. If effective thrust is the same for both types at $.7v_{TO}$, then the ground runs should be about equal. However, at take-off speed and in climb the propeller-driven airplane would have a lower thrust and hence require a greater distance for the climb.

The analytical estimation of landing distance over a 50-ft obstacle involves the estimation of air distance and ground run. Reasonable accuracy is obtained in this analysis by assuming the airplane glides at a constant angle until it reaches the ground and then decelerates at a constant rate along the ground. The accuracy with which flare distance can be estimated does not warrant a detailed calculation. By making the above assumptions an analytical expression is developed which gives landing distances that agree well with experimental data.

Consider first the airplane gliding at a constant angle and decelerating from a speed at the 50-ft altitude, v_{50}, to a landing speed, v_L, the instant it touches the ground (Figure 4–28). At the 50-ft altitude the airplane has kinetic and potential energy which must be dissipated before reaching the ground. This dissipation of energy is assumed to be due to a retarding force, F, acting over the air distance, s_A, which is also the distance along the ground for normal angles of glide. Choosing consistent units,

$$\frac{Wv_{50}^2}{2g} - \frac{Wv_L^2}{2g} + 50W = Fs_A$$

and

$$s_A = \frac{W}{F}\left(\frac{v_{50}^2 - v_L^2}{2g} + 50\right) \tag{4–51}$$

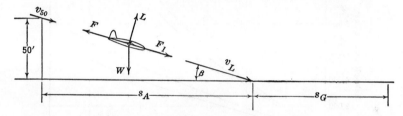

FIGURE 4–28. Landing over 50-ft obstacle.

The term W/F is called an average resistance coefficient and is actually an average L/D during descent. Because of ground effect and departure from a parabolic drag polar at maximum lift coefficient the estimation of this average L/D during the glide and flare is difficult. Good correlation with flight test data is obtained by assuming that these two effects cancel each other, and the drag polar may be expressed as

$$C_D = C_{Df}' + \frac{C_{L\mathrm{max}}^2}{\pi A} \tag{4–52}$$

where C_{Df}' is estimated for the landing configuration. An average value of W/F calculated from equation (4–52) for several airplanes is approximately 8.

The calculation of ground run may be made directly by use of equation (4–47) in a slightly different form.

$$s_G = \int_{v_L}^0 \frac{v}{-a}\, dv \tag{4–53}$$

where $-a$ is the deceleration. The accuracy of the assumptions required to evaluate $v/-a$ in equation (4–53) does not warrant a detailed integration to obtain ground roll. Using an average deceleration during ground run leads to a simplified form of this equation which gives good correlation with experimental tests.

$$s_G = \frac{v_L{}^2}{-2a} \tag{4–54}$$

A value of $-a$ equal to 7 ft/sec^2 is an approximate average for modern airplanes using adequate brake capacity and operating on concrete runways. Combining equations (4–51) and (4–54) with the assumptions noted above and assuming further that

$$v_{50} \cong 1.3v_S$$

and

$$v_L \cong 1.15v_S$$

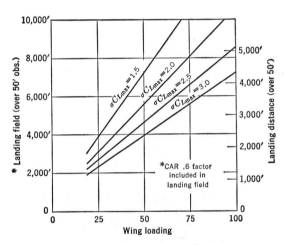

FIGURE 4–29. Landing field chart.

where v_S is stalling speed in feet per second based on flaps fully deflected, the following simplified equation for total distance in feet results:

$$s_L = s_A + s_G = \frac{118\,W/S}{\sigma C_{L\text{max}}} + 400 \tag{4–55}$$

Equation (4–55) is plotted in Figure 4–29 in terms of $\sigma C_{L\text{max}}$ and wing loading. The landing field length required by the Civil Aeronautics Authority requires that the landing distance as obtained from equation (4–55) be divided by a factor of .6 to account for a reasonable margin above the 50-ft altitude at which a pilot may begin an approach.

Acceleration in Climb

Methods of calculating rate of climb based on equilibrium flight, as illustrated in Figure 4–1, no longer apply for high-power interceptors and fighters which accelerate along a steep climb path. In Figure 4–30 the inertia force due to an acceleration, a, is expressed as

$$F_I = \frac{Wa}{g}$$

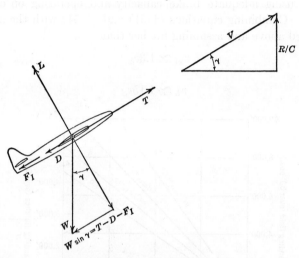

FIGURE 4–30. Acceleration in climb.

Since W is no longer assumed approximately equal to L for high angles of climb,

$$D = \frac{\sigma f V^2}{391} + \frac{124.5 W^2 \cos^2 \gamma}{\sigma e b^2 V^2}$$

and

$$W \sin \gamma = \frac{W \, R/C}{V} = T - \frac{\sigma f V^2}{391} - \frac{124.5 W^2 \cos^2 \gamma}{\sigma e b^2 V^2} - \frac{Wa}{g}$$

then

$$R/C \text{ in ft/min} = \frac{88V}{W} \left(T - \frac{\sigma f V^2}{391} - \frac{124.5 W^2 \cos^2 \gamma}{\sigma e b^2 V^2} - \frac{Wa}{g} \right) \quad (4\text{–}56)$$

If a and γ are very small, equation (4–56) reduces to a more familiar form

$$R/C \text{ in ft/min} = \frac{88V}{W} \left(T - \frac{\sigma f V^2}{391} - \frac{124.5 W^2}{\sigma e b^2 V^2} \right) \quad (4\text{–}57)$$

To solve equation (4–56) an approximate acceleration is assumed at any altitude and rate of climb determined by a method of successive approximation. If acceleration is not accounted for in determining climb performance of a typical rocket interceptor, the time to climb to a high altitude may be low by as much as 25 to 30 per cent. The effect of angle of climb on rate of climb is very small. In equation (4–56) γ affects only the induced drag term, which is very small at the high climbing speeds associated with high-performance aircraft.

Turning Performance

One of the special performance problems to be discussed is that of turning performance. Actually, of course, the performance of an airplane in a steady turn is considerably more of a general problem than those previously discussed. Level flight performance with zero angle of bank is in reality a special problem of the general case. The estimation of turning performance can be accomplished by many short-cut methods, but each in turn is derived from the fundamental relationships of thrust required and available or power required and available for an airplane at various angles of bank. Only the fundamental power curves will be discussed herein. From Figure 4–1d it can be seen that

$$L \cos \phi = W = C_L \frac{\sigma V^2}{391} S \cos \phi \qquad (4\text{–}58)$$

At a constant C_L for any given altitude, then, the speed increases as the angle of bank increases, and this increase is inversely proportional to the square root of the cosine of the angle of bank. Stalling speed in a turn, $V_{S\phi}$, is expressed as

$$V_{S\phi} = \frac{V_S}{\sqrt{\cos \phi}} \qquad (4\text{–}59)$$

To plot the power-required curves shown in Figure 4–31, it is more convenient to compute the P_R for each angle of bank for a constant value of speed. By selecting several values of speed the entire envelope of curves may be drawn. From equation (4–58), C_L at constant V and altitude is inversely proportional to $\cos \phi$. By referring to equations (4–2) and (4–10) only the effective induced thrust and power required are changed by angle of bank, because C_L appears only in these terms. In a more general form, these equations may be written:

$$D = T_R = \frac{\sigma f V^2}{391} + \frac{124.5}{\sigma e} \left(\frac{W}{b}\right)^2 \frac{1}{V^2 \cos^2 \phi} \qquad (4\text{–}60)$$

and

$$P_R = \frac{\sigma f V^3}{146,600} + \frac{.332}{\sigma e}\left(\frac{W}{b}\right)^2 \frac{1}{V\cos^2\phi} \qquad (4\text{-}61)$$

Figure 4–31 is a typical plot of equation (4–61). If the polar of the airplane is known from wind-tunnel tests and varies from a parabolic form, the following technique may be used to obtain a point on the power-required curves for a given angle of bank:

1. Assume a V.
2. At a given ϕ calculate C_L from equation (4–58).
3. Determine C_D from the airplane polar.
4. Compute P_R from $P_R = DV/375$.

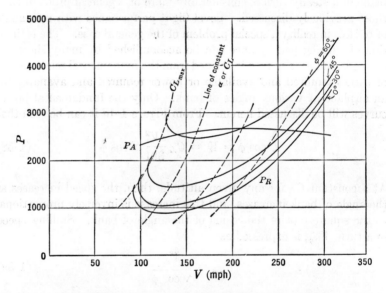

FIGURE 4–31. Power required at various angles of bank.

From the system of curves in Figure 4–31 the speed, rate of climb, and rate of descent performance may be determined in the usual manner, depending on the angle of bank and power requirements. For level flight the power required must be equal to the power available.

A performance item that is very important for fighter aircraft is minimum radius of turn. It is shown in Figure 4–1d that minimum radius in level flight occurs at the $C_{L\max}$ for an airplane, provided the $P_A = > P_R$. If the airplane cannot fly in a level turn at $C_{L\max}$, then the following technique is used to determine the minimum radius:

1. Assume a V.

2. At a value of $P_R = P_A$ compute C_D:

$$P_A = P_R = \frac{C_{D\sigma}V^3S}{146,600}$$

3. Determine C_L from experimental polar or assumed parabolic variation:

$$C_D = C_{D_f} + \frac{C_L^2}{\pi Ae}$$

4. Compute L:

$$L = C_L q S$$

5. Compute $\cos \phi$:

$$L \cos \phi = W$$

6. Determine $\tan \phi$.

7. Compute radius of turn, R. From Figure 4-1d

$$\tan \phi = \frac{\text{C.F.}}{W} = \frac{Wv^2}{gRW} = \frac{v^2}{gR} \qquad (4\text{-}62)$$

8. For several values of V plot

R versus V and determine minimum R

For a spiral descent or climb the above procedure may be used to obtain minimum radius of turn provided the excess or deficiency in power for a given R/C or R/S is calculated and specified in Item 2 above. The corresponding P_R for this condition may then be used to calculate C_D. The rest of the procedure is then the same to obtain minimum radius of turn.

The time in a 360° turn may be obtained as a function of v and ϕ.

$$t = \frac{2\pi R}{v} \qquad (4\text{-}63)$$

From equation (4-62)

$$R = \frac{v^2}{g \tan \phi}$$

Thus

$$t = \frac{2\pi v}{g \tan \phi} \qquad (4\text{-}64)$$

The ϕ and v for minimum time in a turn is not the same as that for minimum R. To obtain a minimum, plot time as obtained for a selected value of v in the same manner as previously discussed.

Design Performance

As was previously discussed, any change in the basic airplane design variables of W, b, e, and f will affect the airplane drag at any given speed and altitude. . Of these variables, f and W can be controlled to a considerable degree by the designer and manufacturer. Lack of control often results in serious performance losses.

Airplane span is determined by design requirements and is usually a compromise between structural and aerodynamic considerations. A large span or low ratio of span loading, W/b, is desirable if take-off climb performance is a critical requirement. Airplane efficiency factor, e, is also important for this climb condition because a low value of e adversely affects a relatively high value of induced drag associated with low speeds. For extreme ranges at relatively low cruising speeds the span should be as large as is structurally practical. Altitude, σ, of course, affects performance but is not an airplane characteristic and as such cannot be controlled. The operational altitude of an airplane is largely determined by power-plant characteristics.

Equivalent parasite area, f, is dependent on the cleanness as well as size of an airplane and may seriously affect the cruising performance of high-speed airplanes. For example, suppose excrescences such as turrets, blisters, and rough surfaces were to increase f by 10 per cent on a given series of airplanes with widely different cruising speeds. From Figure 4–32 it can be seen that the loss in cruising speed due to a 10 per cent change in f or minimum parasite drag becomes very serious for high-speed airplanes like the B-47. For constant power available associated with propeller-driven airplanes, the speed loss for the C-47, B-17, and B-29 amounts to approximately 3 per cent. For constant thrust available the loss on the B-47 is approximately 4.3 per cent. For turbojet airplanes the large speed loss is due not only to the high cruising speed but also to the fact that for a constant cruising thrust available the actual power available is less as the speed is reduced. (See Figure 4–32.) It is also important to note that the addition of the same radome, for example, on a B-47 reduces the cruising or high speed to even a greater extent than that of the B-29 because the percentage reduction in parasite drag is greater for the cleaner airplane.

The weight change on an airplane has even more far-reaching effects on performance than a change in parasite drag. All performance items suffer by an increase in weight. As far as speed is concerned, an increment change in weight reduces the cruising speed by a greater amount than the high speed. The reason is that induced drag is a greater proportion of the total drag at low speeds than at high speeds.

Rate of climb is materially decreased because of both an increase in airplane drag and an increase in weight. See equation (4-6). Take-off and landing distances are unfavorably changed by an increase in weight as indicated by the variables affecting these performance items in Figures 4-27 and 4-29.

Range is the performance item that takes the greatest penalty by an increase in weight. For example, consider the plots of miles per pound of fuel versus weight of a representative bomber as shown in Figure

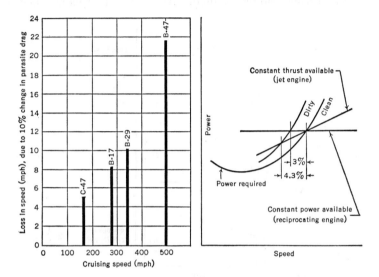

FIGURE 4-32.   Effect of change in parasite drag on speed.

4-33.   The area under each curve represents range as is also indicated in Figure 4-21.   In cases a, b, and c, the weight empty of the airplane is increased 1510 lb by overweights of certain items of equipment. Long-range operations are usually based on maximum allowable take-off weight for critical take-off performance requirements.   In case a, the weight empty is increased 1510 lb, and since take-off weight and bomb load are held constant, fuel is reduced by 1510 lb, which reduces the maximum range approximately 285 miles.   In case b, take-off weight is not considered critical so that fuel weight is increased by the same amount as weight empty.   However, the range is reduced by about 150 miles, as the miles per pound of fuel consumed for the light-weight condition is nearly double that for the heavy take-off condition. In Figure 4-33c is illustrated the increase in take-off weight necessary if the range and bomb load are to remain unaffected by the overweight items.   For every pound of overweight that affects the weight empty, approximately 1 lb of additional fuel is required, resulting in about 2 lb increase in take-off weight.   On long-range missions this effect is extremely important, as the difference in initial and final weights is great.

In the design of an optimum airplane to meet certain performance requirements airplane variables of wing loading and thrust loading are very useful and fundamental.   In propeller-driven airplanes power loading and wing loading are used.   An optimum airplane is one that

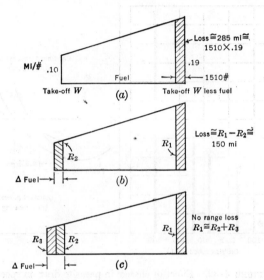

FIGURE 4–33.   Effect of weight increase on range.

will meet all the requirements with as small a gross weight as possible. In terms of transports this airplane will be designed for minimum operating cost.   This corresponds to the highest possible wing loading consistent with Civil Aeronautics Authority landing field requirements and the highest possible thrust or power loading consistent with take-off, climb, cruising, or ceiling requirements.   Landing weight is usually assumed to equal the gross weight minus one-half the fuel weight for climb, maneuver, and cruise.

In Figure 4–34 is presented a typical generalized design chart for a swept-wing turbojet commercial transport in terms of thrust loading, W/T_{0max}, and wing loading, W/S. W is defined as maximum gross weight and T_{0max} as maximum static take-off thrust.   On this chart are presented critical performance requirements dictating the design

W/S and $W/T_{0\text{max}}$ based on data from a number of generalized airplane studies. Assuming a ground deceleration of 7 ft/sec,[2] the landing field distance is a function of W/S. For a given field length requirement, then, the maximum design $W/T_{0\text{max}}$ can be determined by following up a constant field length line until a critical performance cutoff is reached. Engine-out take-off, climb, and ceiling requirements are based on thrusts obtainable under temperature conditions above those for a standard day. Constant cruising speed lines represent maximum

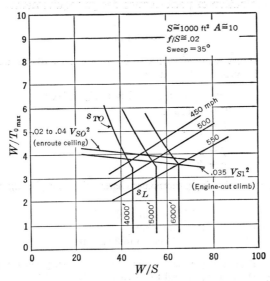

FIGURE 4–34. Generalized design chart.

speeds with normal thrust at 35,000 ft altitude. If field length is a variable, an optimum set of the parameters W/S and $W/T_{0\text{max}}$ would occur at the intersection of a critical climb condition ($.035V_{S1}{}^2$) and the maximum cruising speed desired. This speed would be somewhat less than the critical speed of the airplane. V_{S0} is defined as stalling speed and V_{S1} as take-off speed by the Civil Air Regulations. The weight and size of an optimum airplane are determined from generalized equations set up in terms of any combination of W/S and $W/T_{0\text{max}}$. The cruising speed is established by the engine size required for the critical climb or take-off requirement. Equivalent parasite area, f, is also set up in terms of W/S and $W/T_{0\text{max}}$ and the known variables. For any desired field length, passenger capacity, and range (including reserves) a generalized weight equation, in terms of W/S and $W/T_{0\text{max}}$, can be balanced by successive approximation.

By first assuming a fuel weight a total weight can be estimated. Then, by computing the weight of fuel for climb, maneuver, cruise, and reserve conditions, a second gross weight can be more accurately determined, which in turn will permit the evaluation of direct operating costs. It is possible to generalize Figure 4–34 further by cross-plotting lines of constant direct operating costs. A generalized design chart represented in Figure 4–34 is very helpful for making a rapid selection of an optimum aircraft in a particular class of airplanes to conform with any combination of performance design requirements.

PROBLEMS

4–1. *a.* Plot thrust required versus speed at sea level, 20,000, and 40,000 ft altitude for the following airplane characteristics:

$$\text{Power plant: single Nene (Figure 3–5)}$$
$$W = 15,000 \text{ lb}$$
$$b = 40 \text{ ft}$$
$$A = 6$$
$$f = 4$$
$$e = .8$$

b. Determine the following performance characteristics, neglecting compressibility effects:

High speed at each altitude
Service ceiling
Time to climb to 30,000 ft

c. Determine the cruising speed at 15,000 ft altitude at 90 per cent normal thrust.

4–2. Express the power required to move a wing forward through the air at a constant angle of attack as a function of (C_L, C_D, σ). $W/S = 15$, $S = 200$ ft.²

4–3. Given an airplane of weight W_0, flying at sea level so that $W_0 = C_L q_0 S$. If part of the fuel is used up so that the weight is W at some new altitude, find the relationship between the power required, P_R, for the new condition as a function of P_{R_0}, W_0, W, and σ, assuming the angle of attack of the wing remains constant.

4–4. Derive equation (4–18).

4–5. Airplane, $W = 6000$ lb, climbs at $V = 75$ mph at sea level, $L/D = 10$, $T = 1800$ lb. Find the rate of climb in feet per minute, and the lift:

(1) Assuming $L = W$.
(2) Assuming $L \neq W$.

4–6. A pursuit plane, weighing 6000 lb, is in a terminal velocity dive with zero thrust at an altitude of 10,000 ft (standard conditions).

Given: $C_{m_{ac}} = -.05$, $C_D = .025$, $S = 200$ ft² (rectangular planform). a.c. and c.g. of plane coincide. Tail length, l_t, to c.g. = 21 ft, $A = 8$.

Find (assume no compressibility effects):

(1) C_L
(2) D_i

4–7. Find the thrust horsepower needed to give a rate of climb of 3500 ft/min at 190 mph air speed at sea level for the following monoplane:

$$W = 6000 \text{ lb} \qquad C_{D_p} - C_{D_0} = .018 \left.\vphantom{\begin{matrix}1\\1\\1\end{matrix}}\right\} \begin{array}{l} \text{For this flight condition,} \\ \text{assume angle of climb} \\ \text{small.} \end{array}$$
$$b = 42 \text{ ft} \qquad C_{D_0} = .008$$
$$S = 240 \text{ ft}^2$$

4–8. Plane of Problem 4–7 cruises at 350 thp. At what altitude will maximum possible speed in level flight be reached for this condition?

$$e = .75 \qquad \text{At max. } L/D, \ C_D = 2C_{D_f}$$
$$C_{D_f} = .023$$

4–9. If R/C versus altitude is a linear variation to the absolute ceiling, H, derive an expression for time to climb, t/c, to any altitude, h, as a function of the initial rate of climb, R/C_0, at sea level, h, and H.

4–10. Given the following engine and airplane characteristics:

$W = 3000$ lb, including 150 lb gasoline
$S = 200$ ft^2
$A = 6, \ e = .80$
Minimum V_S at sea level (no flaps) = 55 mph
$(L/D)_{\max}$ for airplane = 9.71

Polar for airplane: $C_D = C_{D_f} + \dfrac{C_L{}^2}{\pi A e}$ (Assume equation is valid up to $C_{L_{\max}}$ of

airplane.)

The engine is supercharged to keep a constant output of 125 bhp to the absolute ceiling of the airplane. Specific fuel consumption is constant and = .6 lb/bhp-hr. The propeller is designed to give a constant propulsive efficiency of 80 per cent.
Find:

    (1) Maximum velocity in miles per hour (fully loaded)
    (2) Maximum R/C in feet per minute ("   ")
    (3) Minimum R/S in feet per minute ("   ")
    (4) Absolute ceiling in feet   ("   ")
    (5) Minimum radius of turn in 42° bank at sea level (fully loaded)
    (6) Maximum endurance in hours
    (7) Maximum range in miles

4–11. Given the following airplane characteristics:

$$W = 75,000 \text{ lb} \qquad f = 30$$
$$S = 1100 \text{ ft}^2 \qquad e = .85$$
$$A = 12$$

Two P & W Wasp Major engines (Figure 3–13)

    a. From Figure 4–10 determine cruising speed, V_c, at 10,000 ft altitude. Engine $N = 2400$. What is the specific fuel consumption, c, for this speed (auto lean)?

    b. Determine time to climb from 10,000 ft to 20,000 ft, using normal rated power. Indicated climb speed = 160 mph. How much fuel is used for this climb?

    c. Determine time to descend from 20,000 ft to 10,000 ft if one engine fails at 20,000 ft. Assume operating engine $N = 2200$. Indicated descent speed = 150 mph.

4–12. If airplane in Problem 2–22 is powered by a single Nene engine (Figure 3–5), determine the following performance characteristics, using generalized performance charts:

(1) High speed at sea level, 20,000, and 35,000 ft.
(2) Minimum time to climb to 35,000 ft.
(3) Maximum range at 35,000 ft, assuming $W_0 = 10,000$ lb and $W_1 = 6000$ lb.
(4) Maximum endurance.
(5) Take-off distance over 50 ft at sea level and at W_0, assuming $C_{L_{TO}} = 2.0$; landing distance over 50 ft at sea level and at W_1, assuming $C_{L_{max}} = 2.5$.

4–13. Given the following characteristics for a rocket interceptor:

$$W = 7500 \text{ lb}$$
$$T = 5000 \text{ lb and is constant with altitude and speed}$$
$$f = 2$$
$$e = .8$$
$$b = 22.4 \text{ ft}$$

Determine the time to climb to 35,000 ft:

(1) Neglecting γ and a.
(2) Accounting for γ and a.

Part 2

AIRPLANE STABILITY AND CONTROL

CHAPTER 5

STATIC LONGITUDINAL STABILITY AND CONTROL
STICK-FIXED

5-1 General

It was pointed out in Chapter 1 that the problems of airplane longitudinal equilibrium, static stability, and control can be developed from the very important static equation of moments about the Y axis through the airplane's center of gravity. As these problems are probably the most important in design of the airplane for adequate stability and control characteristics, considerable emphasis is placed on them in this section. A study of the static equation of moments about the Y axis is a very complex one when the effects of power, component interference, and free controls are taken into account. However, the equations can be handled even in their most complicated forms if use is made of experimental evidence gained in many wind-tunnel and flight tests.

The longitudinal stability theory will first be developed for the simplified case of an airplane in gliding flight with controls locked (stick-fixed) and propellers windmilling. The theories thus obtained will be expanded later to account for the effects of power and free controls.

The study of the longitudinal equilibrium and static stability of the airplane, then, requires an investigation into the moments about the airplane's Y axis through the c.g. and their variation with the airplane's lift coefficient. Equilibrium demands that the summation of these moments equal zero, and static stability demands that a diving moment accompany an increase in lift coefficient and a stalling moment accompany a decrease in lift coefficient from equilibrium.

Throughout this section, it will be assumed that any wing or tail surface can be represented by a mean aerodynamic chord, the forces and moments on which represent all the forces and moments operating on the surface. It will also be assumed that there exists an aerodynamic center on this mean aerodynamic chord about which the wing pitching moment coefficient is invariant with lift coefficient. The forces and moments acting on any wing or tail surface can be represented by a lift and drag force acting at the aerodynamic center, together with a pitching moment about the aerodynamic center whose coefficient is invariant with lift coefficient. The aerodynamic center of any airfoil section is located very close to the 25 per cent chord point.

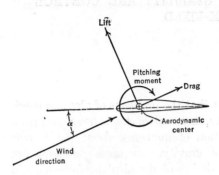

The aerodynamic center of a typical airfoil section is shown in Figure 5–1. Methods for predicting the mean aerodynamic chord and the aerodynamic center location for the three-dimensional wing planform are discussed in Section 2–2.

FIGURE 5–1. Aerodynamic forces on an airfoil section.

The forces and moments operating on a normal airplane configuration are shown below in detail for gliding flight with propeller off. The angles i_w and i_t are the wing and tail incidence with respect to the airplane reference line, and ϵ is the average angle of downwash behind the wing at the horizontal tail. All angles and derivatives with respect to angles will be given in degrees to escape confusion.

The lift and drag are by definition always perpendicular and parallel to the wind, respectively. It is therefore inconvenient to use these forces to obtain moments, for their arms to the center of gravity vary with angle of attack. For this reason all forces are resolved into normal and chordwise forces whose axes remain fixed with the airplane and whose arms are therefore constant.

Resolving the wing forces perpendicular and parallel to the airplane reference:

$$N = L \cos (\alpha - i_w) + D \sin (\alpha - i_w)$$

$$C = D \cos (\alpha - i_w) - L \sin (\alpha - i_w)$$

$$(5-1)$$

Resolving the tail force along the same axes as shown in Figure 5–2, the following summation of moments about the airplane's c.g. is obtained:

$$M_{cg} = Nx_a + Cz_a + M_{ac} + M_{Fus} + M_{Nac} + M_{act}$$
$$+ C_t h_t - N_t l_t \quad (5-2)$$

For equilibrium it is necessary for M_{cg} to equal zero. It has been found convenient to place this equation in coefficient form by dividing through by qS_wc, where q is the dynamic pressure in pounds per square foot, S_w the wing area in square feet, and c the mean aerodynamic chord in feet. The ratio q_t/q, is called the tail efficiency, η_t, and for power-off flight is less than unity, because of the loss of energy as the air interacts with parts of the wing wake and fuselage boundary layer.

$$C_{mcg} = C_N \frac{x_a}{c} + C_c \frac{z_a}{c} + C_{mac} + C_{m_{Fus}\atop Nac} + C_{mact} \frac{S_t}{S_w} \frac{c_t}{c} \eta_t$$

$$+ C_{ct} \frac{S_t}{S_w} \frac{h_t}{c} \eta_t - C_{Nt} \frac{S_t}{S_w} \frac{l_t}{c} \eta_t \quad (5\text{–}3)$$

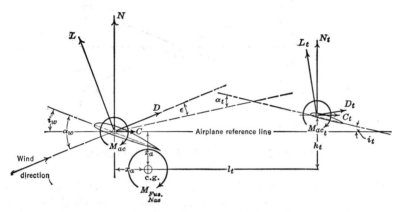

FIGURE 5–2.   Forces and moments in plane of symmetry.

The fifth and sixth terms of equation (5–3) have been found to be negligible when compared with the other terms, and will therefore be eliminated from further consideration. This leaves

$$C_{mcg} = C_N \frac{x_a}{c} + C_c \frac{z_a}{c} + C_{mac} + C_{m_{Fus}\atop Nac} - C_{Nt} \frac{S_t}{S_w} \frac{l_t}{c} \eta_t \quad (5\text{–}4)$$

This is the equilibrium equation in pitch which must sum up to $C_{mcg} = 0$ for equilibrium in a given flight condition. The pitching moment coefficient will be shown to be a function of the lift coefficient, and the slope of the curve of pitching moment coefficient plotted against lift coefficient is used for evaluating the static stability of the airplane in pitch. Typical plots of this curve are shown in Figure 5–3.

For the airplane to be stable the slope of the C_m versus C_L curve must be negative. If, as is shown in Figure 5–3, the airplane is in equilibrium at $C_L = 1.0$ ($C_{mcg} = 0$), an increase in airplane lift coefficient along curve A will create a nose-down pitching moment tending to reduce the lift coefficient back to equilibrium, whereas a decrease in lift coefficient will create a nose-up moment tending to increase the lift coefficient. A positive slope of this curve, as shown by curve B, will result in static instability, for, although in equilibrium at the same lift coefficient, an increase in airplane lift coefficient will create nose-up moments tending to increase the lift coefficient still further.

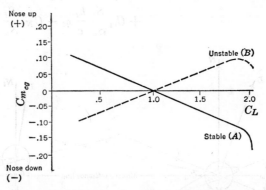

FIGURE 5-3. Typical pitching moment curves.

The slope of this curve of C_m versus C_L is given by the derivative dC_m/dC_L, and a negative sign for this derivative is required for static longitudinal stability. The slope of the pitching moment curve dC_m/dC_L can be obtained analytically by differentiating equation (5-4) with respect to C_L:

$$\frac{dC_m}{dC_L} = \underbrace{\frac{dC_N}{dC_L}\frac{x_a}{c} + \frac{dC_c}{dC_L}\frac{z_a}{c} + \frac{dC_{mac}}{dC_L}}_{\text{Contr. of wing}} + \underbrace{\left(\frac{dC_m}{dC_L}\right)_{Nac}^{Fus}}_{\substack{\text{Contr. of} \\ \text{fuselage and} \\ \text{nacelles}}} - \underbrace{\frac{dC_{Nt}}{dC_L}\frac{S_t}{S_w}\frac{l_t}{c}\eta_t}_{\substack{\text{Contr. of} \\ \text{horizontal tail}}} \quad (5\text{-}5)$$

The contribution of the various parts of the airplane to the total airplane stability can be broken down as shown in equation (5-5). These various contributions will be studied separately.

5-2 Wing Contribution

The wing contribution to the static longitudinal stability of the airplane can be analyzed by studying the first three terms of equation (5-5). In order to express these in terms of the familiar lift coefficient, use can be made of equations (5–1) as follows:

$$C_N = C_L \cos (\alpha - i_w) + C_D \sin (\alpha - i_w)$$

$$C_c = C_D \cos (\alpha - i_w) - C_L \sin (\alpha - i_w)$$

(5–6)

The slopes dC_N/dC_L and dC_c/dC_L can be obtained by differentiating the above equations with respect to C_L:

$$\frac{dC_N}{dC_L} = \cos (\alpha - i_w) - C_L \sin (\alpha - i_w) \frac{d\alpha}{dC_L}$$

$$+ \frac{dC_D}{dC_L} \sin (\alpha - i_w) + C_D \cos (\alpha - i_w) \frac{d\alpha}{dC_L}$$

(5–7)

$$\frac{dC_c}{dC_L} = \frac{dC_D}{dC_L} \cos (\alpha - i_w) - C_D \sin (\alpha - i_w) \frac{d\alpha}{dC_L}$$

$$- C_L \cos (\alpha - i_w) \frac{d\alpha}{dC_L} - \sin (\alpha - i_w)$$

As the angle of attack is usually a small angle, it is permissible to make approximations to the terms $\sin (\alpha - i_w)$ and $\cos (\alpha - i_w)$. Making these assumptions and noting that the drag coefficient is very small with respect to unity, one can reduce (5–7) to the following form:

$$\frac{dC_N}{dC_L} = 1$$

$$\frac{dC_c}{dC_L} = \frac{dC_D}{dC_L} - C_L \frac{d\alpha}{dC_L} = C_L \left(\frac{2}{\pi e A} - \frac{.035}{(dC_L/d\alpha)^\circ} \right) \quad (5–8)$$

with $(dC_L/d\alpha)^\circ$, per degree.

The derivative $dC_{m_{ac}}/dC_L = 0$ by definition of the aerodynamic center.

With these substitutions the wing contribution of equation (5–5) becomes

$$\left(\frac{dC_m}{dC_L} \right)_{Wing} = \frac{x_a}{c} + C_L \left(\frac{2}{\pi e A} - \frac{.035}{dC_L/d\alpha} \right) \frac{z_a}{c} \quad (5–9)$$

The wing contribution to stability is therefore a function of the c.g. position with respect to the wing aerodynamic center. If the c.g. is ahead of the aerodynamic center, x_a will be negative and the first term will be stabilizing, whereas if the c.g. is aft of the aerodynamic center, x_a will be positive and the first term will be destabilizing. To study the effects of the second term, average values substituted for the terms in the brackets give $-\dfrac{C_L}{10}\dfrac{z_a}{c}$. If the c.g. position is under the wing aerodynamic center, z_a will be negative and the term will be stabilizing, whereas if the c.g. is over the wing aerodynamic center, the term will be unstable. The second term is usually very small with respect to the first term except at high values of the lift coefficient. At a $C_L = 1$

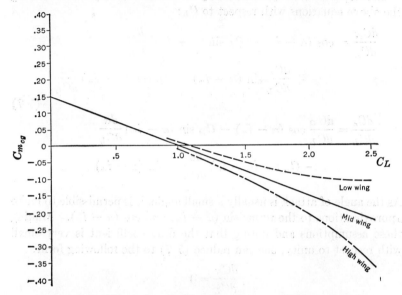

FIGURE 5–4.   Effect of vertical location of c.g. on pitching moments.

the second term is only one-tenth as powerful as the first term for equal horizontal and vertical shifts of the center of gravity. The effects of the second term, or the drag term, as it is often referred to, can usually be neglected. It will increase the stability of the high wing airplane at high lift coefficient and destabilize the low wing airplane at high lift coefficient. At a low C_L the term is negligible. These effects are shown in Figure 5–4.

In the rest of this study the drag term will be neglected. This leaves for the wing contribution the very simple relationship

$$\left(\frac{dC_m}{dC_L}\right)_{Wing} = \frac{x_a}{c} \tag{5-10}$$

which can be written as

$$\left(\frac{dC_m}{dC_L}\right)_{Wing} = x_{cg} - x_{ac} \tag{5-11}$$

where x_{cg} and x_{ac} are the locations of the airplane center of gravity and the wing aerodynamic center in percentage of the wing mean aerodynamic chord.

A study of this expression shows that if the airplane's c.g. is at the wing aerodynamic center, the stability contribution of the wing will be zero, and that the numerical value of the wing contribution is equal to the per cent aerodynamic chord between the c.g. and the aerodynamic center. If the c.g. is 1 per cent aft of the wing aerodynamic center, the slope of the pitching moment curve dC_m/dC_L will be .01.

5-3 Tail Contribution

The tail contribution to the static longitudinal stability of the airplane can be analyzed by studying the last term of equation (5-5). The derivative dC_{Nt}/dC_L is not approximately equal to unity as was assumed for dC_N/dC_L because of the effects of the wing downwash. The evaluation of this derivative can be developed as follows:

$$C_{Nt} = \left(\frac{dC_N}{d\alpha}\right)_t \alpha_t \qquad (5\text{-}12)$$

The angle of attack of the horizontal tail can be seen from Figure 5-2 to be

$$\alpha_t = \alpha_w - \epsilon + i_t - i_w \qquad (5\text{-}13)$$

The angle of attack of the horizontal tail is modified by the downwash from the wing and any incidence between the wing and the horizontal tail.

Equation (5-12) therefore becomes

$$C_{Nt} = \left(\frac{dC_N}{d\alpha}\right)_t (\alpha_w - \epsilon + i_t - i_w) \qquad (5\text{-}14)$$

and the derivative dC_{Nt}/dC_L becomes:

$$\frac{dC_{Nt}}{dC_L} = \left(\frac{dC_N}{d\alpha}\right)_t \left(\frac{d\alpha_w}{dC_L} - \frac{d\epsilon}{dC_L}\right) \qquad (5\text{-}15)$$

which can be reduced to

$$\frac{dC_{Nt}}{dC_L} = \frac{(dC_N/d\alpha)_t}{(dC_N/d\alpha)_w}\left(1 - \frac{d\epsilon}{d\alpha}\right) \qquad (5\text{-}16)$$

The tail contribution to stability becomes:

$$\left(\frac{dC_m}{dC_L}\right)_{Tail} = -\frac{(dC_N/d\alpha)_t}{(dC_N/d\alpha)_w}\frac{S_t}{S_w}\frac{l_t}{c}\eta_t\left(1 - \frac{d\epsilon}{d\alpha}\right) \qquad (5\text{-}17)$$

A short-hand notation is introduced in order to simplify the writing of this expression, which appears many times in the course of the developments that follow. This notation is:

$$\left(\frac{dC_N}{d\alpha}\right)_t = a_t$$

$$\left(\frac{dC_N}{d\alpha}\right)_w = a_w$$

$$\frac{S_t}{S}\frac{l_t}{c} = \bar{V} \tag{5-18}$$

With this new notation equation (5-17) becomes

$$\left(\frac{dC_m}{dC_L}\right)_{Tail} = -\frac{a_t}{a_w}\bar{V}\eta_t\left(1 - \frac{d\epsilon}{d\alpha}\right) \tag{5-19}$$

In order to evaluate the tail term of the stability equation, it is necessary to evaluate the slopes of the two lift curves a_t and a_w. The slope of the lift curve (C_N or C_L versus α) is a function of the wing aspect ratio, A, and the infinite aspect ratio or section slope of the lift curve a_0.

$$a_w = \frac{a_0}{1 + \frac{57.3ra_0}{\pi A}} \tag{5-20}$$

Curves of the slope of the lift curve versus aspect ratio are given in Figure 5-5, corrected at low aspect ratio from experimental data. A correction, r, for end plates is given in Figure 5-6 for use in equation (5-20).

The contributions of the horizontal tail surface to the airplane's equilibrium and static longitudinal stability are affected seriously by the downwash from the wing due to the wing vortex system, and therefore, in order to determine these contributions, it is necessary to evaluate the downwash at the horizontal tail for a given airplane design.

A wing's vortex system can be represented as shown in Figure 5-7, consisting of the "bound" vortex located at the wing quarter chord and a vortex sheet streaming from the wing trailing edge, rolling up to form the familiar two trailing vortices.

The wing wake centerline, which is coincident with the vortex sheet, is displaced downward and deformed by the influence of the bound vortex and the powerful trailing vortices. The whole flow field behind the wing is determined by this vortex system, with the major influence coming from the trailing vortices and the flow of air into the wake. The strength of the vortex system is proportional to the lift coefficient,

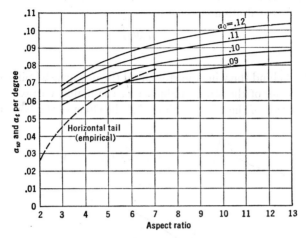

FIGURE 5-5. Slope of lift curve versus aspect ratio.

and therefore the downwash at any particular point will be proportional to the lift coefficient.

$$\epsilon = f\,(C_L) \qquad\qquad (5\text{-}21)$$

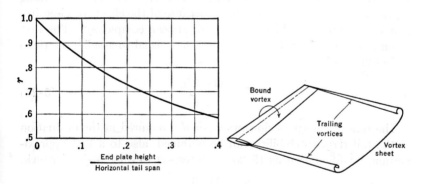

FIGURE 5-6. End plate correction to   FIGURE 5-7. The trailing vortex
slope of lift curve.   sheet.

To complete the disturbed flow regime due to the wing, it is necessary to consider the flow in front of the wing. In this region the bound vortex carries the predominant influence and creates a strong upwash just ahead of the wing, which decreases rapidly the farther forward from the wing the upwash is measured.   A sketch of the wing upwash and downwash distribution is shown in Figure 5–8.

The theoretical downwash at the wing aerodynamic center for an elliptical lift distribution is given as

$$\epsilon_{ac}^{\circ} = \frac{57.3C_L}{\pi A} \tag{5-22}$$

Directly behind the wing quarter chord the downwash builds up very rapidly because of the strong influence of the bound vortex; for positions farther aft of the wing trailing edge the effect of the bound vortex

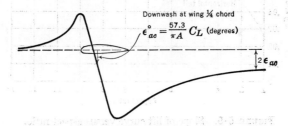

FIGURE 5–8.   Downwash distribution in front of and behind a finite wing.

rapidly diminishes and the downwash becomes almost entirely a function of the two trailing vortices.   At infinity the downwash theoretically becomes equal to $2\epsilon_{ac}$.   Actually, however, the trailing vortices are dissipated quite rapidly because of the effects of viscosity, and the downwash will eventually disappear completely.   For first estimates of the downwash at the tail, it is usual to assume the theoretical value of the downwash at infinity:

$$\epsilon^{\circ} = \frac{114.6C_L}{\pi A} \tag{5-23}$$

The rate of change of downwash, $d\epsilon/d\alpha$, required in the evaluation of the tail term to stability can be obtained, also to a first approximation, by differentiating (5–23) with respect to wing angle of attack.

$$\frac{d\epsilon}{d\alpha} = \frac{114.6}{\pi A} a_w \tag{5-24}$$

The downwash at the tail can be calculated with good accuracy by referring to NACA TR 648, in which design charts are given for predicting downwash angles behind plain and flapped wings.   This report is recommended to the reader for a precise evaluation for downwash angles or rates of change of downwash.

The actual downwash at the tail can vary considerably from the theoretical downwash given in (5-24), and depends largely on the location of the tail with respect to the wing and the position of the horizontal tail with respect to the wing wake. If the horizontal tail is positioned so that it lies either close to or inside the wing wake, large changes in downwash occur, as well as reduced tail efficiency and unpleasant tail buffeting. In the usual design the horizontal tail is

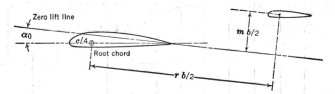

FIGURE 5-9. Dimensions for determination of downwash.

kept high enough to avoid the wing wake at all lift coefficients. If this is done, a simplified empirical scheme developed from the original work of TR 648 can be used to estimate the rate of change of downwash at the horizontal tail. For this purpose, the a.c. of the horizontal tail is positioned with respect to the wing, and zero lift line by the parameters r and m defined as the percentage semispan that the horizontal tail aerodynamic center is behind the wing root quarter chord and above the wing's zero lift line respectively (Figure 5-9). With these parameters the curves given in Figure 5-10 may be used to estimate $(d\epsilon/d\alpha)$. This estimate will be satisfactory for most airplane configurations in which the horizontal tail surface is well away from the wing wake.

The value of $(d\epsilon/d\alpha)$ obtained will be that at the airplane centerline. To account for the variation of $(d\epsilon/d\alpha)$ across the span of the horizontal tail, the correction to $(d\epsilon/d\alpha)$ for the ratio of tail span to wing span is given in Figure 5-11. To use these curves, choose the group aspect ratio and taper ratio that are closest to the actual airplane under investigation. If the airplane is so designed that the horizontal tail is close to or within the wing wake, the methods given in NACA TR 648 should be followed.

5-4 The Fuselage Contribution

The contributions of the fuselage and nacelles to the static longitudinal stability of the airplane are nearly always destabilizing, and in many cases the destabilizing effects are quite large in magnitude.

They must, therefore, be taken into account in the equilibrium and stability equations.

The theory accounting for the effects of the fuselage and nacelle on the airplane stability is rather a complex one, especially when the

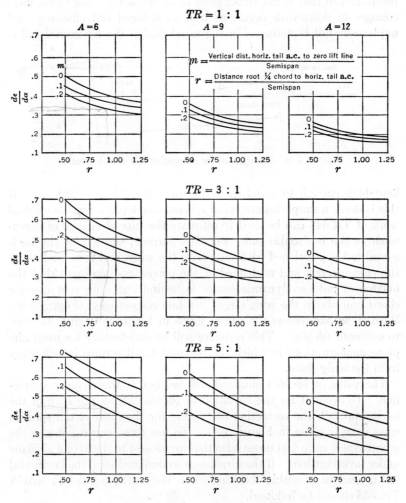

FIGURE 5–10.   Downwash charts.

interference effects of the wing and its flow pattern are taken into account. Two methods for predicting these contributions are included herein that have been used with good success on many occasions.

If the fuselage is considered operating at some angle of attack to the

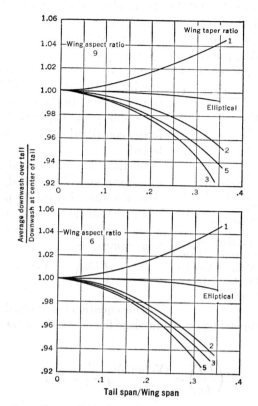

FIGURE 5-11. Correction to $d\epsilon/d\alpha$ for variations across span. From NACA TR 648, "Design Charts for Predicting Downwash Angles and Wake Characteristics behind Plain and Flapped Wings," by Silverstein and Katzoff.

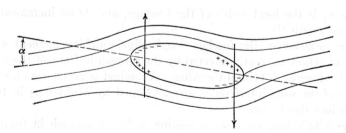

FIGURE 5-12. Potential flow about a body of revolution.

free stream in an ideal fluid, the resulting pressure distribution over the fuselage yields only a pure couple, with no resultant force, the center of pressure being at infinity. (See Figure 5-12.)

For real fluids, the center of pressure moves to a position somewhat ahead of the nose of the fuselage, and the location of the center of gravity on the fuselage will be of slight consequence. In the case of longitudinal stability the frictional lift and drag force contributions are usually neglected.

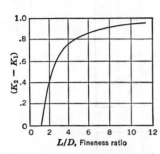

FIGURE 5-13. Fuselage correction for fineness ratio. From NACA TM 1036, "Aerodynamics of the Fuselage," by H. Multhopp.

In 1923 Max Munk* demonstrated that, for a very slender body of revolution, the variation of the pitching moment with angle of attack in degrees is a function of the volume and the dynamic pressure.

$$\left(\frac{dM}{d\alpha}\right) = \frac{(\text{Volume})}{28.7} q \qquad (5\text{-}25)$$

This equation is corrected by a factor $(K_2 - K_1)$, depending on the fuselage fineness ratio (L/D) as given in Figure 5-13.

$$\left(\frac{dM}{d\alpha}\right) = \frac{(\text{Volume})}{28.7} q(K_2 - K_1) \qquad (5\text{-}26)$$

For axially unsymmetric bodies equation (5-26) can be written as approximately

$$\left(\frac{dM}{d\alpha}\right) = \frac{q(K_2 - K_1)}{36.5} \int_0^l w_f^2 \, dx \qquad (5\text{-}27)$$

where w_f is the local width of the fuselage, and dx an increment of fuselage length (l).

The variation of the fuselage longitudinal pitching moment with angle of attack is greatly affected by the upwash in front of the wing and the downwash behind the wing. A method of accounting for the effects of the wing interference was developed by Multhopp† in 1941 and is included herein.

The wing's induced flow, consisting of heavy upwash in front of the wing due to the bound vortex, has a heavy destabilizing influence

* NACA TR 184.
† NACA TM 1036.

on the fuselage or nacelle sections ahead of the wing, whereas the downwash behind the wing reduces the unstable contribution of the fuselage or nacelle segments behind the wing. For this reason the location of the wing on the longitudinal axis of the fuselage is of considerable

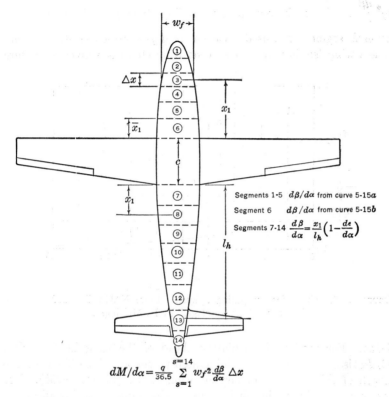

Segments 1-5 $d\beta/d\alpha$ from curve 5-15a

Segment 6 $d\beta/d\alpha$ from curve 5-15b

Segments 7-14 $\dfrac{d\beta}{d\alpha} = \dfrac{x_1}{l_h}\left(1 - \dfrac{d\epsilon}{d\alpha}\right)$

$$dM/d\alpha = \frac{q}{36.5}\sum_{s=1}^{s=14} w_f{}^2 \frac{d\beta}{d\alpha}\,\Delta x$$

FIGURE 5–14. Typical layout for computing fuselage moments.

importance to its destabilizing influence. Multhopp proposed the following formula to account for this phenomenon:

$$\frac{dM}{d\alpha} = \frac{q}{36.5}\int_0^l w_f{}^2 \frac{d\beta}{d\alpha}\,dx \tag{5–28}$$

In equation (5–28), β is the angle of the local flow and is equal to the free-stream angle of attack plus the angle of the induced flow due to the wing. Ahead of the wing the induced upwash adds to the angle of the free stream, making $d\beta/d\alpha$ greater than unity, while behind the wing the induced downwash subtracts from the free-stream angle and $d\beta/d\alpha$ is less than unity and becomes the now familiar $(1 - d\epsilon/d\alpha)$, at the tail. In the region between the wing leading and trailing edge, $d\beta/d\alpha$ is considered zero.

The integral in equation (5–28) is evaluated by dividing the fuselage, or nacelle, into segments (see Figure 5–14), computing the value of $w_f^2 \dfrac{d\beta}{d\alpha} \Delta x$ for each segment and adding them up. Average values for each segment are used. Curves of $d\beta/d\alpha$ versus positions ahead of the wing leading edge in percent wing chord are given in Figure

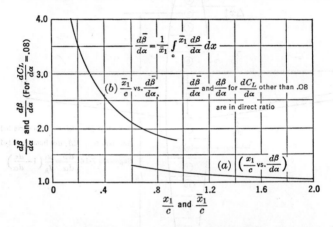

FIGURE 5–15. Chart for computing upwash. From NACA TM 1036, "Aerodynamics of the Fuselage," by H. Multhopp.

5–15a. For sections immediately ahead of the wing leading edge, $d\beta/d\alpha$ rises so abruptly that integrated values are given based on the length of this segment in percent wing chord (Figure 5–15b). For segments aft of the wing, it is assumed that $d\beta/d\alpha$ rises linearly from zero at the root trailing edge to $(1 - d\epsilon/d\alpha)$ at the horizontal tail aerodynamic center.

An additional factor is included that takes into account the effect of the fuselage or nacelle on the wing pitching moments. This may be computed from the following formula:

$$\frac{dM}{d\alpha} = \frac{qc^2}{290} (w_{LE} + 2w_{Mid} - 3w_{TE}) \tag{5–29}$$

where w_{LE}, w_{Mid}, and w_{TE} are the widths of the fuselage at the wing leading edge, midchord, and trailing edge, respectively.

The stability contribution of the fuselage or nacelle in terms of

dC_m/dC_L is obtained by dividing by qS_wca_w:

$$\left(\frac{dC_m}{dC_L}\right)_{\substack{Fus \\ Nac}} = \frac{(dM/d\alpha)_{Fus,\,Nac}}{qS_wca_w} \tag{5-30}$$

A simpler, but less accurate, method for estimating the fuselage and nacelle contributions to equilibrium and stability is to use the following formula:

$$\left(\frac{dC_m}{dC_L}\right)_{\substack{Fus \\ or\ Nac}} = \frac{K_f w_f^2 L_f}{S_w ca_w} \tag{5-31}$$

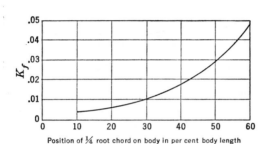

FIGURE 5-16. Fuselage stability coefficient. From NACA TR 711, "Analysis and Prediction of Longitudinal Stability of Airplanes," by R. R. Gilruth.

where L_f is the over-all fuselage length, w_f is the maximum width of the fuselage, and K_f is an empirical factor developed from experimental evidence. This factor was first introduced by Gilruth and White[*] in 1941. It depends entirely on the wing root chord's position on the fuselage or nacelle. The variation of K_f with wing position is given in Figure 5-16.

For normal fuselage arrangements, the shorter method will usually give reliable results, but for careful analysis and for abnormal fuselage or nacelle arrangements, the longer method is recommended.

5-5 Neutral Point (Stick-fixed)

From the studies just completed, the final stability equation can be written for the airplane in gliding flight with fixed controls with no propeller:

$$\frac{dC_m}{dC_L} = \frac{x_a}{c} + \left(\frac{dC_m}{dC_L}\right)_{\substack{Fus \\ Nac}} - \frac{a_t}{a_w}\bar{V}\eta_t\left(1 - \frac{d\epsilon}{d\alpha}\right) \tag{5-32}$$

[*] NACA TR 711.

and the equilibrium equation:

$$C_{m_{cg}} = C_L \frac{x_a}{c} + C_{m_{ac}} + C_{m_{\substack{Fus \\ Nac}}} - \frac{a_t}{a_w} C_L \bar{V} \eta_t \left(1 - \frac{d\epsilon}{d\alpha}\right)$$

$$- a_t(\alpha_0 - i_w + i_t)\bar{V}\eta_t \quad (5\text{–}33)$$

The contributions of the various airplane components to its static longitudinal stability (dC_m/dC_L) are shown in Figure 5–17.

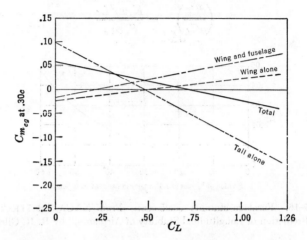

FIGURE 5–17. Typical longitudinal stability breakdown.

For any given airplane, the stability equation (5–32) is fixed except for movement of the center of gravity, which affects the wing term through the variation of x_a. A shift of center of gravity has very small influence on the tail term, a negligible influence on the fuselage or nacelle term, and a very strong influence on the wing term. The wing contribution can be written as in equation (5–11):

$$\left(\frac{dC_m}{dC_L}\right)_{Wing} = x_{cg} - x_{ac} \quad (5\text{–}34)$$

This indicates that for every per cent of the m.a.c. that the c.g. is moved aft, $(dC_m/dC_L)_{Wing}$ will increase positively 1 per cent. Center of gravity movement, therefore, has a powerful influence on the airplane's static longitudinal stability, and as will be shown in the following pages, is probably the most important variable in airplane design for stability. The effect of c.g. shift on the pitching moment coefficient of an assumed airplane is shown in Figure 5–18. It will be noted that all the curves rotate about a fixed C_m at $C_L = 0$.

An examination of Figure 5–18 shows that as the c.g. is moved aft the slope of the pitching moment curve (dC_m/dC_L) for the assumed airplane becomes more positive. When a c.g. position of 30 per cent is reached, the slope becomes zero. At this point the airplane is neutrally stable $(dC_m/dC_L = 0)$, and this c.g. position is called the stick-fixed neutral point and is denoted by the symbol (N_0). Equations (5–32) and (5–34) can be used to solve for this neutral point.

$$N_0 = x_{cg(dC_m/dC_L=0)} = x_{ac} - \left(\frac{dC_m}{dC_L}\right)_{\substack{Fus \\ Nac}} + \frac{a_t}{a_w}\bar{V}\eta_t\left(1 - \frac{d\epsilon}{d\alpha}\right) \quad (5\text{–}35)$$

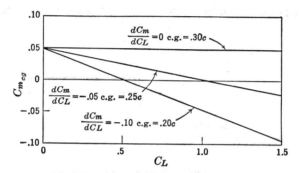

FIGURE 5–18.   Typical effect of c.g. shift on pitching moments.

The neutral point gives the most aft location at which the c.g. can be placed before making the airplane unstable. It therefore places one limit on the airplane's permissible c.g. travel, for the airplane should never be balanced aft of this point if a stable airplane is desired.

The neutral point is also very convenient to obtain, for once the neutral point is known, the stability at any other c.g. position may be obtained with good accuracy from the following relation:

$$\frac{dC_m}{dC_L} = x_{cg} - N_0 \quad (5\text{–}36)$$

In other words, the slope of the pitching moment coefficient versus C_L is numerically equal to the difference between the c.g. location and the neutral point, both expressed in percentage of the mean aerodynamic chord.

5–6   Power Effects

The development of the criteria for static longitudinal stability (dC_m/dC_L) has not taken into account the effects of the propulsive unit. Actually the contributions of the running propellers or jet units can have profound effects on both the equilibrium equation and the stability equation. Unfortunately it is possible to account for only parts of these effects analytically with any degree of accuracy, so that complete analysis of the power or propeller effects is done more or less qualitatively at the present time. On any new design the procedure that is usually followed in the early stages is to predict the contribution of the propellers at low thrust coefficients and then to allow a certain margin based on experience to account for the destabilizing effects of power at the higher thrust coefficients and at the lower speeds. The final design is nearly always analyzed in the wind tunnel by carefully testing a powered model on which the propeller characteristics are reproduced.

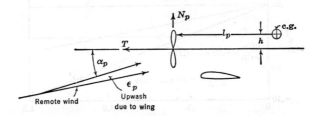

FIGURE 5–19.   Direct power components.

It is beyond the scope of this book to go into the theoretical detail required for a complete analysis of these power effects, even if this were possible. However, they will be discussed to show where they spring from and their relative magnitude. The effects of running propellers will be discussed first, and the effects of the jet propulsion unit discussed second.

The contribution of the running propeller to the airplane's equilibrium and static longitudinal stability can be broken down into two main effects. The first of these is the direct propeller contribution, arising as the result of the forces created by the propeller itself, and the second includes the indirect effects which arise as the result of the slipstream from the propeller and its interaction with the wing and tail surfaces. The direct propeller effects will be discussed first.

The components of the forces due to the running propeller at some angle to the wind include a thrust force, T, along the thrust axis and a normal force, N_p, perpendicular to the thrust axis, in the plane of the propeller disc. The summation of the moments about the airplane's c.g. due to the forces shown in Figure 5–19 is as follows:

$$M_{cg_p} = T \times h + N_p \times l_p \qquad (5–37)$$

Using the thrust coefficient defined as $T_c = T/\rho V^2 D^2$ and defining a normal force coefficient $C_{N_p} = N_p/qS_p$, where S_p is the propeller disc area $(\pi D^2/4)$, and letting N be the number of propellers, equation (5-37) expressed in coefficient form becomes

$$C_{m_p} = T_c \frac{2D^2}{S_w} \frac{h}{c} N + C_{N_p} \frac{l_p}{S_w} \frac{S_p}{c} N \qquad (5\text{-}38)$$

The stability contribution of these direct propeller forces may be obtained by differentiating equation (5-38) with respect to lift coefficient:

$$\frac{dC_{m_p}}{dC_L} = \frac{dT_c}{dC_L} \frac{2D^2}{S_w} \frac{h}{c} N + \frac{dC_{N_p}}{dC_L} \frac{l_p}{S_w} \frac{S_p}{c} N \qquad (5\text{-}39)$$

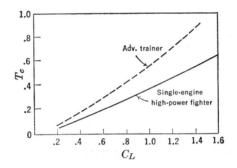

FIGURE 5-20. Typical variations of T_c with C_L.

The rate of change of thrust coefficient with airplane lift coefficient (dT_c/dC_L) must be determined for each particular case. The thrust coefficient may be written as the following function of the lift coefficient for unaccelerated flight:

$$T_c = \frac{Bhp\ \eta_p\ 550 C_L^{3/2} \rho^{1/2}}{(2\ W/S)^{3/2} D^2} \qquad (5\text{-}40)$$

or

$$T_c = K\eta_p C_L^{3/2} \qquad (5\text{-}41)$$

The propeller efficiency falls off rapidly at the higher lift coefficients, tending to linearize the variation of T_c versus C_L. Several examples of the variation of T_c with C_L are given in Figure 5-20. The slope of the variation of T_c with C_L can be obtained by differentiation of equation (5-41).

$$\frac{dT_c}{dC_L} = \frac{3}{2} K\eta_p C_L^{1/2} \qquad (5\text{-}42)$$

If an average figure of $dT_c/dC_L = .5$ is assumed as a typical value of this slope, then the stability contribution of the thrust component will be:

$$\left(\frac{dC_m}{dC_L}\right)_T = \frac{D^2}{S_w}\frac{h}{c} \tag{5-43}$$

As the term (D^2/S_w) will have a value close to .25 for fighter-type airplanes, an approximation to this term is

$$\left(\frac{dC_m}{dC_L}\right)_T = .25\,\frac{h}{c} \tag{5-44}$$

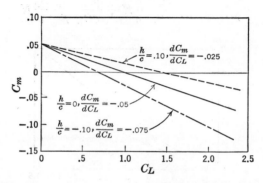

FIGURE 5-21.   Effect of vertical displacement of the thrust line.

This demonstrates that, for the c.g. above the thrust line (h/c positive), the slope of the pitching moment curve will be increased positively one-quarter of 1 per cent for every 1 per cent of the mean aerodynamic chord that the thrust line is below the c.g. Conversely, for every per cent m.a.c. that the thrust line is over the c.g. (h/c negative) the slope of the pitching moment curve will be increased negatively one-quarter of 1 per cent.   A high thrust line, then, with respect to the c.g. will be an important stabilizing influence, whereas a low thrust line will give an unstable contribution.   The calculated effect of the vertical shift of the thrust line on a typical airplane is shown in Figure 5-21.

The effect of the vertical location of the thrust line is a very important factor, and a high thrust line or a thrust line rotated to bring its line of action higher over the c.g. is a means of combating the very severe destabilizing effects of power to be discussed later.

The derivative dC_{Np}/dC_L can be evaluated as follows:

$$\frac{dC_{Np}}{dC_L} = \frac{dC_{Np}}{d\alpha_p}\frac{d\alpha_p}{dC_L} \tag{5-45}$$

The angle of attack of the propeller (α_p) is somewhat greater than the wing angle of attack for tractor airplanes, because of its location in the upwash field ahead of the wing.

$$\alpha_p = \alpha_w + \epsilon \qquad (5\text{-}46)$$

Therefore
$$\frac{d\alpha_p}{dC_L} = \frac{d\alpha_w}{dC_L} + \frac{d\epsilon}{dC_L}$$

or
$$\frac{d\alpha_p}{dC_L} = \frac{(1 + d\epsilon/d\alpha)}{a_w}$$

Finally, equation (5-45), the stability contribution of the propeller normal force, becomes:

$$\left(\frac{dC_m}{dC_L}\right)_{N_p} = \frac{\left(\dfrac{dC_N}{d\alpha}\right)_p \left(1 + \dfrac{d\epsilon}{d\alpha}\right) l_p S_p}{S_w c a_w} \qquad (5\text{-}47)$$

The value of the rate of change of propeller normal force coefficient with angle of attack of the thrust line $(dC_N/d\alpha)_p$ can be obtained from charts developed by Ribner* for the particular propeller involved. This derivative is a function of the propeller thrust coefficient and the projected side area; the destabilizing effect of this factor is sometimes referred to as propeller fin effect, for it acts in somewhat the same manner as a horizontal fin. Its influence increases with number of blades, blade angle, and blade width.

The stability contribution of the propeller normal force for a typical single-engined fighter is

$$\left(\frac{dC_m}{dC_L}\right)_{N_p} = .02\,\frac{l_p}{c} \qquad (5\text{-}48)$$

This demonstrates that for a propeller positioned one m.a.c. length ahead of the c.g. $(l_p/c$ positive), the propeller will destabilize the airplane by 2 per cent. For a pusher airplane $(l_p/c$ negative) the normal force contribution will be stabilizing by the same magnitude. The effect of the horizontal shift of the propeller disc is shown in Figure 5-22 for a typical single-engined airplane.

The normal force contribution is a large one and has considerable influence even for cases where $T_c = 0$.

The contributions of the running propellers to the static longitudinal stability just studied are those due to the components of forces at the propeller. Although these effects are sometimes quite large, they are

* NACA WR L-217.

no more important than the indirect effects due to the slipstream interaction with the wing and the horizontal tail. These indirect effects are very complex and do not lend themselves to an accurate analytical treatment. Several attempts have been made to account for slipstream effects theoretically, but all have lacked both simplicity and accuracy. No attempt will be made herein to offer a complete theoretical treatment of these effects, as it would be beyond the scope of this book even if it were possible. The major contributions of the slipstream will be pointed out, and an approximate method for allowing for these effects will be described.

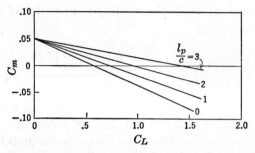

FIGURE 5–22.   Effect of horizontal displacement of the propeller disc.

There are four major contributions making up the indirect effects of the running propellers on the static longitudinal stability. They are as follows:

    a. Effect of slipstream on wing-fuselage moments.

    b. Effect of slipstream on wing lift coefficient.

    c. Effect of slipstream downwash at the horizontal tail.

    d. Effect of increased slipstream dynamic pressure on the tail.

In order to study the effects of the slipstream, it is necessary to return to the equilibrium equation, including the direct propeller force terms.

$$C_{m_{cg}} = C_L \frac{x_a}{c} + C_{m_{ac}} + C_{m_{Fus}} - C_{L_t}\bar{V}\eta_t + T_c \frac{2D^2}{S_w}\frac{h}{c} N$$

$$+ C_{N_p} \frac{l_p}{S_w} \frac{S_p}{c} N \qquad (5\text{–}49)$$

The direct effects of the running propeller are included in the last two terms of (5–49). The slipstream effects on the wing and fuselage moments must be obtained from the second and third terms, while the slipstream effect on the horizontal tail must be developed from the tail term $(-C_{L_t}\bar{V}\eta_t)$.

The change in pitching moments introduced by the interaction of the propeller slipstream on wing and fuselage will be neglected, as they are very difficult to analyze and because these terms are quite small in comparison to the other slipstream or power effects. The effects of the slipstream interaction with the horizontal tail cannot be neglected, because this interaction is in many cases one of the major factors making up the equilibrium or stability picture.

The tail contribution to the equilibrium equation at a given lift coefficient will be altered when the effects of the propeller are considered, first because of the fact that the wing will operate at a new angle of attack for the same C_L, on account of the partial immersion in the slipstream. This will place the horizontal tail at a new angle of attack and alter its pitching moment contribution. Second, the angle of attack of the horizontal tail will be changed by the influence of the downwash from the propeller created as a result of the normal force in the plane of the propeller disc, and finally, the tail contribution will be changed because of the increased velocity in the slipstream.

If it is assumed that the horizontal tail lies completely in the propeller slipstream, the tail pitching moment about the center of gravity can be expressed as follows:

$$C_{mt} = -a_t(\alpha_w - \epsilon_w - \epsilon_p - i_w + i_t)\,\bar{V}\left(\frac{v_s}{v}\right)^2 \qquad (5\text{-}50)$$

where ϵ_w is the downwash from the wing as calculated before, ϵ_p the downwash from the propeller, and v_s and v the velocity of the slipstream and the free stream, respectively. Equation (5-50) may be expanded in terms of the zero lift angle α_0 and C_L as follows:

$$C_{mt} =$$

$$-a_t\left(\alpha_0 + \frac{d\alpha}{dC_L}C_L - \frac{d\epsilon_w}{dC_L}C_L - \frac{d\epsilon_p}{dC_L}C_L - i_w + i_t\right)\bar{V}\left(\frac{v_s}{v}\right)^2 \qquad (5\text{-}51)$$

which may be further simplified to

$$C_{mt} = -\frac{a_t}{a_w}C_L\left(1 - \frac{d\epsilon_w}{d\alpha} - \frac{d\epsilon_p}{d\alpha}\right)\bar{V}\left(\frac{v_s}{v}\right)^2$$

$$-a_t(\alpha_0 + i_t - i_w)\,\bar{V}\left(\frac{v_s}{v}\right)^2 \qquad (5\text{-}52)$$

The velocity in the slipstream, v_s, is the following function of the thrust coefficient, T_c, as developed from the simple momentum theory

$$v_s{}^2 = v^2\left(1 + \frac{8T_c}{\pi}\right) \qquad (5\text{-}53)$$

or

$$\left(\frac{v_s}{v}\right)^2 = \left(1 + \frac{8T_c}{\pi}\right) \tag{5-54}$$

The velocity of the slipstream is therefore a function of the airplane's thrust coefficient, with $v_s = v$ when $T_c = 0$.

The increment of downwash at the tail because of the slipstream, ϵ_p, is a function of the normal force and thrust coefficient at the propeller. The derivative $(d\epsilon_p/d\alpha)$ has been expressed by Ribner* as the following function of the rate of change of propeller normal force coefficient with angle of attack for $T_c = 0$:

$$\frac{d\epsilon_p}{d\alpha} = A + B\left(\frac{dC_N}{d\alpha}\right)_{pT_c=0} \tag{5-55}$$

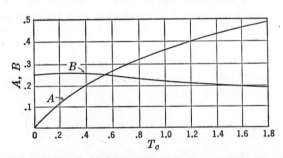

FIGURE 5–23. Propeller downwash factors. From NACA WR L–25, "Notes on the Propeller and Slipstream in Relation to Stability," by H. S. Ribner.

where A and B are functions of the thrust coefficient, T_c, as shown in Figure 5–23.

From a knowledge of the variation of the thrust coefficient with lift coefficient, together with calculated values of the derivative $(dC_N/d\alpha)_p$, it is possible to compute to a first approximation the tail pitching moment, including the slipstream effects. It must be understood that these calculations are approximate because of the lack of reliable information on the serious interference factors.

The slipstream effect on the horizontal tail moments varies with airplane lift coefficient and therefore also affects the tail contribution to the static longitudinal stability. The generalized tail term may be written as follows:

$$C_{mt} = -C_{Lt}\,\bar{V}\left(\frac{v_s}{v}\right)^2 \tag{5-56}$$

* NACA WR L-25.

Equation (5–56) may be differentiated with respect to C_L to obtain the tail stability term power-on.

$$\left(\frac{dC_m}{dC_L}\right)_t = -\frac{dC_{Lt}}{dC_L}\,\bar{V}\left(\frac{v_s}{v}\right)^2 - C_{Lt}\bar{V}\,\frac{d(v_s/v)^2}{dC_L} \tag{5-57}$$

The first term is similar to the power-off tail term, but the second term is a new one, required to account for the variation of the slip-stream velocity with C_L through the variation of T_c with C_L, and indicates the very important effect that the stability contribution of the tail, power-on, is dependent on the magnitude and sign of the load the tail is carrying.

Equation (5–57) may be rewritten as follows:

$$\left(\frac{dC_m}{dC_L}\right)_t = -\frac{a_t}{a_w}\,\bar{V}\left(1 - \frac{d\epsilon}{d\alpha} - \frac{d\epsilon_p}{d\alpha}\right)\left(\frac{v_s}{v}\right)^2 - C_{Lt}\bar{V}\,\frac{d(v_s/v)^2}{dC_L} \tag{5-58}$$

The first term is the same as the power-off tail term given in equation (5–35) except for the added propeller downwash factor ($d\epsilon_p/d\alpha$). As this factor is nearly always positive, the contribution is destabilizing and is of importance even for a windmilling propeller ($T_c = 0$). The second term arises as a result of the variation of the slipstream velocity with airplane lift coefficient. The derivative $\dfrac{d(v_s/v)^2}{dC_L}$ is positive and varies with propeller T_c. If the tail is carrying an upload (C_{Lt} positive), this term will be stabilizing, whereas if it is carrying a download, it will be destabilizing. Its contribution is, of course, zero when $T_c = 0$.

The magnitude of these effects can be very large, especially for high power at low air speeds, and contribute large changes from the propeller-off stability. The value of $d\epsilon_p/d\alpha$ can reach magnitudes as high as .5 for large values of the thrust coefficient, which, when combined with values of $d\epsilon/d\alpha$ from the wing, can obviate the whole tail term and shift the neutral point by 15 to 20 per cent of the mean aerodynamic chord. Even at $T_c = 0$ this term can cause forward shifts of the neutral point as high as 5 per cent m.a.c.

The second term of equation (5–58) can contribute largely to the stability, again depending on the magnitude and sign of the tail lift coefficient (C_{Lt}) and the thrust coefficient. Its effect is to reduce the variation in stability as the c.g. moves forward or aft. This comes about as a result of the fact that as the c.g. is moved aft, rapidly destabilizing the wing term (x_a/c), it calls for a more positive C_{Lt} for trim, giving a stabilizing effect through this term and thereby reducing the net destabilization of the airplane for a given aft c.g. movement. Conversely, as the c.g. is moved forward, a more negative C_{L_t} is required for trim, and the consequent destabilizing effect of the horizontal tail offsets somewhat the added wing stability due to the forward c.g. movement. An airplane that is made stable in spite of a rearward c.g. position by use of a large horizontal tail will carry an upload on the tail through the high C_L range and obtain a stabilizing effect from the rate of change of $(v_s/v)^2$ with C_L. When the propellers are windmilling $(T_c = 0)$, this term, of course, disappears.

The indirect effects of the slipstream on the airplane's longitudinal equilibrium and stability can be largely charged to the increase in downwash at the tail due to the added downwash from the propeller, and the increase in the dynamic pressure at the tail due to its immersion in the slipstream. These effects, when combined with the direct propeller effects discussed earlier in this section, are the major effects of the running propellers on the equilibrium and static longitudinal stability of the airplane. The power effects on static longitudinal stability just discussed are presented only as an indication of where these effects come from and their orders of magnitude. They are only considered qualitative and should be used with great care. In a more detailed study of these effects, account should be taken of the slipstream interaction with the fuselage and wing and its effects on their moments. Account should also be taken of the partial immersion of the tail in the slipstream and the more complicated effects such as propeller rotation on multiengine airplanes. There have been several attempts to account analytically for the effects of the running propellers on the airplane's stability, but their accuracy is not very good and they are not used extensively.

The procedure used by most airplane designers* is to estimate the static longitudinal stability of the airplane with propellers windmilling $(T_c = 0)$ and then to allow certain margins for the effects of high power at low lift coefficients. These margins are obtained from a great many flight test data on a great many different airplanes. This system has proven quite successful and yields answers that give good indications of the power-on stick-fixed neutral points.

At $T_c = 0$ only two of the propeller contributions to stability remain. One of these is the direct effect of the normal force, and the other is the effect of the downwash from the propeller on the horizontal tail. These two terms come from equations (5–47) and (5–48) for the case of zero thrust.

* NACA WR L-116.

A direct propeller contribution at $T_c = 0$ is given as follows:

$$\left(\frac{dC_m}{dC_L}\right)_{N_{pTc=0}} = \frac{(dC_N/d\alpha)_{pT=0}(1 + d\epsilon/d\alpha)l_p S_p\, N}{S_w c a_w} \qquad (5\text{–}59)$$

where $(dC_N/d\alpha)_{pT=0}$ must be obtained from propeller data for the particular installation, or may be taken from the following table for first estimates:

$(dC_N/d\alpha)_{pT=0}$*

.00165	Two-bladed propellers
.00235	Three-bladed propellers
.00296	Four-bladed propellers
.00510	Six-bladed counter-rotating propellers

The term $(1 + d\epsilon/d\alpha)$ is the factor $d\beta/d\alpha$ used in computing the fuselage contribution to stability and may be obtained from Figure 5–15.

The effect of the propeller downwash on the horizontal tail may be evaluated from the following equation:

$$\left(\frac{dC_m}{dC_L}\right)_{\epsilon_{pT=0}} = \frac{a_t}{a_w}\frac{\bar{V}\eta_t}{.07}\left(\frac{dC_N}{d\alpha}\right)_{pT=0}\frac{d\beta}{d\alpha} \qquad (5\text{–}60)$$

The static longitudinal equation for the airplane in flaps-up, gliding flight with propellers windmilling is given in equation (5–61):

$$\frac{dC_m}{dC_L} = \frac{x_a}{c} + \left(\frac{dC_m}{dC_L}\right)_{\substack{Fus \\ Nac}} - \frac{a_t}{a_w}\bar{V}\eta_t(1 - d\epsilon/d\alpha)$$

$$+ N\frac{(dC_N/d\alpha)_{pT=0}\, d\beta/d\alpha\, l_p S_p}{S_w c a_w} + \frac{a_t}{a_w}\frac{\bar{V}\eta_t}{.07}\left(\frac{dC_N}{d\alpha}\right)_{pT=0}\frac{d\beta}{d\alpha} \qquad (5\text{–}61)$$

and the stick-fixed neutral point with propellers windmilling is given by equation (5–62)

$$N_{0\text{wind}} = x_{ac} - \left(\frac{dC_m}{dC_L}\right)_{\substack{Fus \\ Nac}} + \frac{a_t}{a_w}\bar{V}\eta_t\left(1 - \frac{d\epsilon}{d\alpha}\right)$$

$$- N\frac{(dC_N/d\alpha)_{pT=0}\, d\beta/d\alpha\, l_p S_p}{S_w c a_w} - \frac{a_t}{a_w}\frac{\bar{V}\eta_t}{.07}\frac{d\beta}{d\alpha}\left(\frac{dC_N}{d\alpha}\right)_{pT=0} \qquad (5\text{–}62)$$

The shift in the stick-fixed neutral point, from propeller windmilling to critical power on flight configuration, is given in the following empirically determined table:

* NACA WR L-25.

Type Airplane	ΔN_0
Single-engine fighters	--.04
Twin-engine medium bombers	--.06
Multiengined heavy bombers	−.10
Twin-engine cargo	−.08
Four-engine cargo	−.10
Pusher-type airplane	0

Typical flight test evaluations of the stick-fixed neutral point for a single-engine fighter and a twin-engine medium bomber are presented

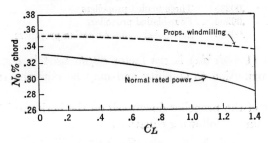

FIGURE 5–24. Typical variation of stick-fixed neutral point with power (single-engine fighter).

in Figures 5–24 and 5–25. The stick-fixed neutral points for propeller windmilling ($T_c = 0$) gliding flight are shown, together with the stick-fixed neutral points for flight with normal rated power.

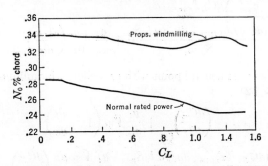

FIGURE 5–25. Typical variation of stick-fixed neutral point with power (twin-engined medium bomber).

5–7 Power Effects on the Jet Airplane

The effects of power on the equilibrium and static longitudinal stability of the jet propelled airplane are somewhat simpler to analyze than were the effects of power on an airplane equipped with the engine-propeller combination. This is due to the elimination of the propeller and the very complicated effects of its slipstream on the horizontal tail surface.

The jet blast from either a rocket or turbojet installation is kept at a safe distance from the horizontal tail because of the heat of this blast, and as a consequence the slipstream problems are simpler than those for the propeller slipstream which either fully or partially engulfs the horizontal tail.

There are three major contributions of a jet unit to the equilibrium and static longitudinal stability of the airplane. These are the direct thrust effects, the direct normal force effects at the air-duct inlet, if

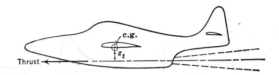

FIGURE 5-26.   Jet thrust moment.

any, and the effect of the induced flow at the tail due to the inflow to the jet blast. The rocket installation which requires no air intake will encounter only the first and last of these, while the turbojet will encounter all three.

The direct effect of the jet thrust can be obtained quite readily by obtaining the moment coefficient due to the thrust acting at some vertical arm to the airplane's center of gravity. (See Figure 5-26.)

$$C_m = \frac{T \times z_t}{q S_w c} \tag{5-63}$$

If the airplane is considered in unaccelerated flight, the dynamic pressure, q, will be the following function of the lift coefficient: $q = \frac{W/S_w}{C_L}$. Equation (5-63) will therefore become:

$$C_m = \frac{T \times z_t \times C_L}{W \times c} \tag{5-64}$$

In most cases the thrust, T, is nearly independent of airplane speed and can be considered a constant. The stability contribution can be obtained by differentiating (5-64) with respect to C_L.

$$\left(\frac{dC_m}{dC_L}\right)_{th} = \frac{T \times z_t}{W \times c} \tag{5-65}$$

For airplanes in which the thrust variation with speed is large, the pitching moment coefficient must be calculated for several values of the airplane's lift coefficient and the results plotted against C_L. In most cases this refinement is not necessary. Jet-propelled airplanes with thrust lines below the c.g. (z_t positive) will be destabilized by the thrust contribution, whereas for thrust lines above the c.g. this term will be stabilizing.

It should be noted that there is an effect on the airplane's stability due to a constant thrust when only the speed is varied. For unaccelerated flight, the condition $C_L V^2 = K$ must be satisfied. In accelerated flight where the lift coefficient is changed at constant air

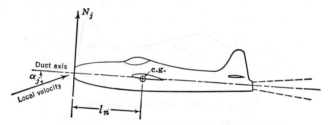

FIGURE 5-27.   Duct inlet normal force moment.

speed, the thrust will contribute nothing to the airplane's stability. This will be discussed at greater length in a subsequent chapter.

The second equilibrium and stability contribution is that due to the normal force created at the air-duct inlet to the turbojet unit. This normal force is the force created as a result of the momentum change incurred as the free stream is bent to flow along the duct axis (Figure 5-27).

The normal force N_j can be roughly determined from momentum considerations:

$$N_j = \frac{W_a}{g} v \frac{\alpha_j{}^\circ}{57.3} \qquad (5\text{--}66)$$

where N_j is the normal force at the duct entrance in pounds, W_a is the weight of air flowing through this duct in pounds per second, and α_j is the angle between the local flow at the duct entrance and the duct axis in degrees. The local flow at the duct entrance is affected by the upwash from the induced flow field of the wing, as was the local flow at the propeller, and can be expressed as

$$\alpha_j = \alpha \frac{d\beta}{d\alpha} \qquad (5\text{--}67)$$

where $d\beta/d\alpha$ is given in Figure 5-15.

The moment coefficient about the center of gravity due to this normal force is equal to

$$C_{mj} = \frac{2W_a(d\beta/d\alpha)l_n}{57.3gc\rho S_w v}\, \alpha \tag{5-68}$$

The weight of air flowing through the duct per second varies slightly with speed, and so for a careful analysis of this contribution the pitching moment coefficient at several values of C_L should be obtained. To a first approximation, however, it is possible and fairly accurate to consider W_a a constant. In this case equation (5–68) can be differentiated readily with respect to C_L and the stability contribution obtained:

$$\left(\frac{dC_m}{dC_L}\right)_j = \frac{W_a}{g}\frac{d\beta/d\alpha}{a_w}\frac{l_n}{c}\frac{.05}{\rho S_w v} \tag{5-69}$$

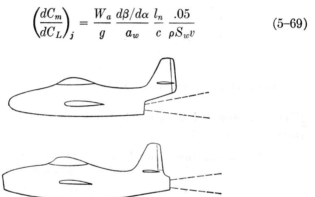

FIGURE 5–28.   Typical jet exhausts.

The third contribution of the jet unit to the equilibrium and static longitudinal stability is the effect of the jet-induced downwash at the horizontal tail. The only configurations for which this term need be considered are for those installations where the jet blast passes under or over the horizontal tail surface. On installations where the jet blast leaves the airplane behind the horizontal tail, this term may be neglected. Typical examples of these configurations are shown in Figure 5–28.

The jet spreading out as it leaves the airplane will, because of turbulent mixing, draw the outer flow into the jet, thereby creating an inclined flow pattern around the jet. If the horizontal tail is located in this inclined flow, it will operate at a different angle of attack from what it would at zero power, and thereby creates moments about the c.g. that affect both the equilibrium and stability of the airplane.

The change in pitching moment coefficient due to the change in flow inclination at the horizontal tail can be given as follows:

$$\Delta C_m = -a_t \bar{V} \eta_t \, \Delta \alpha_t \tag{5–70}$$

The change in angle of attack of the tail due to the flow inclination must be determined at each value of the lift coefficient and the result plotted against C_L. The methods of computing this angle of attack change are in the literature,[*] and because of their length will not be included herein. A final plot of C_m versus C_L will give the required information.

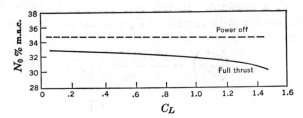

FIGURE 5–29. Typical effect of jet power on stick-fixed neutral point.

The jet-propelled airplanes tested to date indicate the following changes in stability:

a. Because of direct thrust effects   Depends entirely on vertical location of c.g. with respect to thrust axis

b. Because of normal force $\Delta \, dC_m/dC_L = .01 - .03$

c. Because of flow inclination   Jet under tail $\Delta \, dC_m/dC_L = .04$

Jet behind tail $\Delta \, dC_m/dC_L = 0$

The effects of the jet on stability, then, can be readily computed with a good degree of accuracy. A typical flight test variation between power-on and power-off stability for a jet airplane is shown in Figure 5–29.

5–8 The Most Aft Center of Gravity

Establishment of the airplane's stick-fixed neutral point, N_0, immediately places a restriction on the possible location for the airplane's center of gravity. If it is required that the airplane possess some static

* NACA WR L-213.

longitudinal stability, then it should never be possible to load the airplane so that its center of gravity lies aft of the neutral point.   In almost all airplane designs, the necessary dispersal of the disposable load means that the airplane's center of gravity will move somewhat because of the burning of fuel, expenditure of armament, change in loading arrangement, etc.   In the initial stages of the airplane design the anticipated c.g. range must be established, and then, from an analysis of the neutral points, this c.g. range must be placed on the mean aerodynamic chord so that the most aft c.g. lies in front of the neutral point.   If, for example, an airplane's c.g. range extends over 10 per cent of the mean aerodynamic chord and the neutral point is calculated to be at 30 per cent m.a.c., the range must be placed so that the most aft location does not lie behind 30 per cent m.a.c.   A forward limit on the c.g. range will be established later, but for the present the extreme importance of the airplane's c.g. location should be understood and the aft limit of the stick-fixed neutral point recognized.

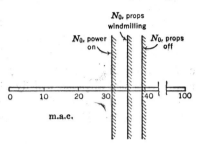

FIGURE 5-30.   Typical aft c.g. limits.

As a preknowledge of the location of the airplane's neutral point is of extreme importance to the designer, the analytical treatment just completed is used especially during the first phase of the airplane's design. The stability of the airplane is more thoroughly investigated in the wind tunnel by testing a geometrically similar scale model equipped with electric driving motor and propellers, so adjusted as to simulate carefully the full-scale slipstream conditions.   The results of these tests are presented in the form of curves of C_m versus C_L, and from these curves the airplane's neutral point is determined.   The wind-tunnel results are considered more reliable than the results obtained from the analytical development and should always be obtained on any important airplane design.

5-9 Longitudinal Control

In the previous section, the static longitudinal stick-fixed stability of the airplane was discussed for both power-off and power-on flight conditions, and it was demonstrated that the static stability level of the airplane is measured by the slope of the pitching moment coefficient versus lift coefficient (dC_m/dC_L).   In this same chapter the airplane equilibrium equation (5-33) was given with the statement that the airplane components must be designed to satisfy the equilibrium condition ($C_{m_{cg}} = 0$) at the desired lift coefficient. This section will deal with this equilibrium equation and the means provided the pilot for satisfying it for any desired lift coefficient, thereby giving him control of the speed of the airplane in unaccelerated flight, and the curvature of the flight path in accelerated flight.

For a particular airplane, the curve of pitching moment coefficient versus lift coefficient may be as shown in Figure 5-31.

The airplane, in the condition shown, is stable ($dC_m/dC_L = -.10$) and is in equilibrium ($C_{m_{cg}} = 0$) at a lift coefficient $C_L = .5$. This means that the airplane will fly at a speed corresponding to this lift coefficient, its stability resisting any tendency of the airplane to deviate from this speed. If the pilot, however, desires to slow the airplane

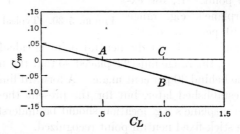

FIGURE 5-31.   Typical variations of C_m versus C_L.

down and fly at $C_L = 1.0$, he must be equipped with some control that can overpower the stability diving moment ($C_{m_{cg}} = -.05$) as shown at B-C in Figure 5-31 in order to obtain equilibrium at $C_L = 1.0$. Obviously the more stable the airplane becomes, the more powerful will be the control required to change the airplane lift coefficient. It is this condition that actually limits the amount of stability permissible in any airplane.

To find a means for producing control moments that can effectively change the equilibrium lift coefficient, it is necessary to return to the equilibrium equation. The propeller-off equations will be used for simplicity.

$$C_{m_{cg}} = C_{m_{ac}} + C_L \frac{x_a}{c} + C_{m_{F_{us} \atop Nac}} - \left(\frac{dC_L}{d\alpha}\right)_t \alpha_t \bar{V} \eta_t \qquad (5\text{-}71)$$

The three factors of this equation that might be used to control the trim lift coefficient are $C_{m_{ac}}$, x_a/c, and α_t. The first of these terms is the wing pitching moment about its aerodynamic center ($C_{m_{ac}}$), whose magnitude is a function of the wing section camber and the aerodynamic twist of the wing. This moment can be controlled by a flap at the wing trailing edge, the major control used by the tailless airplane. On an airplane equipped with a horizontal tail to the rear, the effectiveness of such a device is reduced, for as the flap is deflected to change $C_{m_{ac}}$ negatively, there is an accompanying increase of lift on the wing and a consequent increase in downwash, giving a more negative tail angle of attack which produces a moment in the opposite direction. Control can be obtained by this means, but it is small and used only on tailless airplanes.

The term x_a/c is a pure function of c.g. position. The equilibrium lift coefficient can be altered by shifting the c.g. As the c.g. moves forward the equilibrium C_L is reduced, and as the c.g. moves aft the equilibrium C_L is increased. Unfortunately, there is a change in stability accompanying this c.g. shift, which, together with its mechanical complexity, rules it out as a possibility of control.

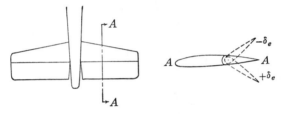

FIGURE 5–32. The elevator.

The final term is the angle of attack of the tail, α_t. Changing the angle of attack of the tail can produce large changes in pitching moment and can be made into a very powerful control. If the horizontal tail were all one piece, rotatable about some axis near its leading edge, it would prove to be the most powerful control available. It will be shown in subsequent chapters that the reason tail surfaces are not usually operated in this manner is the very difficult problem of designing for proper control forces with this type of control. Some experimental airplanes have been tested in the past that have had all-moving horizontal tails, but up to the present they have never been used on any airplanes produced in quantity.

In the usual case, the longitudinal control is an elevator, essentially a flap on the aft section of the horizontal tail. This flap is hinged near its leading edge and can be rotated by the pilot to alter the tail moment and thereby the equilibrium lift coefficient. A typical horizontal tail is shown in Figure 5–32.

5–10 Elevator Power

Deflecting the elevator effectively changes the angle of attack of the whole horizontal tail, thereby changing its lift and producing a control moment about the airplane's center of gravity. The magnitude of the moment coefficient obtained per degree deflection of the elevator is termed the elevator power and can be obtained analytically from equation (5–71) by differentiating C_m with respect to elevator deflection, δ_e. It should be noted again that, for simplicity, equation (5–70) contains no propeller or power terms. This omission will introduce no error for the propeller-windmilling condition, but will give results somewhat low for high-power, low-speed flight, because of the increase in dynamic pressure in the slipstream. It shall be assumed that a downward deflection of the elevator is positive. The elevator power criterion from equation (5–71) is as given below:

$$\frac{dC_m}{d\delta_e} = -\left(\frac{dC_L}{d\alpha}\right)_t \bar{V}\eta_t \frac{d\alpha_t}{d\delta_e} \qquad (5–72)$$

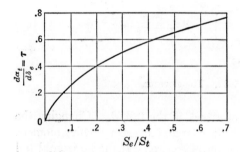

FIGURE 5–33. Elevator effectiveness.   FIGURE 5–34. Effect of elevator deflection on horizontal tail lift.

All the terms of this equation have been introduced before except the derivative $d\alpha_t/d\delta_e$. The rate of change of effective tail angle of attack with elevator deflection is a function of the ratio of the elevator area, S_e, to the whole horizontal tail area, S_t, but is not a linear function of this percentage. An empirically determined curve of the control effectiveness parameter $d\alpha_t/d\delta_e$ versus the area ratio S_e/S_t is given in Figure 5–33.

The effective change in horizontal tail angle of attack can be demonstrated from simple tail lift curves as shown in Figure 5–34. The cross-plot of α_t versus δ_e obtained at constant C_{Lt} yields the negative of this gradient. The cross-plot is usually taken at zero lift, and the elevator effectiveness parameter, $d\alpha_t/d\delta_e$, is sometimes thought of as the control's ability to rotate the zero lift line of the horizontal tail. The parameter $d\alpha_t/d\delta_e$ is usually given the symbol τ. Using τ and the short-hand derivative forms $(dC_m/d\delta_e = C_{m\delta})$, equation (5–72) may be written as:

$$C_{m\delta} = -a_t \bar{V} \eta_t \tau \qquad (5\text{–}73)$$

$C_{m\delta}$, then, will be used as the measure of elevator control power throughout the remainder of this book.

5–11   Elevator Angle versus Equilibrium Lift Coefficient

The change in equilibrium lift coefficient due to deflecting the elevator may be studied by again referring to equation (5–71). The only term affected by the elevator deflection is the tail angle of attack, α_t, which, for this study, can be broken down in terms of wing angle of attack, downwash, tail and wing incidence, and change in angle of attack due to elevator deflection:

$$\alpha_t = (\alpha_w - \epsilon - i_w + i_t + \tau\delta_e) \qquad (5\text{–}74)$$

The propeller-off equilibrium equation can be rewritten

$$C_{m_{cg}} = C_{m_{ac}} + C_L \frac{x_a}{c} + C_{m_{\substack{Fus \\ Nac}}}$$
$$- a_t(\alpha_w - \epsilon - i_w + i_t + \tau\delta_e)\bar{V}\eta_t \qquad (5\text{–}75)$$

The control of the equilibrium lift coefficient is effected through the influence of the term $\tau\delta_e$ of equation (5–75). It is of interest to note, at this point, that a change in elevator deflection and lift coefficient will not change the slope of the pitching moment curve (dC_m/dC_L), for τ is independent of the lift coefficient and the term vanishes when the derivative of (5–75) is taken with respect to C_L. Deflecting the elevator will change the equilibrium lift coefficient, but will not change the static longitudinal stability for the propeller-off condition. This is not the case for the constant-power flight condition, as the change in tail lift accompanying the change in elevator deflection will give small changes in the slope of the pitching moment curve. In spite of the errors introduced for high-power flight, it is safe to assume that the airplane's stability is independent of elevator angle.

An example of the curves of pitching moment coefficient versus lift coefficient is given in Figure 5–35.

It can be seen from Figure 5–35 that the elevator control is powerful enough to bring the airplane into equilibrium at any C_L throughout the unstalled range. A cross-plot of these curves, to show the elevator angle required for equilibrium versus lift coefficient, can readily be made and is shown in Figure 5-36.

The elevator deflection required for equilibrium may be obtained analytically from a knowledge of the slope of the pitching moment

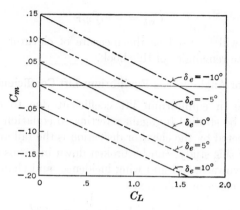

FIGURE 5-35. C_m versus C_L for various elevator angles.

curve, dC_m/dC_L, and the elevator power, $C_{m\delta}$. The equation for the elevator angle required versus equilibrium lift coefficient can be written

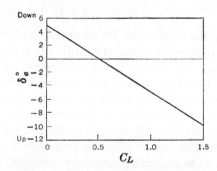

FIGURE 5-36. Elevator angle versus C_L.

as follows, with δ_{e0} the elevator angle at zero lift:

$$\delta_e = \delta_{e0} + \frac{d\delta_e}{dC_L} C_L \qquad (5\text{-}76)$$

Equation (5-76) may be placed in a more usable form by determining the value of the derivative $d\delta_e/dC_L$ in terms of familiar quantities. The elevator angle required for equilibrium may be obtained from equation (5–75) by equating it to zero and solving for δ_e.

$$\delta_e = \frac{C_{m_{ac}} + C_L \dfrac{x_a}{c} + C_{m_{Fus}} - a_t(\alpha_w - \epsilon - i_w + i_t)\bar{V}\eta_t}{a_{t\tau}\bar{V}\eta_t} \tag{5–77}$$

The rate of change of elevator angle with equilibrium lift coefficient may be obtained from (5–77) by differentiating with respect to C_L.

$$\frac{d\delta_e}{dC_L} = \frac{\dfrac{x_a}{c} + (dC_m/dC_L)_{Fus} - (a_t/a_w)\bar{V}\eta_t(1 - d\epsilon/d\alpha)}{a_{t\tau}\bar{V}\eta_t} \tag{5–78}$$

The numerator of equation (5–78) can be recognized as the propeller-off stick-fixed stability equation, while the denominator is the expression for elevator power, $-C_{m\delta}$. Equation (5–78) may therefore be rewritten as follows:

$$\frac{d\delta_e}{dC_L} = -\frac{dC_m/dC_L}{C_{m\delta}} \tag{5–79}$$

Equation (5–76) for the elevator angle required versus equilibrium lift coefficient can therefore be written as follows:

$$\delta_e = \delta_{e0} - \frac{dC_m/dC_L}{C_{m\delta}} C_L \tag{5–80}$$

The expression for δ_{e0} can be determined for any given airplane by solving (5–77) for the condition that $C_L = 0$, remembering that

$$\alpha_w = \alpha_{0L} + \frac{d\alpha}{dC_L} C_L \quad \text{and} \quad \epsilon = \frac{d\epsilon}{dC_L} C_L$$

$$\delta_{e0} = \frac{C_{m_{Fus}} + C_{m_{ac}} - a_t\bar{V}\eta_t\,(\alpha_0 - i_w + i_t)}{-C_{m\delta}} \tag{5–81}$$

The pitching moment of the fuselage and nacelle at zero lift is approximately zero and for an approximate evaluation can be neglected. If this is done (5–81) will simplify to

$$\delta_{e0} = -\frac{C_{m_{ac}}}{C_{m\delta}} - \frac{(\alpha_0 - i_w + i_t)}{\tau} \tag{5–82}$$

The elevator angle required to vary the equilibrium lift coefficient therefore varies inversely with the elevator power, $C_{m\delta}$, and directly with the static stick-fixed longitudinal stability, dC_m/dC_L. For a given elevator power, then, the slope of the elevator angle versus equilibrium lift coefficient curve is dependent on the slope of the pitching moment curve and therefore on the c.g. location with respect to the neutral point, N_0. When the c.g. is at the neutral point $(dC_m/dC_L = 0)$, this slope will also be zero. As the c.g. is moved forward, both criteria will have increasingly negative slopes; whereas as the c.g. moves aft, they will become more positive. The slope of the elevator angle versus lift coefficient curve $(d\delta_e/dC_L)$ can therefore be used as a criterion for static stick-fixed longitudinal stability as well as dC_m/dC_L. The elevator angle versus lift coefficient criteria is used in flight test work to estimate the airplane's stick-fixed neutral point. The airplane's c.g. is moved progressively aft until the slope $d\delta_e/dC_L$ vanishes. This c.g. will then be the neutral point sought, and the slope dC_m/dC_L will also be zero.

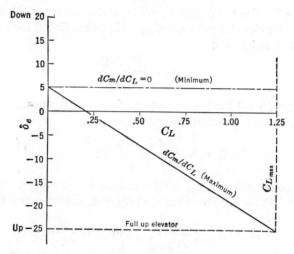

FIGURE 5-37. Limitation on forward center of gravity.

5-12 The Most Forward c.g. for Free Flight

All the developments completed so far indicate how critical the center of gravity location is to the longitudinal motion of the airplane. An indication of the most aft c.g. has been obtained with the requirement that the airplane possess stick-fixed longitudinal stability with power-on. An indication of the most forward c.g. permissible comes from the elevator theory just completed: the requirement that the elevator must always be capable of bringing the airplane into equilibrium at the maximum lift coefficient attainable by the airplane.

As the c.g. moves forward, the airplane becomes more stable, and more up elevator is required to trim out $C_{L\max}$. Obviously, there is some forward c.g. at which the elevator will be just powerful enough to attain equilibrium at the maximum lift coefficient. If the airplane is flown when balanced at any c.g. ahead of this, maximum lift will not be obtained with full up elevator. These c.g. limits are demonstrated by Figure 5–37.

The maximum allowable forward c.g. is that for which maximum up elevator will just balance out the maximum lift coefficient. An expression for this limiting stability can be obtained analytically by substituting δ_{emax} and C_{Lmax} in equation (5–80) and solving for $(dC_m/dC_L)_{max}$.

$$\left(\frac{dC_m}{dC_L}\right)_{max} = (\delta_{e0} - \delta_{emax})\frac{C_{m\delta}}{C_{Lmax}} \qquad (5\text{–}83)$$

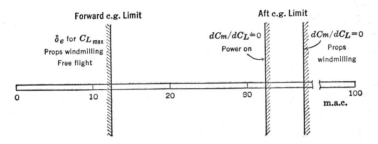

FIGURE 5–38. Typical center of gravity limits.

As the airplane is usually more stable with props windmilling than it is with high power, the critical forward c.g. will be for props windmilling, and equation (5–83) will indicate how far forward the c.g. may be moved from the propeller-windmilling neutral point. The maximum allowable stability, and consequently the most forward permissible c.g. position, therefore vary directly with the elevator power, $C_{m\delta}$, and inversely with the airplane's maximum lift coefficient.

The first approximation to the airplane's permissible c.g. range is therefore obtained and is depicted in Figure 5–38.

The extent of the airplane's c.g. range depends almost entirely on the elevator power term, $C_{m\delta}$. An airplane whose loading schedules require a large c.g. range must have high elevator power. Equation (5–73) indicates that elevator power varies directly with the slope of the tail lift curve, a_t, tail area, S_t, and tail length, l_t, and the elevator effectiveness term, τ. In other words, for large c.g. range, the airplane must be equipped with a large horizontal tail of high aspect ratio with a long arm to the c.g. Actually the c.g. range anticipated for the airplane dictates to a large degree the size of the horizontal tail.

5–13 Elevator Required for Landing

It has been demonstrated that the magnitude of the elevator control power limits the amount of static longitudinal stability allowable and thereby limits the forward c.g. position. This development was carried out under the assumption that the airplane was being flown at a high enough altitude to be free from the confining effects of the ground.

For all normal airplane configurations (horizontal tail aft of the wing) there is a large effect on the equilibrium and stability because of the effects of the ground on the induced flow field from the wing. The major effect of the ground is to reduce the downwash at the tail, thereby increasing the tail angle of attack positively, requiring more up elevator to maintain equilibrium at the same lift coefficient. The reduction in downwash also increases the slope of the wing and tail lift curves. As the limitation on the airplane's forward center of gravity is the requirement that full up elevator will just trim out $C_{L\mathrm{max}}$, it is apparent from this discussion that the critical condition on the elevator will be to trim out $C_{L\mathrm{max}}$ in the presence of the ground. This requirement will be in general a more severe limitation on the forward c.g.

The reduction in downwash at the tail and the increase in wing and tail slopes of the lift curve, due to the ground effect, all have an effect on the tail contribution to stability. The downwash at the tail is a major factor, and as it is reduced for a given C_L because of the ground, the term $d\epsilon/d\alpha$ is also reduced materially, increasing the tail contribution to stability, or, in other words, moving the neutral points aft. A convenient method for obtaining the most forward c.g. for trimming out $C_{L\mathrm{max}}$ in the ground effect is developed from the equilibrium equation including the pitching moment contribution of the wind-milling propeller, assumed zero at zero lift, and expressed as follows:

$$C_{m_p} = \left[\left(\frac{dC_m}{dC_L}\right)_{N_pT_{c=0}} + \left(\frac{dC_m}{dC_L}\right)_{\epsilon_p}\right] C_L \qquad (5\text{--}84)$$

or

$$C_{m_p} = \left(\frac{dC_m}{dC_L}\right)_p C_L$$

where $(dC_m/dC_L)_p$ will include both contributions of the windmilling propeller.

Equation (5–77) can be rewritten thus:

$$\delta_e = \frac{C_L \dfrac{x_a}{c} + C_{m_{ac}} + C_{m_{Fus}} + C_{m_p} - a_t(\alpha_w - \epsilon - i_w + i_t)\bar{V}\eta_t}{a_t\bar{V}\eta_t\tau} \qquad (5\text{--}85)$$

As $x_a/c = x_{cg} - x_{ac}$, equation (5-85) can be rearranged to yield the most forward permissible center of gravity position ($x_{cg\text{frwd}}$) if the maximum up elevator deflection ($\delta_{e\max}$) and the maximum lift coefficient ($C_{L\max}$) are substituted for δ_e and C_L, respectively, and all other variables correspond to those at $C_{L\max}$ and in the ground effect.

$$x_{cg\text{frwd}} = x_{ac} - \frac{C_{m\delta}}{C_{L\max}} \left[\delta_{e\max} + \frac{(\alpha_w - \epsilon - i_w + i_t)}{\tau} \right.$$

$$\left. + \frac{C_{mac} + C_{mFus} + C_{mp}}{C_{m\delta}} \right] \quad (5\text{-}86)$$

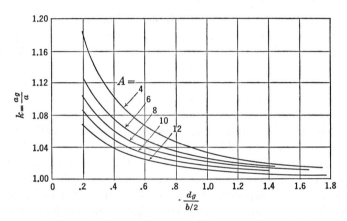

FIGURE 5–39. Ground effect on slope of the lift curve. From NACA WR L-95, "A Method for Predicting the Elevator Deflection Required to Land," by R. F. Goranson.

Equation (5-86) will yield the limiting forward c.g. location for any condition of flight, including flight in close proximity to the ground. To study this forward limit with ground effect, it is necessary to obtain information on the change in the variables of (5-86) due to this influence of the ground on the downwash field.

The variables that will be altered by the ground effect will be the wing angle of attack, α_w, the elevator power, $C_{m\delta}$, due to the increase in the slope of the tail lift curve, a_t, and the downwash angle, ϵ. The change in wing angle of attack is due to an increase in the slope of the lift curve resulting from the reduction in the wing downwash pattern. The change in slope of the lift curve for wing or tail can be obtained by referring to the curves given in Figure 5–39. These curves, developed by the NACA,* give the ratio of the slope of the lift curve with

* NACA WR L-95.

ground effect to the slope in free flight as a function of the height of the root quarter chord of wing or tail surface in semispans, $d_g/(b/2)$. Curves are given for values of constant aspect ratio.

The change in wing angle of attack due to the ground effect may be approximated as follows:

$$\Delta\alpha = \frac{C_L}{a}\left(\frac{1}{k} - 1\right) \tag{5-87}$$

The slope of the lift curve of the tail, a_t, may be obtained directly from Figures 5-5 and 5-39.

The major effect of the ground is brought about through the change in the downwash at the tail, ϵ. A very complete method for computing the downwash with ground effect is given in NACA TR 738. The reader is referred to this report for computing the downwash for any particular case. The reduction in downwash at the tail due to the ground is large, and therefore considerably more up elevator is required to achieve C_{Lmax} in landings.

The wing pitching moment coefficient, C_{mac}, used in equation (5-86) must be for flaps deflected in their landing configuration, and the term x_a/c must be the distance in per cent chord between the c.g. and the a.c. positions for flaps deflected. The wing a.c. usually moves aft a few per cent with flap deflection, depending on the type of flap, the flap configuration, and the flap deflection. The pitching moment coefficient, C_{mac}, with flap deflected can be estimated by the methods presented in Chapter 2.

The section moments, c_{mac}, must be obtained from airfoil data for the particular wing under study. The change in pitching moment due to flap deflection for several flap types can be estimated from the curves given in Figure 5-40.

A very approximate method for allowing for the effects of the ground on the elevator angle required to trim out C_{Lmax} is to assume that all the effects can be approximated by assuming that the downwash at the tail with ground effect is just half that in free flight. In other words, $\epsilon_g = .5\epsilon$. Under this assumption the change in elevator angle due to ground effect becomes simply

$$\Delta\delta_{e_{g-0}} = \frac{\Delta\epsilon}{\tau} \tag{5-88}$$

where $$\Delta\epsilon = -\frac{C_L}{\pi A} \times 57.3 \tag{5-89}$$

giving $$\Delta \delta_{e_{g-0}} = - \frac{C_L}{\pi A_T} \times 57.3 \qquad (5\text{-}90)$$

Typical values substituted into equation (5–90) show changes in elevator angles of $-8°$ to $-12°$. The rearward shift of the limiting forward c.g. location can be given as follows:

$$(\Delta x_{cg})_{\text{frwd}} = \Delta \delta_{e_{g-0}} \frac{C_{m\delta}}{C_{L_{\max}}} \qquad (5\text{-}91)$$

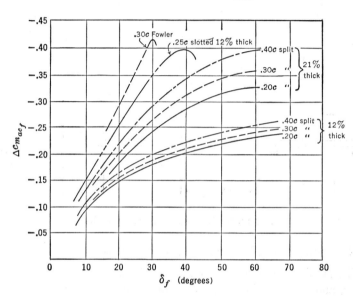

FIGURE 5–40.   Effect of flap deflection on section pitching moments.

For a case where $\Delta \delta_{e_{g-0}} = -12°$, $C_{m\delta} = -.015$, and $C_{L_{\max}} = 2.0$, equation (5–91) gives $\Delta x_{cg} = .09$, or the most forward c.g. limit is moved aft by 9 per cent of the m.a.c.

Some examples of the differences in elevator angles required to trim out $C_{L_{\max}}$ away from the ground and in actual landings are given in Table 2.   It should be pointed out that, in flight measurement of elevator angles during landings, considerable variation can be encountered because of variation in pilot technique.   For example, if the airplane is brought in at only a few miles per hour over the stalling speed, the flight path will be very steep and the airplane will make ground contact while the pilot is applying a lot of up elevator to flare the flight path rapidly.   However, if the approach is made with a comfortable margin of speed over the stalling speed, the airplane can be flared off and held in the attitude for $C_{L\max}$ without undue difficulty. The elevator angles measured during this type of landing agree fairly well with those predicted by the theory.

TABLE 2

Airplane	$-\delta_e{}^\circ$(free flight)	$-\delta_e{}^\circ$(ground)	$\Delta\delta_e{}^\circ{}_{g-0}$
P-40	10.4	22.0	−11.6
F-4U	8.0	17.8	− 9.8
F-4F	12.5	16.0	− 3.5
B-17	5.4	19.8	−14.4
B-25	3.0	14.0	−11.0
Stinson 105	16.0	23.6	− 7.6

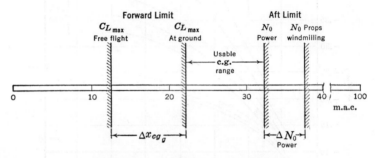

FIGURE 5–41.   Center of gravity limits.

5–14   Restriction of Forward Limit of c.g. Range

The landing requirement on the elevator is more severe than any other, and therefore it becomes the design condition for the elevator power term, $C_{m\delta}$. For a given airplane the per cent m.a.c. that the most forward c.g. can be placed ahead of the propeller windmilling or power-off neutral point is a direct function of $C_{m\delta}$. The elevator power therefore must be made great enough so that, if the most aft c.g. of the airplane's c.g. range is placed at the power-on neutral point, the elevator will still be powerful enough to achieve trim at $C_{L\max}$ in landings when the airplane is balanced at the forward end of the c.g. range. The more c.g. range required, the more powerful will the elevator need to be.   It can be said that the anticipated c.g. travel of the airplane designs the elevator and the horizontal tail.

In using equations (5–90) and (5–92) to calculate the forward c.g. limit, it is advisable to leave a margin of up elevator deflection available to flare the airplane during the landing approach. Usually 5° of up elevator margin is allowed for this.

The balance restrictions imposed on a particular airplane are given in Figure 5–41 to illustrate how severely the allowable c.g. range is restricted in practice.

On any given airplane the stability requirements limit the most aft location of the c.g. The design of the horizontal tail and elevator, which determines the elevator power, must always be carried out with the anticipated balance limits of the airplane in mind. The more range allowed in actual c.g. travel, the more difficult becomes the design problem. It is therefore the responsibility of the designers laying out the airplane to restrict the c.g. range in every way possible, in order to keep the aerodynamic design problem from becoming impossible.

SUGGESTED READING

1. NACA WR L-116, "Estimation of Stick-fixed Neutral Points of Airplanes," by M. D. White, March 1945.
2. NACA TR 711, "Analysis and Prediction of Longitudinal Stability of Airplanes," by Gilruth and White, 1941.
3. AAF TR 5167, "Some Theoretical Developments in Airplane Static Longitudinal Stability and Control," by C. D. Perkins, 1944.
4. NACA TM 1036, "Aerodynamics of the Fuselage," by H. Multhopp, 1942.
5. NACA TR 648, "Design Charts for Predicting Downwash Angles and Wake Characteristics behind Plain and Flapped Wings," by Silverstein and Katzoff, 1940.
6. NACA WR L-25, "Notes on the Propeller and Slipstream in Relation to Stability," by H. S. Ribner, 1944.
7. NACA WR L-761, "Effect of Propeller Operation on the Pitching Moments of Single-engine Monoplanes," by Goett and Pass, 1941.
8. NACA WR L-24, "Effect of Power on the Stick-fixed Neutral Points of Several Single-engine Monoplanes," by M. D. White, 1944.
9. *Jour. Aero Sciences*, "The Influence of Running Propellers on Airplane Characteristics," by C. B. Millikan, January 1940.
10. NACA TN 823, "Methods of Analyzing Wind-tunnel Data for Dynamic Flight Conditions," by Donlan and Recant, 1941.
11. NACA TR 738, "Ground Effect on Downwash Angles and Wake Location," by Katzoff and Sweberg, 1943.
12. NACA WR L-248, "Flight Measurements of the Elevator Deflections Used in Landings of Several Airplanes," by J. R. Vensel, 1941.
13. NACA WR L-95, "A Method for Predicting the Elevator Deflection Required to Land," by R. F. Goranson, 1944.
14. NACA WR L-344, "Some Notes on the Determination of the Stick-fixed Neutral Point from Wind-tunnel Data," by M. Schuldenfrei, 1943.
15. RAF Report BA 1531, "An Analysis of Static Longitudinal Stability in Relation to Trim and Control Forces. Part I. Gliding," by S. B. Gates, 1939.
16. RAF Report BA 1549, "An Analysis of Static Longitudinal Stability in Relation to Trim and Control Force. Part II. Engine On," by S. B. Gates, 1939.
17. NACA WR L-322, "Analysis of Wind-tunnel Stability and Control Tests in Terms of Flying Qualities of the Full-scale Airplane," by G. G. Kayten, 1943.

18. AAF TR 5242, "Stability and Control Flight Test Methods," by C. D. Perkins. and T. F. Walkowicz, 1945.
19. NACA WR L-710, "Determination of the Stability and Control Characteristics of Airplanes from Tests of Powered Models," by Recant and Swanson, 1942.
20. NACA WR L-213, "Field of Flow about a Jet and Effect of Jets on Stability of Jet Propelled Airplanes," by H. S. Ribner, 1946.

PROBLEMS

5-1. Given a rectangular wing of aspect ratio 6 and area of 600 sq ft. The wing section employed is an NACA-4412 airfoil with an aerodynamic center at $.24c$ and $C_{m_{ac}} = -.088$. If the wing is balanced so that the c.g. lies on the wing chord but 6 in. ahead of the a.c., calculate the lift coefficient for which the wing would be in equilibrium $(C_{m_{cg}} = 0)$. Is this lift coefficient useful? Is the equilibrium statically stable? Calculate the position of the c.g. for equilibrium at $C_L = .4$. Is this equilibrium statically stable?

5-2. If the wing given in Problem 5-1 is rebuilt maintaining the same plan form, but using a reflexed airfoil section such that $C_{m_{ac}} = +.02$, with the a.c. at $.24c$, calculate the c.g. position for equilibrium at $C_L = .4$. Is this equilibrium statically stable?

5-3. An airplane is equipped with a wing of aspect ratio 6 $(a_0 = .095)$ and a span efficiency factor, e, of .768, with an airfoil section giving $C_{m_{ac}} = +.02$. Calculate the wings' pitching moment coefficient about the airplane's c.g. located $.05c$ ahead of the a.c. and $.06c$ under the a.c. Make this calculation for every $.2C_L$ from $C_L = 0$ to $C_L = 2.0$, first neglecting the chordwise force component, and then repeating the calculation with the chordwise component included. Finally, repeat the calculation for a c.g. located $.12c$ under the a.c., but with the same horizontal c.g. location. Plot the results as curves of C_m versus C_L for each condition given. (Assume $i_w = 0$, $C_{D_f Wing} = .008$, $\alpha_0 = 0$.)

5-4. Given a glider consisting of a rectangular wing with a slender boom carrying vertical and horizontal tail surfaces. If the moment contribution of the boom is negligible, calculate the center of gravity position for which the combination will be neutrally stable. If the incidence of the wing is 0° and the incidence of the horizontal tail is $-2°$, at what C_L will the glider trim when the c.g. is moved to a position 10% ahead of the neutral point? (Neglect chordwise force components except when specifically asked for.)

The geometric characteristics of the glider are as follows:

$$S_w = 360 \text{ sq ft} \qquad S_t = 72 \text{ sq ft}$$
$$A = 10 \qquad A_t = 4$$
$$c = 6 \text{ ft} \qquad \eta_t = .9$$
$$l_t = 18 \text{ ft} \qquad a_{0_t} = .10$$
$$a_{0_w} = .10 \qquad i_t = -2°$$
$$C_{m_{acw}} = -.02 \qquad i_w = 0°$$
$$a.c._w = .24c \qquad \alpha_{0_t} = 0°$$
$$\alpha_{0_w} = -3° \qquad \text{T.R.} = 1:1$$
$$m = .10 \qquad r = .60$$

5-5. A cargo glider, as shown in the sketch, is ballasted to a c.g. position of $.30c$. What will be the glider's stability margin in this condition? Calculate the downwash from the charts given in Figure 5-10 and the fuselage contribution by both methods referred to in the text.

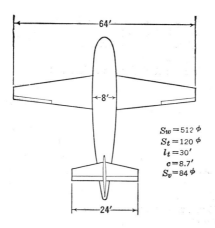

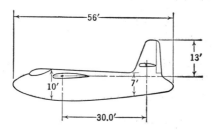

PROBLEM 5-5. Cargo glider.

b_w = 64 ft	i_w = 2°
S_w = 512 sq ft	i_t = 0°
b_t = 24 ft	a_{0w} = .10
S_t = 120 sq ft	C_{mac} = −.01
l_t = 30 ft	W/S_w = 20
c = 8.7 ft	T.R. = 3 :1
η_t = .9	C_{Df} = .025
$a.c._w$ = .24c	e = .80
α_{0w} = −3.0°	

5-6. Calculate the incidence setting of the tail for the glider in Problem 5-5 required to bring the airplane into trim at C_L = .5, with the c.g. moved to .35c. Assume that the moment coefficient of the fuselage is zero at C_L = 0.

5-7. Calculate the horizontal tail load in pounds, required to achieve equilibrium for the glider in Problem 5-5 through a lift coefficient range from C_L = −1.4 to C_L = 2.0 in steps of C_L = .2 and for c.g. locations of .10c, .25c, and .40c. Plot the results as a curve of horizontal tail lift coefficient and tail load in pounds versus airplane lift coefficient. Repeat the above calculation for tail lift coefficient for the flaps-down condition when the flaps are full-span flaps giving a ΔC_{mac} = −.30 and a shift of the wing zero lift angle to −11°. If the maximum lift coefficient of the horizontal tail is C_{Ltmax} = ± 1.0, is there any possibility of the horizontal tail stalling for any of the trim conditions investigated?

5–8. Given a single-engine fighter-type airplane. This airplane is equipped with an internal-combustion engine-propeller combination using a four-bladed propeller 11 ft in diameter. Calculate the airplane's stick-fixed neutral point for the propeller-windmilling condition.

The major geometric characteristics are as follows:

Wing area	= 220 sq ft
Wing span	= 35 ft
Wing airfoil NACA	= 65–1–212
m.a.c.	= 6.5 ft
T.R.	= 2:1
Horizontal tail area	= 40 sq ft
Horizontal tail span	= 13 ft
Tail length	= 16.2 ft
Vertical distance extended wing zero lift line to elevator hinge line	= 3 ft
Horizontal distance wing root ¼ chord to elevator hinge	= 16 ft
Fuselage over-all length	= 30 ft
Maximum width fuselage	= 3 ft
Dist. propeller disc to wing leading edge	= 9 ft
Dist. propeller disc to c.g.	= 11 ft
Incidence of the wing	= 2°
Incidence of the horizontal tail	= 1°

5–9. For the airplane given in Problem 5–8 estimate the stick-fixed neutral point, power-on.

5–10. Given a multiengined tailless bomber as shown in the figure. This airplane is powered by internal-combustion engines and propeller combinations using counterrotating propellers 15 ft in diameter. Calculate the propeller-windmilling and power-on stick-fixed neutral points for this airplane.

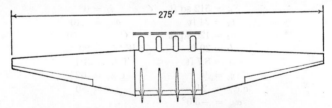

PROBLEM 5–10. Tailless airplane.

The major geometric characteristics are as follows:

Wing area	= 7129 sq ft
Wing span	= 275 ft
Aspect ratio	= 10.6
T.R.	= 5:1
m.a.c.	= 28 ft
Nacelle widths	= 5 ft
Nacelle lengths	= 12 ft
Propeller discs to c.g.	= 25 ft
Propeller discs to wing leading edge	= 13 ft

5–11. Given a turbojet propelled fighter of identical geometric shape as that given in Problem 5–8. The air intake is located at the fuselage nose 11 ft ahead of the c.g., and the jet exhaust is emitted from the very rear of the fuselage behind the tail surfaces. The line of action of the thrust is 1 ft under the airplane's c.g. The jet unit can be considered to deliver a thrust of 3000 lb and the airflow through the duct, W_a, is 70 lb/sec. Calculate the airplane's neutral points for the power-off and power-on conditions. ($W = 12,000$ lb, $V = 400$ mph, Nose intake to c.g. $= 11$ ft, $\rho = \rho_0$.)

5–12. If the slope of the pitching moment curve for a given airplane is $dC_m/dC_L = -.15$, and the pitching moment coefficient at zero lift is $C_m = .10$, at what C_L will the airplane be in trim, and how much pitching moment coefficient must be supplied to achieve trim at $C_L = 1.5$?

5–13. An airplane designer attempts to equip an airplane with adjustable sliding weight to control the trim lift coefficient. If the airplane's pitching moment coefficient at zero lift is $C_m = +.05$ and if its stick-fixed neutral point is at $.30c$, calculate and plot the variation of c.g. location required to achieve trim from $C_L = .2$ to 1.4. Also, plot the variation of dC_m/dC_L with C_L for this arrangement.

5–14. A wind-tunnel test of a horizontal surface yields curves of lift coefficient versus angle of attack for various elevator angles as shown in the accompanying figure. From these data, determine the elevator effectiveness parameter, τ.

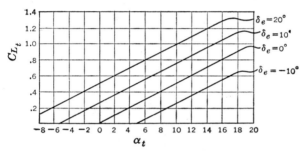

PROBLEM 5–14.   Tail lift characteristics.

5–15. If an airplane is equipped with the horizontal tail surface whose characteristics are given in Problem 5–14, calculate the elevator power, $C_{m\delta}$, for the following airplane characteristics:

Wing area, S_w $= 300$ sq ft
m.a.c., c $= 8$ ft
Tail area, S_t $= 60$ sq ft
Tail span, b_t $= 15$ ft
Tail length, l_t $= 20$ ft
Tail efficiency, $\eta_t = .90$

5–16. An airplane has an elevator control power of $C_{m\delta} = -.010$. If the c.g. is placed so that the static margin is 10 per cent m.a.c., and if the stabilizer is set for trim at zero elevator at $C_L = .5$, plot curves of C_m versus C_L for constant elevator angles of $\delta_e° = -20, -15, -10, -5, 0, 5, 10,$ and 15. Cross-plot these curves for elevator angle for trim versus C_L. ($C_{Lmax} = 2.0$.)

5–17. If the fighter airplane of Problem 5–8 is equipped with an elevator of area ratio $S_e/S_t = .30$, calculate and plot curves of elevator angle required for trim versus C_L for the propeller-windmilling condition for c.g.'s of .20c, .30c, and .40c.

5–18. The airplane of Problem 5–17 has a maximum lift coefficient with flaps retracted of $C_{Lmax} = 1.5$. Calculate the most forward c.g. for which full up elevator of 30° will just trim out the maximum lift coefficient, with the propeller windmilling, for flight out of the ground effect.

5–19. If the airplane of Problem 5–17 is equipped with .25c split flaps extending from .15 to .60 of the wing semispan, and if this wing is a straight wing with a taper ratio of 2:1, calculate the increment to C_{mac} due to a 50° flap deflection. Also, calculate the most forward c.g. at which 25° of up elevator will just trim out this maximum lift coefficient in the presence of the ground and also when flying out of the ground effect, for the propeller-windmilling condition, if the ground effect is such that the downwash at the tail is reduced from that in free flight by 6°.

C_{Lmax} flaps down $= 2.0$, $\left(\dfrac{dg}{b/2}\right)_w = .4$, $\left(\dfrac{dg}{b/2}\right)_t = .6$.

5–20. Make a plot similar to that shown in Figure 5–41 of the forward and aft c.g. limits for the airplane of Problem 5–17, showing the forward limits in and out of the ground effect, flaps down, propeller windmilling, and the power-on and power-off stick-fixed neutral points.

CHAPTER 6

STATIC LONGITUDINAL STABILITY AND CONTROL STICK-FREE

6-1 General Discussion

In Chapter 5 it was pointed out that one of the ways that the static longitudinal stability of the airplane is felt by the pilot is through the variation required of the elevator angle and, thereby, the control position, with change in air speed. This variation was demonstrated to be a function of the stability criterion, dC_m/dC_L, with controls locked and is usually referred to as the static longitudinal stability stick-fixed.

A second way in which the pilot "feels" the static longitudinal stability of the airplane is through the stick force required to change the speed of the airplane from a given trim condition. The magnitude of the control force, or stick force, F_s, required of the pilot to change the speed from any given trim condition ($F_s = 0$) cannot be related directly to the stability criterion, dC_m/dC_L, with fixed elevator as developed, but can be shown to be closely related to a new criterion, dC_m/dC_L, with control left free to float with the wind. This second type of static longitudinal stability is commonly referred to as stick-free stability, and, as will be demonstrated, is of extreme importance to the pilot in controlling the airplane.

If the elevator is mounted on frictionless bearings, it will float with or against the wind, depending on the pressure distribution over the elevator. This pressure distribution will create a moment about the control surface hinge referred to as the hinge moment, causing the elevator to rotate to some new floating angle, thereby altering the stability contribution of the horizontal tail. If, for a certain airplane configuration, an increase in the airplane's lift coefficient from equilibrium yields a restoring moment due to the horizontal tail, this restoring moment may be reduced if the elevator floats up with increase in airplane lift coefficient. The airplane will therefore be less stable with elevator free than with elevator locked. On the other hand, if the elevator is balanced so that it will float down with increase in airplane lift coefficient, it will produce a higher restoring moment than was obtained with the elevator fixed and will consequently be more stable.

The change in stability on freeing the elevators is all contained in the tail contribution and is a function of the floating characteristics of the elevator and therefore of the elevator hinge moments. Any study, therefore, of the airplane's stick-free stability or the variation of control force with speed necessarily requires some estimate of the characteristics of the hinge moments created by the pressure distribution over the control surface as the airplane's attitude is changed.

The pressure distribution over flapped airfoils operating at a given angle of attack and flap deflection can be estimated from the normal potential flow theory, and therefore the hinge moments can be estimated theoretically. It has been found, however, that the values of the control surface hinge moments, as given by theory, differ widely from those found in practice. The reason for this lies in the fact that the simple potential theory neglects the boundary layer, which in reality has a large influence on the flap hinge moments. No attempt will be made herein to develop the potential theory of hinge moments because of its inaccuracy and because of lack of space. The subject of the control hinge moment will be approached from a more qualitative and empirical point of view.

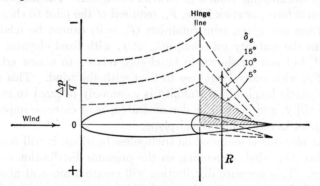

FIGURE 6–1.   Pressure distribution due to control deflection.

6–2 The Hinge Moment Parameters

The control surface, in almost all cases, is a plain flap making up the rear portion of an airfoil section. The pressure distribution over this section and, therefore, over the flap is governed by two major variables. The first of these is the angle of attack of the airfoil, and the second is the deflection of the flap chord with respect to the airfoil chord. Changes in angle of attack and flap deflection alter the pressure distribution over the entire airfoil and consequently change the magnitude of the flap hinge moment.

A typical pressure distribution over a horizontal tail at zero angle of attack, but with varying elevator angles, is shown in Figure 6–1. The suction created on the upper surface as the elevator is deflected downward can be represented as the reaction, R, acting at the centroid of the pressure area, creating the hinge moment, HM, about the hinge line. For up elevator deflections, the picture is the same, but the suction will then be on the bottom surface. The variation of the hinge moment with elevator deflection for zero angle of attack is shown in

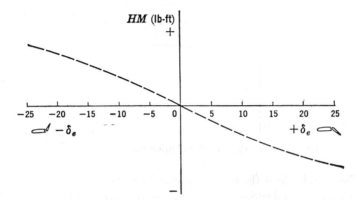

FIGURE 6–2.   Hinge moment versus control deflection.

Figure 6–2, showing that for the plain flap of normal section with the hinge at the nose radius, the slope of the hinge moment versus control surface deflection will be negative, under the assumption that a positive hinge moment tends to rotate the control surface trailing edge down.

The pressure distribution picture shown in Figure 6–1 is quite simple, but unfortunately it is complicated by the effect of angle of attack. As any airfoil section is given an increment in angle of attack, the pressure distribution over it changes rapidly with large suction pressures appearing close to the leading edge of the section. These distributions affect the pressure distribution over the whole section and therefore affect the control surface hinge moments. A typical variation of the pressure distribution with angle of attack is shown in Figure 6–3.

The suction over the upper surface, which increases with angle of attack, again creates a hinge moment which tends to make the elevator "float" up as the angle of attack is increased. The variation of hinge moment with angle of attack for a plain flap of normal section with the hinge line at the nose radius is shown in Figure 6–4. It should be noted that the slope of control surface hinge moment versus angle of attack is negative for this type of control.

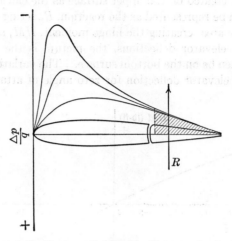

FIGURE 6-3.   Pressure distribution due to angle of attack.

The magnitude of the hinge moment for any combination of angle of attack and control surface deflection is developed from the complicated

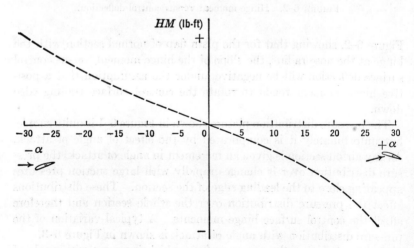

FIGURE 6-4.   Hinge moment versus angle of attack.

pressure distribution created by both the angle of attack and control surface deflection operating together. It has been assumed, and the assumption is well justified, that the total hinge moment, HM, can be thought of as the addition of the two effects taken separately. In other words the hinge moment due to a control surface deflection of 5° at a 5° angle of attack is the sum of the hinge moment due to a 5° deflection at zero angle of attack plus a 5° angle of attack at zero deflection. Mathematically, this can be thought of as a total differential being made up of partial derivatives.

$$dHM = \left(\frac{\partial HM}{\partial \alpha}\right)_{\delta} d\alpha + \left(\frac{\partial HM}{\partial \delta}\right)_{\alpha} d\delta \qquad (6\text{-}1)$$

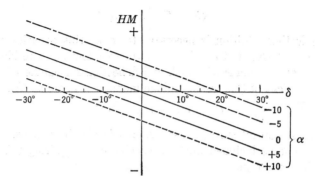

FIGURE 6-5.   Typical hinge moment curves.

Under the assumption that the variation of hinge moment with angle of attack and deflection is linear, the absolute value of the hinge moment, angle of attack, and deflection can replace the differentials.

$$HM = \left(\frac{\partial HM}{\partial \alpha}\right)_{\delta} \alpha + \left(\frac{\partial HM}{\partial \delta}\right)_{\alpha} \delta \qquad (6\text{-}2)$$

The variation of the hinge moment with both control surface deflection and angle of attack represented by equation (6-2) is shown as the family of curves in Figure 6-5.

It has been found very convenient to non-dimensionalize the hinge moment, as was done with the lift, drag, and pitching moment. The hinge moment coefficient, C_h, is obtained by dividing the hinge moment in pound-feet by the dynamic pressure, q, in pounds per square foot, the area of the control surface back of the hinge line, S_c, in square feet, and the root mean square chord of the control surface aft of the hinge line, $\bar{c}_c$, in feet.

$$C_h = \frac{HM}{qS_c\bar{c}_c} \qquad (6\text{-}3)$$

The hinge moment coefficient can be expressed in terms of partial derivatives as was done in equation (6–2).

$$C_h = \left(\frac{\partial C_h}{\partial \alpha}\right)_\delta \alpha + \left(\frac{\partial C_h}{\partial \delta}\right)_\alpha \delta \qquad (6\text{–}4)$$

The partial derivatives $(\partial C_h/\partial \alpha)_\delta$ and $(\partial C_h/\partial \delta)_\alpha$ are usually written in the short-hand form, $C_{h\alpha}$ and $C_{h\delta}$, and will be written in this form throughout the remainder of this book. In this notation equation (6–4) is written as follows:

$$C_h = C_{h\alpha} \alpha + C_{h\delta} \delta \qquad (6\text{–}5)$$

One further addition is necessary to make this equation perfectly general, and that is to include a term to allow for a residual hinge moment at zero angle of attack and deflection for cambered surfaces. The coefficient of this residual hinge moment is given as C_{h0}.

$$C_h = C_{h0} + C_{h\alpha} \alpha + C_{h\delta} \delta \qquad (6\text{–}6)$$

As most control surfaces use symmetric airfoil sections, the term C_{h0} will not be included unless a specific case requiring this term is under investigation.

6–3 Control Surface Floating Characteristics and Aerodynamic Balance

One of the important characteristics of any control surface is its so-called floating characteristics. If the control surface is mounted on frictionless bearings and the whole surface given an angle of attack, the hinge moment developed for the unbalanced flap of Figures 6–1 and 6–3 will tend to make the control float up. This tendency is a function of the parameter $C_{h\alpha}$. However, as soon as the control surface starts to float up, a new pressure distribution due to the parameter $C_{h\delta}$ is added which tends to prevent the elevator from floating up. An equilibrium floating angle is reached when the "floating tendency," $C_{h\alpha}$, is just opposed by the "restoring tendency," $C_{h\delta}$. The floating angle can be solved mathematically from equation (6–5) by equating C_h to zero and solving for the control surface angle, δ.

$$\delta_{\text{Float}} = -\frac{C_{h\alpha}}{C_{h\delta}} \alpha \qquad (6\text{–}7)$$

The control surface floating angle, therefore, varies with angle of attack, and if $C_{h\alpha}$ and $C_{h\delta}$ both have negative signs, the elevator will float up more and more with increasing angle of attack. The rate that the elevator floats with angle of attack can be obtained readily by differentiating equation (6–7) with respect to α, obtaining:

$$\frac{d\delta_{\text{Float}}}{d\alpha} = -\frac{C_{h\alpha}}{C_{h\delta}} \tag{6-8}$$

The floating characteristics of the control surface as well as the magnitude of the hinge moment coefficient depend on the magnitudes of the parameters $C_{h\alpha}$ and $C_{h\delta}$. On most control surfaces it is im-

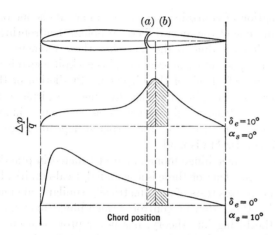

FIGURE 6–6. Set-back hinge line.

portant to reduce the floating tendency to as near zero as practicable. This means $C_{h\alpha}$ should be made to have a zero slope. At the same time $C_{h\delta}$ must be made to have a low slope if the pilot is to be able to move the control at all, especially for large, very fast airplanes. Methods for controlling the parameters $C_{h\alpha}$ and $C_{h\delta}$ are therefore of prime interest and are referred to as methods of aerodynamic balancing control surface.

One of the most frequently used methods of aerodynamic balancing control surfaces is the set-back hinge. This is a very powerful method, and its mechanics can be demonstrated by restudying Figures 6–1 and 6–3. The pressure distribution over the surface can be redrawn as in Figure 6–6.

With the hinge line at the nose radius, the summation of moments due to the pressure distribution will create a resultant moment tending to raise the control surface in each case. However, as the hinge line is moved aft, the summation of moments about the hinge line will yield a total moment somewhat smaller, and finally, if the hinge line is moved far enough aft, the sign of the resultant moment can be reversed. When the hinge line is moved aft far enough so that the resultant moment is reversed, the control is said to be overbalanced. The hinge moment parameters $C_{h\alpha}$ and $C_{h\delta}$ are therefore very sensitive to hinge line position, becoming more positive as the hinge line is moved aft.

6–4   Estimation of Hinge Moment Parameters

The estimation of control surface hinge moments is one of the most important and one of the most difficult problems confronting the aerodynamicist. This is due to the fact that these moments create the forces which the pilot must overcome at his cockpit controls and which give him a sense of "feel" of the airplane. Prediction of the control forces required of the pilot for maneuvering the airplane within its aerodynamic and structural limitations is based on an estimate of these parameters, and the success of the airplane depends largely on how accurately this estimate is made.

The control surface hinge moments are predicted in practice by two general methods. One of these is a semianalytical-empirical method,* and the other utilizes tests of a geometrically similar scale model in the wind tunnel. Prediction of control surface hinge moments, analytically from the lifting line theory, has never proved accurate enough for practical use, probably because of the strong influence of the boundary layer, neglected in these calculations. All methods for predicting control surface hinge moments are subject to rather large errors and are quite unreliable, except possibly for the method of measuring the hinge moments at large scale in the wind tunnel.

A method for obtaining a first approximation to the hinge moment parameters $C_{h\alpha}$ and $C_{h\delta}$ has been developed for the case of aerodynamic balance by the set-back hinge. This method would be relatively simple if setting back the hinge were the only variable. Unfortunately, these parameters are also affected seriously by the ratio of the control chord to the total airfoil chord; the airfoil section used, the nose shape of the control surface, the gap at the control surface nose, and the aspect ratio of the surface upon which the control acts. In estimating the final hinge moments, each of these variables must be dealt with, and therefore considerable error is accumulated during the process.

It has been found convenient to develop the two-dimensional hinge moment parameters first and then to correct them for the effects of three dimensions by the lifting line theory. Airfoil section (two-dimensional) parameters are always written in the lower case, and the

* NACA WR A-11.

three-dimensional parameters are written in the upper case. As an example, the two-dimensional variation of hinge moment coefficient with angle of attack will be written c_{h_α}, while the three-dimensional parameter will be given as C_{h_α}. This will also apply to the slope of the lift curve and all other airfoil derivatives.

The variation of the section hinge moment parameters c_{h_α} and c_{h_δ} with ratio of flap to airfoil chord c_f/c is given in Figure 6–7 for an NACA 0009 airfoil section.

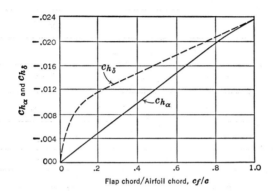

FIGURE 6–7. Section hinge moment parameters versus flap chord/airfoil chord ratio (NACA 0009 airfoil). From NACA WR L–663, "Wind-tunnel Data on the Aerodynamic Characteristics of Airplane Control Surfaces," by R. I. Sears.

The variation of section hinge moment parameters with nose shape, nose gap, and location of hinge axis is shown in Figure 6–8 for a 30 per cent flap on the 0009 section. It should be noted that the sharper the nose shape of the control surface, the farther aft will the hinge axis have to move in order to reduce the hinge moments to zero. These data are taken from the very extensive wind tunnel tests of the NACA 0009 section.*

Although the data presented in Figure 6–8 are for a 30 per cent chord flap, these data may be converted to any other flap chord ratio by using the curves of Figure 6–7. The ratio of the change in c_{h_α} and c_{h_δ} for the new flap chord will be in the same ratio as the change in the curves of Figure 6–7 for a similar variation in flap chord.

The variation in the hinge moment parameters with airfoil section arises mostly as a function of the included angle at the section trailing edge. The hinge moment parameters will be more negative for small included angles and more positive for large included angles. The change in c_{h_α} and c_{h_δ} with variations in the included angle at the trailing

* NACA WR L–663.

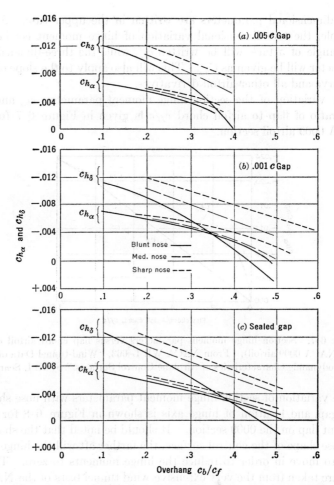

FIGURE 6–8.   Effect of set-back hinge on section hinge moments. From NACA WR L–663, "Wind-tunnel Data on the Aerodynamic Control Surfaces of Airplanes," by R. I. Sears.

edge is given as

$$\Delta c_{h_\alpha} = .005 a_0 \, \Delta \phi \tag{6–9}$$

$$\Delta c_{h_\delta} = .0078 a_0 \tau \, \Delta \phi \tag{6–10}$$

where $\Delta \phi$ is the increment in trailing edge angle from that of the NACA 0009 section, and a_0 and $a_0 \tau$ are the section lift variations with α and δ, respectively.

The correction of the two-dimensional or section data to three-dimensional flow is obtained from the lifting line theory:

$$C_{h\alpha} = c_{h\alpha} \frac{a}{a_0} \tag{6-11}$$

$$C_{h\delta} = c_{h\delta} + \tau(C_{h\alpha} - c_{h\alpha}) \tag{6-12}$$

where the three-dimensional slope of the lift curve, a, may be obtained from Figure 5-5, and the flap effectiveness factor from Figure 5-33.

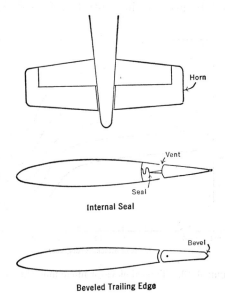

FIGURE 6-9.   Various types of aerodynamic balance.

The method just presented for estimating $C_{h\alpha}$ and $C_{h\delta}$ was outlined to demonstrate the complex treatment required to estimate the hinge moment parameters for an aerodynamically balanced surface by means of a set-back hinge, nose shape, or trailing-edge shape. There are several other means for balancing controls, the analyses of which are just as complex and just as inaccurate; therefore, because of space limitations, they will not be included. The more familiar forms of aerodynamic balance, besides the set-back hinge, are the horn balance, the internal seal, and the beveled trailing edge. Examples of each of these types of aerodynamic balance are given in Figure 6-9.

The linearity of the hinge moment variations with angle of attack and control deflection, $C_{h\alpha}$ and $C_{h\delta}$, holds only for limited angle of attack and control deflection ranges. For large angles of attack and large control deflections, the curves of hinge moment coefficient versus these variables become non-linear, and a more difficult state of affairs exists from the point of view of analysis. At large angles, the suctions created by angle of attack and control deflection start to fall off as a result of incipient stall, and the hinge moments are affected heavily. The usual tendency is for the floating rate of the control surface to increase markedly because of the large negative increase in $C_{h\alpha}$. These effects are very important, and although they are hard to analyze, they must be taken into account in control surface design. A typical

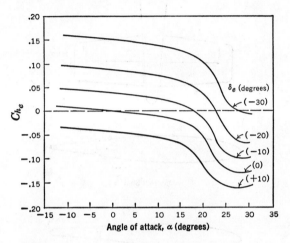

FIGURE 6–10.   Typical elevator hinge moment curves.

variation of hinge moment coefficient with angle of attack and deflection is shown in Figure 6–10 for an elevator balanced very closely by a set-back hinge.

The elevator floating angle may be obtained by cross-plotting the curves of Figure 6–10 for $C_{h_e} = 0$. The result of this cross-plot is shown in Figure 6–11.

The tendency of the control surface to lose its effective balance at high angles is most important and must be carefully kept in mind, as it has considerable bearing on the studies that follow.

The control surface hinge moment parameters, estimated analytically in the first design phases, must always be checked in the wind tunnel on any critical design requiring close aerodynamic balance. This model, as mentioned previously, should represent the actual surface to be installed on the airplane in structure, type of skin, and contour. It is only by carefully testing such a model that variables such as

bulging of the skin, production trailing-edge contour, and skin waviness
can be accounted for in advance.

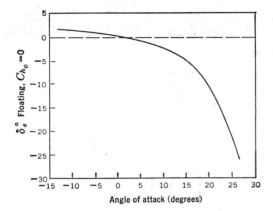

FIGURE 6–11.   Elevator floating angle versus angle of attack.

6–5   The Tab

Before concluding the discussion of control surface hinge moments,
it is necessary to discuss the control tab and to estimate its power.
The control surface tab is a small auxiliary flap usually built into the
trailing-edge portion of the control surface and, because of its location,
can create very powerful moments about the control surface hinge line,
for use as a trimming device, a balancing device, or, in some cases, as
the primary control surface.

There are many types of tabs in use at the present time, all serving
different purposes, but all based on the fundamental ability of the tab
to develop large hinge moments about the control surface hinge line.
Some of the tab types now in use are the trim tab, the balancing tab,
the servo tab, and the spring tab.   The function of each of these
tab types will be discussed later.   In this chapter only the effect of the
tab on the control surface hinge moments will be discussed.

A typical tab installation is shown in Figure 6–12.   Deflecting the
tab down will create a suction on the upper surface of the tab which
creates a resultant upward force whose arm to the main controls hinge
line is so large that a very material moment will be created, tending
to make the surface float up.   Deflecting the tab down then tends to
make the control surface float up, and conversely deflecting the tab
up makes the control surface float down.

As a down (or positive) tab deflection creates a negative hinge
moment, the slope of a curve of control hinge moment versus tab

deflection will be negative. Assuming linearity again, the control
hinge moment coefficient can be written:

$$C_h = C_{h0} + \left(\frac{\partial C_h}{\partial \alpha}\right)_{\delta_c,\delta_t} \alpha + \left(\frac{\partial C_h}{\partial \delta_c}\right)_{\alpha,\delta_t} \delta_c + \left(\frac{\partial C_h}{\partial \delta_t}\right)_{\alpha,\delta_c} \delta_t \quad (6\text{–}13)$$

or, in short-hand notation,

$$C_h = C_{h0} + C_{h\alpha}\alpha + C_{h\delta c}\delta_c + C_{h\delta t}\delta_t \quad (6\text{–}14)$$

The control floating angle will now become

$$\delta_{\text{Float}} = -\frac{C_{h\alpha}}{C_{h\delta c}}\alpha - \frac{C_{h\delta t}}{C_{h\delta c}}\delta_t \quad (6\text{–}15)$$

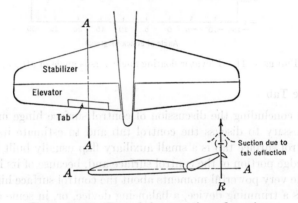

FIGURE 6–12.   Typical elevator tab configuration.

which indicates again that deflecting the tab can change the control
surface floating angle.

In estimating the parameter $C_{h\delta t}$, it is convenient to obtain the two-
dimensional slope, $c_{h\delta t}$, and then correct for the three-dimensional case.
The section parameter $c_{h\delta t}$ depends on the airfoil section, the flap to
airfoil chord ratio, c_f/c, and the tab to airfoil chord ratio, c_t/c.   Figure
6–13* gives experimental data for $c_{h\delta t}$ as functions of c_f/c and c_t/c.

The three-dimensional hinge moment slope, $C_{h\delta t}$, can be estimated,
first assuming the tab is a full span flap, from the following formula:

$$C_{h\delta t} = c_{h\delta t} - c_{h\alpha}\tau_t\left(1 - \frac{a}{a_0}\right) \quad (6\text{–}16)$$

where τ_t is the effectiveness of the tab.

As in most cases, the tab is only a partial span flap; a correction to

* NACA TR 721.

(6–16) is required to allow for this. The simplest way to do this is to assume that the correction will be in proportion to the area-moment ratios of the actual tab to the full span tab of equal chord ratio. The area moment is taken as the tab area times the perpendicular distance of the centroid of this area from the control hinge line. If the area-moment ratio is given as K,

$$K = \frac{(\text{Area moment}) \text{ partial span tab}}{(\text{Area moment}) \text{ full span tab}} \tag{6–17}$$

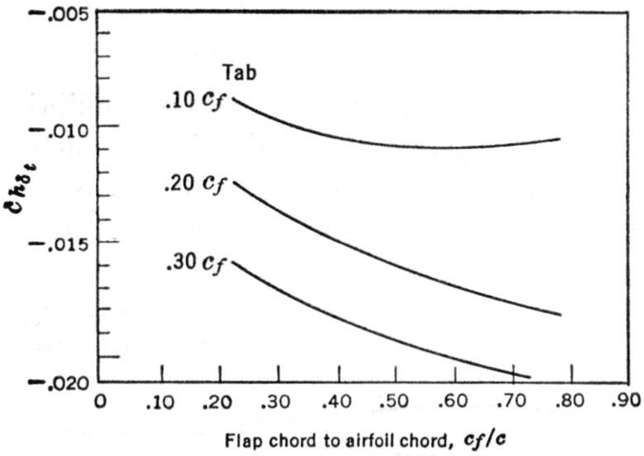

FIGURE 6–13.   Tab hinge moments.

equation (6–16) must be corrected as given in (6–18). It should be noted that the correction factor, K, should only multiply $c_{h\delta t}$, as the tab effectiveness factor, τ_t, is already corrected for proper area.

$$C_{h\delta t} = c_{h\delta t}K - \tau_t c_{h\alpha}\left(1 - \frac{a}{a_0}\right) \tag{6–18}$$

Tab hinge moments are very sensitive to control deflection. For large deflections the flap hinge moment drops off very rapidly. The value of $C_{h\delta t}$ computed from (6–18) is good up to about 10–15° of control deflection, but falls off very rapidly for larger deflections until the value of $C_{h\delta t}$ for 30° deflection is only about one-half its original value.

It must be realized that the estimates of all hinge moment parameters $C_{h\alpha}$, $C_{h\delta c}$, $C_{h\delta t}$ are subject to considerable error, and that the estimating procedure given above is only a first approximation at best. More elaborate analytical treatments of hinge moment parameters are given in the literature, but even these are subject to considerable error. Accurate wind-tunnel tests are required on large-scale models at nearly flight Reynolds numbers for accurate determinations of these characteristics.

6-6   The Stick-free Neutral Point

From the discussion of control surface hinge moments, the total hinge moment coefficient of the elevator may be given as:

$$C_{h_e} = C_{h_0} + C_{h_\alpha}\alpha_s + C_{h_{\delta_e}}\delta_e + C_{h_{\delta_t}}\delta_t \qquad (6\text{-}19)$$

from which the floating angle of the elevator can be obtained by equating C_{h_e} to zero and solving for δ_e. For the case of $C_{h_0} = 0$ and $\delta_t = 0$

$$\delta_{e_{\text{Float}}} = - \frac{C_{h_\alpha}}{C_{h_\delta}}\alpha_s \qquad (6\text{-}20)$$

where α_s is the angle of attack of the stabilizer.

In Chapter 5 the elevator angle required for equilibrium at each lift coefficient was obtained, and the derivative $d\delta_e/dC_L$ given as the criterion for stick-fixed longitudinal stability. Now, if the elevator floating angle at each lift coefficient just happened to be the same as

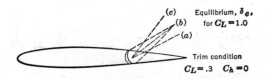

FIGURE 6-14.   Typical floating conditions of the elevator.

the elevator angle required for equilibrium, the pilot would not have to apply any force to hold any other air speed than the original trim speed, and the airplane would be termed neutrally stable stick-free. If for an increase in lift coefficient from trim the elevator floats up, but at a smaller angle than required of the elevator for equilibrium, the pilot will have to supply a pull force at the stick to hold the equilibrium elevator angle, and the airplane would be considered stick-free stable. Finally, if the floating angle exceeded the equilibrium angle required for an increase in C_L from trim, the pilot would have to supply a push force at the stick which would be termed stick-free instability. The three conditions are shown schematically in Figure 6-14.

To evaluate the static longitudinal stability stick-free, it is necessary to determine the value of the derivative $(dC_m/dC_L)_{\text{Free}}$. This is obviously the value of $(dC_m/dC_L)_{\text{Fix}}$ plus the destabilizing effect of freeing the elevators.

$$\left(\frac{dC_m}{dC_L}\right)_{\text{Free}} = \left(\frac{dC_m}{dC_L}\right)_{\text{Fix}} + \left(\frac{dC_m}{dC_L}\right)_{\substack{\text{Free} \\ \text{elevator}}} \qquad (6\text{-}21)$$

The moment coefficient due to the tail was developed in Chapter 5 and can be restated as follows:

$$C_{mt} = - C_{Lt} \bar{V} \eta_t \qquad (6\text{-}22)$$

and

$$C_{m\delta} = - \frac{dC_{Lt}}{d\delta_e} \bar{V} \eta_t \qquad (6\text{-}23)$$

The stability contribution of the free elevator is therefore

$$\left(\frac{dC_m}{dC_L}\right)_{\substack{\text{Free} \\ \text{elevator}}} = \left(\frac{d\delta_e}{dC_L}\right)_{C_{he=0}} \times C_{m\delta} \qquad (6\text{-}24)$$

The value of the floating angle derivative $(d\delta_e/dC_L)_{C_{he=0}}$ comes from equation (6–20).

$$(\delta_e)_{C_{he=0}} = - \frac{C_{h\alpha}}{C_{h\delta}} \alpha_s \qquad (6\text{-}25)$$

$$\left(\frac{d\delta_e}{dC_L}\right)_{C_{he=0}} = - \frac{C_{h\alpha}}{C_{h\delta}} \frac{d\alpha_s}{dC_L} \qquad (6\text{-}26)$$

The angle of attack of the stabilizer, α_s, was given in Chapter 5 as follows:

$$\alpha_s = \alpha_w - \epsilon - i_w + i_t \qquad (6\text{-}27)$$

therefore

$$\frac{d\alpha_s}{dC_L} = \frac{d\alpha}{dC_L}\left(1 - \frac{d\epsilon}{d\alpha}\right) = \frac{(1 - d\epsilon/d\alpha)}{a_w} \qquad (6\text{-}28)$$

also

$$\frac{dC_{Lt}}{d\delta_e} = \left(\frac{dC_L}{d\alpha}\right)_t \tau = a_t\tau \qquad (6\text{-}29)$$

Substituting (6–23), (6–26), (6–28), and (6–29) into (6–24),

$$\left(\frac{dC_m}{dC_L}\right)_{\substack{\text{Free} \\ \text{elevator}}} = \frac{C_{h\alpha}}{C_{h\delta}} \frac{a_t}{a_w} \bar{V} \eta_t \tau \left(1 - \frac{d\epsilon}{d\alpha}\right) \qquad (6\text{-}30)$$

finally

$$\left(\frac{dC_m}{dC_L}\right)_{\text{Free}} = \left(\frac{dC_m}{dC_L}\right)_{\text{Fix}} + \frac{C_{h\alpha}}{C_{h\delta}} \frac{a_t}{a_w} \bar{V} \eta_t \tau \left(1 - \frac{d\epsilon}{d\alpha}\right) \qquad (6\text{-}31)$$

The stability criterion $(dC_m/dC_L)_{\text{Fix}}$ was developed in Chapter 5 as follows for the props-off case:

$$\left(\frac{dC_m}{dC_L}\right)_{\text{Fix}} = \frac{x_a}{c} + \left(\frac{dC_m}{dC_L}\right)_{\substack{Fus \\ Nac}} - \frac{a_t}{a_w} \bar{V} \eta_t \left(1 - \frac{d\epsilon}{d\alpha}\right) \qquad (6\text{-}32)$$

Substituting (6–32) into (6–31) and rearranging,

$$\left(\frac{dC_m}{dC_L}\right)_{\text{Free}} = \frac{x_a}{c} + \left(\frac{dC_m}{dC_L}\right)_{\substack{\text{Fus} \\ \text{Nac}}} - \frac{a_t}{a_w}\bar{V}\eta_t\left(1 - \frac{d\epsilon}{d\alpha}\right)\left(1 - \frac{C_{h_\alpha}}{C_{h_\delta}}\tau\right) \quad (6\text{–}33)$$

This is the equation for static longitudinal stability, stick-free, props off. The stick-free stability, props windmilling, can be written as follows:

$$\left(\frac{dC_m}{dC_L}\right)_{\text{Free}} = \frac{x_a}{c} + \left(\frac{dC_m}{dC_L}\right)_{\substack{\text{Fus} \\ \text{Nac}}} - \frac{a_t}{a_w}\bar{V}\eta_t\left(1 - \frac{d\epsilon}{d\alpha}\right)\left(1 - \frac{C_{h_\alpha}}{C_{h_\delta}}\tau\right)$$

$$+ \frac{N(dC_N/d\alpha)_{pT=0}\,(d\beta/d\alpha)l_pS_p}{S_wca_w} + \frac{a_t}{a_w}\frac{\bar{V}\eta_t}{.07}\left(\frac{d\beta}{d\alpha}\right)\left(\frac{dC_N}{d\alpha}\right)_{pT=0} \quad (6\text{–}34)$$

The effect of freeing the elevator enters the tail term as the multiplying factor $\left(1 - \tau\dfrac{C_{h_\alpha}}{C_{h_\delta}}\right)$. For an airplane equipped with an elevator having no change in hinge moment with angle of attack ($C_{h_\alpha} = 0$), this term becomes unity, and the stick-fixed and stick-free stabilities are equal. However, if the elevator has a large floating tendency (the ratio $C_{h_\alpha}/C_{h_\delta}$ large and positive), the stability contribution of the horizontal tail can be reduced materially. For instance, for a ratio of $C_{h_\alpha}/C_{h_\delta} = 2$ and a normal value of τ equal to .5, the floating of the elevator can obviate the whole tail contribution to stability. From this brief analysis the importance of careful elevator balance design to insure proper hinge moment characteristics and thereby good stick-free stability can be readily appreciated.

The stick-free neutral point, N_0', can be obtained directly from equation (6–33) for the propeller-off condition.

$$N_0' = x_{ac} - \left(\frac{dC_m}{dC_L}\right)_{\substack{\text{Fus} \\ \text{Nac}}} + \frac{a_t}{a_w}\bar{V}\eta_t\left(1 - \frac{d\epsilon}{d\alpha}\right)\left(1 - \frac{C_{h_\alpha}}{C_{h_\delta}}\tau\right) \quad (6\text{–}35)$$

The difference between the stick-fixed and stick-free neutral points is as follows:

$$N_0 - N_0' = \frac{a_t\bar{V}\eta_tC_{h_\alpha}}{a_wC_{h_\delta}}\tau\left(1 - \frac{d\epsilon}{d\alpha}\right) = -\frac{C_{m\delta}}{a_w}\frac{C_{h_\alpha}}{C_{h_\delta}}\left(1 - \frac{d\epsilon}{d\alpha}\right) \quad (6\text{–}36)$$

A typical value of this difference is .02 to .05 per cent m.a.c.

In practice, curves of C_m versus C_L can be obtained in the wind tunnel for the stick-free case by mounting the elevator on ball bearings and allowing it to float freely throughout the test run. Typical curves of wind-tunnel results of elevator free tests in comparison to elevator fixed curves are given in Figure 6-15.

Actually this technique is not usually resorted to. In actual wind-tunnel practice elevator hinge moments are obtained during pitching moment versus lift coefficient runs, with varying fixed values of the

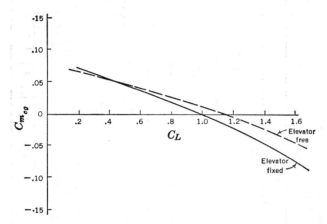

FIGURE 6-15. Typical reduction of stability due to freeing elevator.

elevator angle. From these data the elevator hinge moment coefficients, C_{h_e}, are plotted versus airplane lift coefficient for various elevator angles. The stick-free characteristics are nearly always deduced from such curves. Elevator hinge moment runs are shown in Figure 6-16 for a typical case.

6-7   Stick Force Gradients in Unaccelerated Flight

The criterion $(dC_m/dC_L)_{\text{Free}}$ is important because of its basic influence on the variation of the stick force, F_s, required of the pilot to change the airplane's speed from a given trim speed, $F_s = 0$. For desirable flying qualities it is necessary that the pilot be required to apply pull forces on the stick for flight at speeds below the trim speed and push forces for flight at speeds above the trim speed, and the stick-free stability criterion $(dC_m/dC_L)_{\text{Free}}$ has a determining influence on the gradient of the stick force versus speed curve through trim. The theory of this will be explained in the remainder of this chapter.

The pilot's force at the top of the stick is determined by the hinge moment at the elevator and the gearing between the elevator and the pilot's control. A typical arrangement of stick and elevator is shown in Figure 6-17.

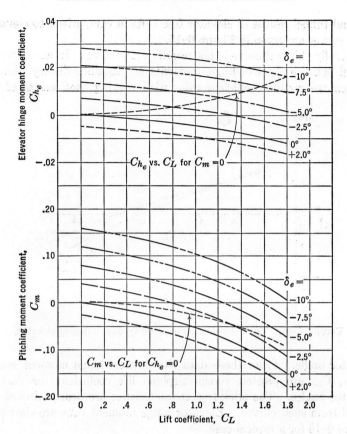

FIGURE 6–16.   Typical wind-tunnel pitching moment—hinge moment runs.

The gearing between the stick and the elevator may be obtained by equating the work done at the top of the stick to the work done at the elevator.

Work done at top of stick = work done at elevator

$$\frac{F_s \times l_s \times \delta_s}{2} = \frac{HM \times \delta_e}{2} \tag{6-37}$$

where l_s is length of the stick in feet, and δ_s and δ_e are the angular rotations of the stick and elevator, respectively, in radians. Equation (6-37) may be reduced as follows:

$$F_s = HM \frac{\delta_e}{l_s \delta_s} \tag{6-38}$$

The term $(\delta_e/l_s\delta_s)$ is called the elevator gearing, G, and is sometimes written as $d\delta_e/ds$, where $s = l_s\delta_s$, the linear movement of the top of the stick.

The stick force required of the pilot can be written as follows, considering a push force as positive:

$$F_s = -G \cdot HM \qquad (6\text{-}39)$$

Equation (6–39) can be written in coefficient form

$$F_s = -GC_{h_e}S_e c_e q \eta_t \qquad (6\text{-}40)$$

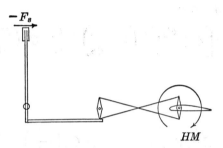

FIGURE 6–17.   Elevator-stick gearing.

where the hinge moment coefficient can be written as

$$C_{h_e} = C_{h0} + C_{h_\alpha}\alpha_s + C_{h\delta_e}\delta_e + C_{h\delta_t}\delta_t \qquad (6\text{-}41)$$

Substituting (6–41) into (6–40),

$$F_s = -GS_e c_e q \eta_t (C_{h0} + C_{h_\alpha}\alpha_s + C_{h\delta_e}\delta_e + C_{h\delta_t}\delta_t) \qquad (6\text{-}42)$$

Now

$$\alpha_s = \alpha_w - \epsilon - i_w + i_t$$

or $\qquad\qquad\qquad\qquad\qquad\qquad\qquad\qquad\qquad\qquad\qquad$ (6–43)

$$\alpha_s = \alpha_0 + \frac{C_L}{a_w}\left(1 - \frac{d\epsilon}{d\alpha}\right) - i_w + i_t$$

The elevator angle, δ_e, can be expressed in terms of the elevator angle at zero lift, the stick-fixed stability parameter $(dC_m/dC_L)_{\text{Fix}}$, and the elevator effectiveness, $C_{m\delta}$, as given in equation (5–72).

$$\delta_e = \delta_{e0} - \left(\frac{dC_m}{dC_L}\right)_{\text{Fix}}\frac{C_L}{C_{m\delta}} \qquad (6\text{-}44)$$

If the values of δ_e and α_s are substituted into equation (6–42), the final stick force equation results.

$$F_s = -GS_ec_e \frac{1}{2}\rho V^2 \eta_t \left[C_{h0} + C_{h\alpha}\left(\alpha_0 + \frac{C_L}{a_w}\left(1 - \frac{d\epsilon}{d\alpha}\right) - i_w + i_t\right)\right.$$

$$\left. + C_{h\delta}\left(\delta_{e0} - \left(\frac{dC_m}{dC_L}\right)_{\text{Fix}}\frac{C_L}{C_{m\delta}}\right) + C_{h\delta t}\delta_t\right] \quad (6\text{–}45)$$

letting $\qquad K = -GS_ec_e\eta_t$

and $\qquad A = C_{h0} + C_{h\alpha}(\alpha_0 - i_w + i_t) + C_{h\delta}\delta_{e0}$

Equation (6–45) becomes

$$F_s = K\frac{1}{2}\rho V^2\left[A + \frac{C_{h\alpha}C_L}{a_w}\left(1 - \frac{d\epsilon}{d\alpha}\right) - \frac{C_{h\delta}}{C_{m\delta}}C_L\left(\frac{dC_m}{dC_L}\right)_{\text{Fix}}\right.$$

$$\left. + C_{h\delta t}\delta_t\right] \quad (6\text{–}46)$$

Rearranging

$$F_s = K\frac{1}{2}\rho V^2\left[A + C_{h\delta t}\delta_t - C_L\left(\frac{dC_m}{dC_L}\right)_{\text{Free}}\frac{C_{h\delta}}{C_{m\delta}}\right] \quad (6\text{–}47)$$

for unaccelerated flight, $C_L = \dfrac{2W/S}{\rho V^2}$, substitution into (6–47) gives

$$F_s = K\frac{1}{2}\rho V^2(A + C_{h\delta t}\delta_t) - K\frac{W}{S}\frac{C_{h\delta}}{C_{m\delta}}\left(\frac{dC_m}{dC_L}\right)_{\text{Free}} \quad (6\text{–}48)$$

Equation (6–48) brings out the interesting fact that the stick force variation with speed is dependent on the first term only and independent in general of the stability level. The slope of the stick force versus speed curve is simply

$$\frac{dF_s}{dV} = K\rho V(A + C_{h\delta t}\delta_t) \quad (6\text{–}49)$$

A plot of elevator stick force, F_s, versus velocity is shown in Figure 6–18 and is made up of a constant force springing from the second or stability term of equation (6–48) plus a variable force proportional to the velocity squared, introduced through the constant A and the tab term $C_{h\delta t}\delta_t$.

For a given center of gravity, then, a stable or negative value of the stability criterion (stick-free) will introduce a constant pull force, while an unstable value will introduce a push force. It can be seen from

Figure 6–18 that an airplane possessing stick-free stability will require
a nose-up tab setting to trim out the stick force ($F_s = 0$) for a given
trim speed, and the resultant variation of stick force with air speed
will be stable.   If $(dC_m/dC_L)_{\text{Free}}$ is unstable, then in order to trim the
airplane out at the given trim speed a nose-down tab is required, giving
an unstable variation of stick force with air speed.   In other words the
tab creates the required slope, but the static stability criterion stick-
free is essential to allow the tab to move in a stable direction for trim.
It is important to notice again that a stable slope is of interest only if
equilibrium or trim is established.

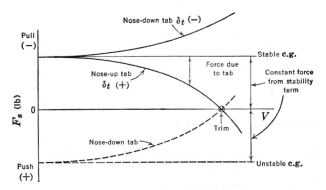

FIGURE 6–18.   Stick force build-up.

From this discussion it can be seen that the stability criterion
$(dC_m/dC_L)_{\text{Free}}$ plays an important but rather complex role in establishment of the flight condition of a stable stick force variation with speed.
It is interesting to note how explicitly $(dC_m/dC_L)_{\text{Free}}$ can be brought
into the picture, if it is required that the trim tab always be deflected
to trim the airplane out ($F_s = 0$) to a given speed (V_{Trim}).
The value of $C_{h_{\delta t}}\delta t$ for this trim condition can be obtained from
equation (6–48) by substituting V_{Trim} for V and equating F_s to zero.

$$C_{h_{\delta t}}\delta t = \frac{2W/S}{\rho V_{\text{Trim}}{}^2} \frac{C_{h_\delta}}{C_{m_\delta}} \left(\frac{dC_m}{dC_L}\right)_{\text{Free}} - A \qquad (6\text{–}50)$$

Substituting (6–50) into (6–48) gives

$$F_s = K \frac{W}{S} \frac{C_{h_\delta}}{C_{m_\delta}} \left(\frac{dC_m}{dC_L}\right)_{\text{Free}} \left(\frac{V^2}{V_{\text{Trim}}{}^2} - 1\right) \qquad (6\text{–}51)$$

and

$$\frac{dF_s}{dV} = 2K \frac{W}{S} \frac{C_{h_\delta}}{C_{m_\delta}} \left(\frac{dC_m}{dC_L}\right)_{\text{Free}} \frac{V}{V_{\text{Trim}}{}^2} \qquad (6\text{–}52)$$

The slope when $V = V_{\text{Trim}}$ will be

$$\frac{dF_s}{dV} = 2K \frac{W}{S} \frac{C_{h\delta}}{C_{m\delta}} \left(\frac{dC_m}{dC_L}\right) \frac{1}{V_{\text{Trim}}} \qquad (6\text{-}53)$$

Equation (6-53) indicates that the slope F_s versus V varies with c.g. position if the tab is rolled to maintain the trim speed (V_{Trim}), the slope becoming more stable as the c.g. is moved forward, and more

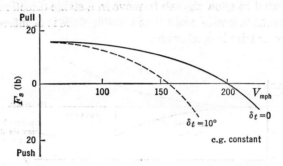

FIGURE 6-19. Stick force versus velocity for different tab angles.

unstable as the c.g. is moved aft. Equation (6-53) also shows that the slope dF_s/dV varies inversely with the trim speed, being higher at the lower speeds. See Figure 6-19.

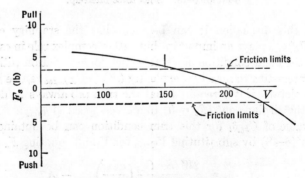

FIGURE 6-20. Friction limits.

The gradient of stick force versus speed, dF_s/dV, is extremely important, as it plays a major role in determining the pilot's feel of the airplane, i.e., the pilot's concept of its stability. A large gradient will tend to keep the airplane flying at constant speed and will resist the influence of disturbances toward changing the speed. It will also enable the pilot to trim the airplane out easily and will not require a lot of pilot attention to hold the given trim speed.   It is also essential that the friction in the longitudinal control system be kept very low so as not to mask the feel completely.   Friction forces much above 1 or 2 lb will be excessive for any except very large airplanes.

Figure 6–20 shows a typical stick force versus velocity curve with 2.5 lb of friction superposed.   If the airplane with its tab set for trim

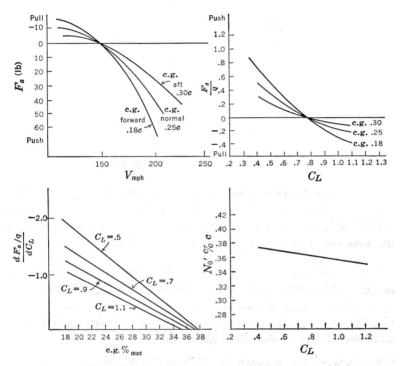

FIGURE 6–21.   Flight test determination of stick-free neutral point.

at 200 mph is slowed down to 100 mph, upon release of the stick, the speed would increase up to 150 mph.   The force gradient inside the friction belt is masked and gives an apparent instability.

There are two methods in common use for obtaining the stick-free neutral point from flight test.   The first of these can be demonstrated by returning to equation (6–48), dividing by the dynamic pressure, $q = \frac{1}{2}\rho V^2$ and substituting $C_L = \dfrac{2W/S}{\rho V^2}$,

$$\frac{F_s}{q} = K(A + C_{h\delta t}\delta_t) - K\frac{C_{h\delta}}{C_{m\delta}}\left(\frac{dC_m}{dC_L}\right)_{\text{Free}} C_L \qquad (6\text{--}54)$$

If (6–54) is differentiated with respect to C_L,

$$\frac{d(F_s/q)}{dC_L} = -K \frac{C_{h\delta}}{C_{m\delta}} \left(\frac{dC_m}{dC_L}\right)_{\text{Free}} \qquad (6\text{–}55)$$

The above demonstrates that the slope of F_s/q versus C_L is a direct function of the stability criterion $(dC_m/dC_L)_{\text{Free}}$. In reducing flight

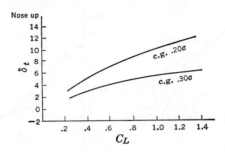

FIGURE 6–22. Tab angle versus C_L.

data, curves of F_s versus V are replotted as F_s/q versus C_L. The slope of these curves is then plotted versus c.g. position, and the c.g. for zero $d(F_s/q)/dC_L$ interpolated or extrapolated. This c.g. position will be the stick-free neutral point for the lift coefficient at which the slopes were taken. See Figure 6–21.

The second way to obtain the stick-free neutral point from flight test is to obtain flight curves of tab angle to trim ($F_s = 0$) versus air speed for varying c.g. locations. The slope of δ_t versus C_L is a function of $(dC_m/dC_L)_{\text{Free}}$, and the c.g. for a zero slope of $d\delta_t/dC_L$ is also the stick-free neutral point. See Figure 6–22.

6–8 Gadgetry for Improving Stick Force Gradients

Because of the very low force gradients found in most modern aircraft at the aft center of gravity, improvement in the longitudinal stability stick-free has been obtained by many different devices, most of which operate so as to produce a constant pull force on the stick independent of speed, which allows a more nose-up tab setting and a steeper stick force gradient. The theory of this operation is shown schematically in Figure 6–23.

There are quite a few ways now in existence for giving the constant pull force. The most common are the downspring or the bobweight. These installations are sketched in Figure 6–24.

The operation of the downspring and bobweight is quite simple and does provide a means for improving stick force versus velocity gradients.   They are not too satisfactory in ground operations, for the heavy pull force required to hold the wheel or stick in neutral is somewhat objectionable.

Another device that shows great promise for accomplishing the same end is the "Vee tab" or "springy tab."   This tab is a free elevator tab

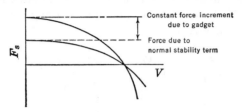

FIGURE 6–23.   Ersatz stick-free stability.

with a spring giving constant torque about its hinge line, which makes the tab ride against its full up stop at zero air speed.   The hinge moment about the tab hinge line can be expressed as follows, assuming

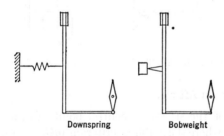

Downspring   Bobweight

FIGURE 6–24.   The downspring and bobweight.

that the hinge moment due to the up spring is K lb-ft, and $C_{ht\delta t}$ is the rate of change of tab hinge moment about its own hinge line with change in tab deflection.

$$HM_{\text{Tab}} = K + C_{ht\delta t}\delta_t \tfrac{1}{2}\rho V^2 S_t \bar{c}_t \qquad (6\text{--}56)$$

The tab floating angle can be determined by equating (6–56) to zero and solving for δ_t.

$$\delta_t = -\frac{2K}{\rho V^2 S_t \bar{c}_t C_{ht\delta t}} \qquad (6\text{--}57)$$

If the spring tends to rotate the tab upwards, K will be negative, and the tab will float up against the stop until the velocity increases to a point where the spring torque is just overcome.   For speeds past this critical speed, the tab will float down proportionally to the velocity squared; the hinge moment due to the tab is also proportional to the velocity squared, and the net effect is to cancel out any hinge moment variation with speed and leave a net constant hinge moment tending to push the stick forward, giving the same effect as a downspring except that the device is more powerful for improving the stick force gradient and has the added advantage of not imposing any static stick loads during ground maneuvers.

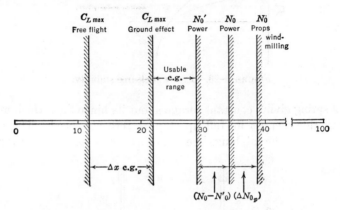

FIGURE 6-25.   Typical center of gravity limits.

6-9   Restriction on Aft c.g.

The concept of the stick-free neutral point, N_0', as that center of gravity location where the stability criterion $(dC_m/dC_L)_{\text{Free}}$ vanishes or where the stick force versus velocity gradient is zero through a trim speed, brings a new restriction on the aft limit to the allowable c.g. range.   At the present writing the Army and Navy specifications call only for the most aft c.g. to be ahead of the stick-free neutral points. It is obvious, however, that zero stick force gradients are undesirable, and any well-designed airplane should always try to maintain a stable stick force gradient even at the most aft c.g. location.   This is difficult to do on modern high-speed airplanes with large variations in c.g.; however, by proper design and making use of aids such as the down-spring, bobweight, or Vee tab, adequate gradients can be obtained. No desirable minimum gradient is suggested herein, but it should be tied up rather closely with airplane type, control friction, etc.

To point up the new aft limit and its further restriction on the c.g. range, the c.g. limitations on a typical mean chord are shown in Figure 6-25.

SUGGESTED READING

1. NACA WR L-663, "Wind-tunnel Data on the Aerodynamic Characteristics of Airplane Control Surfaces," by R. I. Sears, 1943.
2. NACA TR 721, "Determination of Control Surface Characteristics from NACA Plain Flap and Tab Data," by Ames and Sears, 1941.
3. British ARC R & M 1095, "Theoretical Relationships for an Aerofoil with Hinged Flap," by H. Glauert, 1927.
4. NACA Series on "Wind-tunnel Investigation of Control Surface Characteristics," by various authors.
5. NACA WR A-11, "Computation of Hinge Moment Characteristics of Horizontal Tails from Section Data," by R. M. Crane, 1945.
6. NACA TR 528, "Reduction of Hinge Moments of Airplane Control Surfaces by Tabs," by T. A. Harris, 1935.
7. British ARC R & M 1420, "A Collection of Wind-tunnel Data on the Balancing of Controls," by F. B. Bradfield, 1922.
8. NACA TR 675, "Effect of Elevator Nose Shape, Gap, Balance, and Tabs on the Aerodynamic Characteristics of a Horizontal Tail Surface," by Goett and Reeder, 1939.

PROBLEMS

6-1. The hinge moment parameters of an assumed elevator control are as follows: $C_{h\alpha} = -.006$, $C_{h\delta} = -.010$, $C_{h0} = 0$. Plot curves of elevator hinge moment coefficient, C_{h_e}, versus elevator deflection, δ_e, for the following angles of attack: $-15°$, $-10°$, $-5°$, $0°$, $5°$, $10°$, $15°$. Carry these plots for a deflection range of $\pm30°$.

6-2. Cross-plot the curves obtained in Problem 6-1 to obtain the elevator floating angle versus angle of attack. Check the slope obtained analytically.

6-3. If the stick-elevator gearing for a given airplane is such that the force on the stick in pounds is just one-half the elevator hinge moment in pound-feet, calculate the force required of a pilot to change the airplane's speed from 300 mph V_i, where the stick force is zero, tail angle of attack ($\alpha_s = 0$), and elevator deflection ($\delta_e = 0$), to a speed of 100 mph V_i, where the tail angle of attack is $4°$ and the elevator deflection is $-4°$. The horizontal tail area is 200 sq ft with a 24.5 ft span, with an NACA 0009 airfoil section. The elevator is a .30c flap of 60 sq ft area, hinged at the leading edge with a sealed gap. Its root-mean-square chord is 3.0 ft.

6-4. If the elevator given in Problem 6-3 has a medium nose shape, how far back should the elevator hinge line be moved in order to give a stick force for the flight condition of Problem 6-3 of approximately 50 lb?

6-5. Estimate the three-dimensional hinge moment parameters $C_{h\alpha}$ and $C_{h\delta}$ for the following elevator:

Horizontal tail area = 60 sq ft
Horizontal tail span = 15.5 ft
Airfoil section NACA = 0015 ($\Delta\phi = 4°$)
$c_f/c = .4$
$c_b/c_f = .25$
Nose shape: blunt
Nose gap: .001c

6-6. Given a glider with the following geometric and aerodynamic characteristics:

Wing aspect ratio—8 (NACA 23012)
Horizontal tail aspect ratio—4 (NACA 0009)
Tail volume, $\bar{V}$ = .6
Rate of change of downwash, $d\epsilon/d\alpha$ = .5
Tail efficiency, η_t = .9
Aerodynamic center = .24c
$(dC_m/dC_L)_{Fus}$ = .08
Elevator area ratio S_e/S_t = .35
Floating tendency, $C_{h\alpha}$ = −.003
Restoring tendency, $C_{h\delta}$ = −.0055
Residual hinge moment, C_{h0} = 0

Calculate the stick-fixed neutral point, N_0, and the stick-free neutral point, N_0'. If the floating tendency is reversed such that $C_{h\alpha}$ = .003, what would be the stick-free neutral point?

6-7. The horizontal tail of the airplane of Problem 5-17 is equipped with an elevator balanced by a set-back hinge such that c_b/c_f = .30 and with a medium nose shape with sealed gap. Calculate the propeller-windmilling stick-free neutral point and the power-on stick-free neutral point. The airfoil section used for the tail surface is an NACA 0009.

6-8. For the airplane of Problem 6-7 draw a c.g. limit diagram as shown in Figure 6-25, including the power-on stick-free neutral point, N_0'.

6-9. If the airplane of Problem 6-6 has a 20 sq ft elevator with a root-mean-square chord of 2 ft and if the gearing between the elevator and control column is G = .5 rad/ft, calculate and plot the stick force required versus indicated air speed for a constant tab of δ_t = 0 for speeds from 100 mph V_i to 300 mph V_i for stick-fixed stability margins of $(dC_m/dC_L)_{Fix}$ = .05, −.05, and −.15. The geometric characteristics of this glider are as listed below:

$$\alpha_0 = -2°$$
$$i_w = 0°$$
$$i_t = -1°$$
$$\delta_{e0} = -2°$$
$$C_{h\delta t} = -.003$$
$$W/S = 30 \text{ lb per sq ft}$$

6-10. Calculate and plot the variation of stick force with indicated air speed for the airplane of Problem 6-9 for a 5° nose-up (positive) tab setting and then for a 5° nose-down (negative) tab setting, both for a c.g. giving $(dC_m/dC_L)_{Fix}$ = −.05.

6-11. Calculate the tab setting required for the airplane of Problem 6-9 to trim out the stick force (F_s = 0) at a speed of 200 mph V_i for the three c.g. locations given. Plot the value of F_s versus V_i from 100 mph V_i to 300 mph V_i for each c.g. location.

6-12. If the airplane of Problem 6-9 is equipped with a spring giving a constant force tending to move the elevator down equivalent to 20 lb at the stick, recalculate the trim tab angles for trim at 200 mph V_i and replot the F_s versus V_i curves for the three given c.g.'s.

6-13. Given an airplane with the following geometric and aerodynamic characteristics:

$$(dC_L/d\alpha)_w = .08 \qquad \alpha_0 = -2°$$
$$G = .5 \text{ rad/ft} \qquad i_w = 0°$$
$$S_e = 15 \text{ sq ft} \qquad i_t = 1°$$
$$c_e = 1.5 \text{ ft} \qquad d\epsilon/d\alpha = .55$$
$$\eta_t = .9 \qquad \delta_{e_0} = -3°$$
$$C_{h_0} = 0.0 \qquad C_{m\delta} = -.012$$
$$C_{h\alpha} = -.002 \qquad W/S = 40 \text{ lb/sq ft}$$
$$N_0 = .35c$$
$$C_{h\delta} = -.005$$
$$C_{h\delta_t} = -.006$$

Calculate and plot curves of stick force, F_s, versus tab angle, δ_t, for the following constant indicated air speeds, for a c.g. location of .20c ($V_{i\text{mph}}$ = 100, 140, 180. 220, 260, 300, 340), tab angles running from 15° to $-10°$.

Recalculate and plot the same curves for a c.g. location at .40c. (In both cases cut off the stick force at 80 lb.)

6–14. Cross-plot the curves obtained in Problem 6–13 for stick force, F_s, versus speed, V_i, for constant trim tab angles required for trim at 220 mph V_i for the two c.g.'s.

CHAPTER 7

MANEUVERING FLIGHT

7-1 Introduction

The study just completed of the airplane's static longitudinal stability and control has considered the airplane only from the point of view of its flying on equilibrium unaccelerated flight paths with the lift equal to the weight, $C_L V^2 = K$. This, of course, is the major flight consideration, as the airplane flies on straight flight paths over 90 per cent of the time. This even includes fighter-type airplanes which are

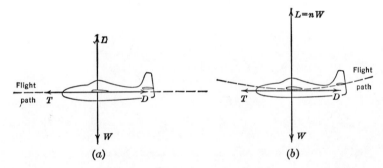

FIGURE 7-1. Forces of unaccelerated and accelerated flight (pull-up).

designed for maneuvering. The static longitudinal stability both stick-fixed and stick-free is therefore of prime importance, and a knowledge of the two neutral points, N_0 and N_0' (stick-fixed and stick-free) is essential for the satisfactory operation of the airplane.

In spite of the fact that the airplane flies on straight flight paths 90 per cent of the time, it is nevertheless essential that the airplane be provided with stability and control in flight along curve flight paths or in maneuvering flight. This is, of course, more true for airplanes that are required to do extensive maneuvering, such as fighter-type airplanes, but as some maneuvering is done by all known airplanes, it is necessary to study these characteristics for all airplanes.

When the airplane is flying on straight, unaccelerated flight paths, all the forces are in static equilibrium (Figure 7-1a). In order to curve

298

the flight path this static equilibrium must be destroyed in such a way that the resulting unbalanced forces operate perpendicular to the flight path, thereby causing it to curve. If, for instance, the pilot changes the airplane's angle of attack, and thereby its lift coefficient, quickly enough so that the speed does not fall off to satisfy the equilibrium condition, $C_L V^2 = K$, the airplane's lift will exceed its weight, creating an unbalanced force in the vertical plane which will curve the flight path upwards (Figure 7–1b). This maneuver is termed a pull-up and is used for pulling airplanes out of dives, looping, and other maneuvers in the plane of symmetry.

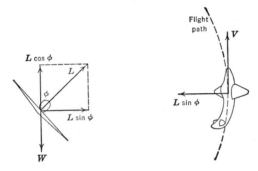

FIGURE 7–2. Accelerated flight (turns).

Another method for creating unbalanced forces perpendicular to the airplane's flight path is that used by the pilot for the normal turn. In order to curve the flight path in the horizontal plane, the airplane's wings are banked over, rotating the resultant lift vector out of the vertical, creating an unbalanced horizontal component perpendicular to the flight path which curves the flight path in the direction of the unbalanced force. (See Figure 7–2.)

The airplane will make climbing or diving turns, depending on whether the vertical component of the lift is equal to or less than the airplane's weight. The subject of the airplane's turning performance was discussed in Part 1 in detail and will not be repeated here.

In both the pull-up maneuver and the turn, the acceleration of the airplane, in response to the unbalanced lift forces perpendicular to the flight path, manifests itself as a curvature of this flight path and a rotation of the airplane about its Y axis. The rotation about the Y axis produces damping moments tending to stop the rotation and must be overcome by the application of more up elevator and more stick force than those required to overcome the stability moments as the lift coefficient is increased. This phenomenon gives rise to a sort of pseudo-stability or apparent stability, for when the airplane is balanced at either the stick-fixed or stick-free neutral point, it will be necessary to deflect the elevator and increase the stick forces, respectively, in order to increase the lift coefficient in accelerated flight.

The airplane's stability in accelerated flight is affected by the airplane's c.g. position, just as it was in unaccelerated flight, the only difference being that for a given c.g. position the airplane usually has a little more stability in accelerated flight because of the damping term. If the c.g. is moved aft behind the stick-fixed and stick-free neutral points, N_0 and N_0', the stability in accelerated flight will be reduced until at some c.g. the change in elevator angle and stick force required to accelerate the airplane will become zero. These c.g.'s are termed the maneuver points. The c.g. where the elevator angle required to accelerate the airplane vanishes is the stick-fixed maneuver point, N_m, and the c.g. where the stick force required to accelerate the airplane vanishes is the stick-free maneuver point, N_m'. This chapter will deal with the elevator angle and stick forces required to accelerate the airplane on curved flight paths both in pull-ups and in turns and will present methods for determining the two maneuver points.

In maneuvering flight, the difference in the stability over unaccelerated flight arises because of the angular velocity of the airplane about its Y axis and also because of the fact that the lift coefficient is changed at constant speed instead of according to the equation $C_L V^2 = K$. Usually the first of these is the most important effect, and to start this investigation the dependence of the angular velocity, q, about the Y axis, on the severity of the maneuver will be determined.

If the airplane is making a pull-up, the unbalanced lift force creates centripetal acceleration satisfying the Newtonian laws of motion ($F = ma$). The unbalanced lift force, L, can be given in terms of the airplane's weight as follows:

$$\Delta L = (nW - W) = W(n - 1) \qquad (7-1)$$

This force is balanced by the resulting acceleration

$$W(n - 1) = \frac{W}{g} a \qquad (7-2)$$

or

$$(n - 1)g = a \qquad (7-3)$$

From the laws of mechanics the centripetal acceleration equals V^2/R, so

$$\frac{V^2}{R} = g(n - 1) \qquad (7-4)$$

The angular velocity, q, in radians per second from the laws of mechanics equals V/R; therefore

$$q = \frac{g(n-1)}{V} \qquad (7\text{-}5)$$

Equation (7-5) gives the relationship between the airplane's angular velocity, q, in radians per second, the forward velocity in feet per second, and the airplane's load factor, n, for pull-ups from level flight.

In steady turning flight, the angular velocity is a different function of the forward velocity and the load factor, because of the fact that the rotation about the Y and Z axes varies with angle of bank. From Figure 7-3 it can be seen that if the angular velocity of the airplane about some vertical axis is given as Ω, the angular velocity in pitch, q, will be $\Omega \sin \phi$, where ϕ is the angle of bank.

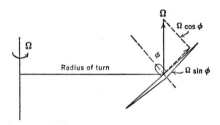

FIGURE 7-3. Angular velocity components in turns.

The angular velocity of the airplane about its Y axis is therefore

$$q = \frac{V}{R} \sin \phi \qquad (7\text{-}6)$$

From the developments given in Part 1 on airplane's turning performance, the following relationships were determined from a balance of forces in the horizontal and vertical planes:

$$R = \frac{V^2}{g \tan \phi} \qquad (7\text{-}7)$$

and

$$n = \frac{1}{\cos \phi} \qquad (7\text{-}8)$$

Combining equations (7-6), (7-7), (7-8) and simplifying yield the following relation:

$$q = \frac{g}{V}\left(n - \frac{1}{n}\right) \qquad (7\text{-}9)$$

Equation (7–9) gives the relationship between the airplane's angular velocity, q, the forward velocity, V, and the airplane's load factor for steady level turns.

As it is the angular velocity in pitch that gives rise to changes in elevator angle and stick force beyond those required by the normal stability moments, it is convenient to express the airplane's load factor, n, as the independent variable for maneuvering flight. This is a direct outcome of equations (7–5) and (7–9), which show the fundamental importance of the load factor in maneuvering flight. The increment of elevator angle and stick force to produce an increment in normal acceleration equal to one g at constant speed, both in pull-ups and in steady turns, is taken as the criterion for the longitudinal stability and control in maneuvering flight.

7–2   Elevator Angle per g

If an airplane is pulled into a steady turn at a given value of the normal acceleration, its lift coefficient must be increased over that

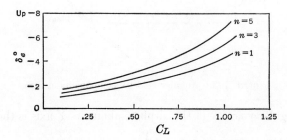

FIGURE 7–4.   Elevator angle versus C_L in pull-ups.

required for straight level flight at the same speed, in order to have a vertical component of lift equal to the weight. The airplane's lift coefficient in general must be expressed as given below:

$$C_L = \frac{2n(W/S)}{\rho V^2} \tag{7–10}$$

The elevator angle required to bring the airplane into equilibrium at any lift coefficient was given in equation (6–44), Chapter 6:

$$\delta_e = \delta_{e0} - \frac{C_L}{C_{m\delta}} \left(\frac{dC_m}{dC_L}\right)_{\text{Fix}} \tag{7–11}$$

If there were no additional requirements on the elevator angle, the curve of δ_e versus C_L would be independent of the normal acceleration.

However, flight test curves of the elevator angle versus airplane lift coefficient in accelerated flight show a decided dependence on the magnitude of the acceleration. Typical curves are given in Figure 7–4.

The reason for the spread of the δ_e versus C_L curves is the elevator requirement to overcome the damping moments arising from the airplane's angular velocity when flying on curved flight paths. The angular velocity, q, is a function of the forward velocity, the normal acceleration, and the type of maneuver as given in equations (7–5) and (7–9).

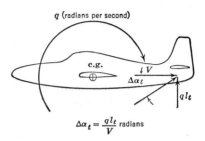

q (radians per second)

c.g.

$\Delta\alpha_t = \dfrac{ql_t}{V}$ radians

FIGURE 7–5. Increment in tail angle of attack due to pitching velocity.

The angular rotation of the airplane creates an increment in tail angle of attack, which in turn creates a moment about the airplane's c.g. tending to stop the rotation. This moment is the airplane's longitudinal damping and must be overcome by application of sufficient up elevator.

The increment in tail angle of attack due to the angular velocity in pitch, q, is shown schematically in Figure 7–5.

For small angles and with q expressed in radians per second

$$\Delta\alpha_s = \frac{ql_t}{V}\text{ radians} \tag{7–12}$$

The change in effective tail angle of attack due to deflecting the elevator is

$$\Delta\alpha_t = \tau\delta_e \tag{7–13}$$

The elevator angle required to offset the effect of the increment in tail angle of attack due to the pitching velocity is obtained by adding (7–12) to (7–13), equating to zero, and solving for δ_e, in consistent units:

$$\delta_e{}^\circ = -\frac{57.3ql_t}{\tau V} \tag{7–14}$$

If the pull-up formula, equation (7–5), is substituted for the angular velocity, q, the elevator required to overcome the increment in tail angle of attack, i.e., the tail damping effect, may be written as follows:

$$\delta_e{}^\circ = -\frac{gl_t}{\tau V^2}(n-1)57.3 \tag{7–15}$$

This elevator angle increment is required to overcome the damping of the horizontal tail.   It is also necessary to make some allowance for the damping effect of the rest of the airplane.   This is usually done by multiplying the tail damping requirement by 1.10.   This gives the total elevator required to overcome the rotational moments as

$$\delta_e = \frac{-63 g l_t}{\tau V^2} (n - 1) \tag{7-16}$$

This term must finally be added to equation (7-11) to obtain the general expression for the elevator angle.

$$\delta_e = \delta_{e0} - \frac{C_L}{C_{m\delta}} \left(\frac{dC_m}{dC_L} \right)_{\text{Fix}} - \frac{63 g l_t}{\tau V^2} (n - 1) \tag{7-17}$$

Equation (7-17) may be rewritten by substituting equation (7-10) for C_L:

$$\delta_{e\text{Pull-up}} = \delta_{e0} - \frac{2n \, (W/S)}{\rho V^2 C_{m\delta}} \left(\frac{dC_m}{dC_L} \right)_{\text{Fix}} - \frac{63 g l_t}{\tau V^2} (n - 1) \tag{7-18}$$

The gradient of elevator angle per g at constant air speed may be obtained directly from equation (7-18) by simple differentiation.

$$\left(\frac{d\delta_e}{dn} \right)_{\text{Pull-up}} = - \frac{1}{V^2} \left[\frac{63 g l_t}{\tau} + \frac{2(W/S)}{\rho C_{m\delta}} \left(\frac{dC_m}{dC_L} \right)_{\text{Fix}} \right] \tag{7-19}$$

Equations (7-18) and (7-19) are for wings-level pull-ups.   Similar expressions may be obtained for steady turns by substituting the steady turn expression for angular velocity, q, equation (7-9), into (7-19) and proceeding as before.   This development yields the following:

$$\delta_{e\text{Turns}} = \delta_{e0} - \frac{2n(W/S)}{\rho V^2 C_{m\delta}} \left(\frac{dC_m}{dC_L} \right)_{\text{Fix}} - \frac{63 g l_t}{\tau V^2} \left(n - \frac{1}{n} \right) \tag{7-20}$$

$$\left(\frac{d\delta_e}{dn} \right)_{\text{Turns}} = - \frac{1}{V^2} \left[\frac{63 g l_t}{\tau} \left(1 + \frac{1}{n^2} \right) + \frac{2(W/S)}{\rho C_{m\delta}} \left(\frac{dC_m}{dC_L} \right)_{\text{Fix}} \right] \tag{7-21}$$

The gradient of elevator angle per g is one of the criteria for judging the stability and control characteristics of the airplane in maneuvering or accelerated flight.   Examination of equation (7-19) shows that this gradient is a linear function of the stick-fixed stability and therefore varies directly with c.g. position, becoming less stable as the c.g. is moved aft.   When the airplane is balanced at the stick-fixed neutral point, N_0, the gradient $d\delta_e/dn$ is only a function of the airplane's damping.   As the airplane's c.g. is moved aft of N_0, the gradient

$d\delta_e/dn$ continues to reduce until at some aft c.g. it will vanish. This c.g. position is termed the stick-fixed maneuver point, N_m. It may be obtained from equation (7-19) by equating it to zero and solving for $(dC_m/dC_L)_{Fix}$.

$$\left(\frac{dC_m}{dC_L}\right)_{Fix} = \frac{-63gl_t\rho C_{m\delta}}{2\tau(W/S)} \tag{7-22}$$

or in terms of the stick-fixed neutral point,

$$N_m = N_0 - \frac{63\rho C_{m\delta}gl_t}{2\tau(W/S)} \tag{7-23}$$

Equation (7-23) demonstrates that the stick-fixed maneuver point, N_m, is aft of the stick-fixed neutral point, N_0. The difference between

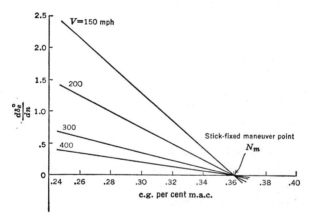

FIGURE 7-6. Typical variation of $d\delta_e/d_n$ with c.g. shift.

these neutral points is the greatest at sea level and for light wing loadings.

The variation of the gradient $d\delta_e/dn$ with c.g. position for a typical airplane is shown in Figure 7-6. It can be seen that the gradient is essentially an inverse function of the velocity squared, as given by equation (7-19).

Examination of equations (7-21) and (7-19) shows that a little more up elevator is required to pull some acceleration increment in steady turns than it is in pull-ups. This difference vanishes at the higher g's.

The elevator control must be made powerful enough to allow the pilot to accelerate the airplane to the stall or its design load factor. As the c.g. is moved forward, more and more up elevator is required, so that a new limit on the forward c.g. is defined. This limit is that c.g. forward of which it is impossible to accelerate the airplane either to its stall or to its limit load factor throughout the airplane's speed range. Another way of saying this is that the elevator should be powerful enough to allow equilibrium at the most forward c.g. around the airplane's flight envelope or V-G diagram. An example of this is given in Figure 7–7.

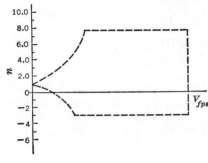

FIGURE 7–7. The V-G diagram.

In almost all cases this requirement is not a critical one, as the landing requirements on the elevator power are much more demanding. The forward limit on the c.g. due to this requirement is seldom worked out, as it is well forward of the limiting c.g. to meet the landing requirement.

A new limit on the rearward c.g. must also be considered in the light of this stability criterion. It is required that the most rearward c.g. be forward of the stick-fixed maneuver point, N_m. However, as this point is usually well aft of the stick-fixed neutral point, it is hardly ever critical, and designers seldom concern themselves with it. The elevator angle required in accelerated flight is of small importance to the pilot, especially when compared to the stick force per g which will be taken up in the next section.

7–3   Stick Force per g

The second criterion in common use for evaluating the stability and control of the airplane in maneuvering flight is the stick force required to produce a one-g increment in normal acceleration in either a pull-up or steady turn. This criterion has been found to be of tremendous importance during the past few years, especially for fighter-type airplanes. It has been well proved that the pilot's feeling of the maneuvering capabilities of his airplane is very closely tied up with the gradient of stick force versus g, and it is therefore necessary to design the airplane very carefully to maintain the gradient within the desirable limits. It will be shown that this gradient varies directly with c.g. position, being heavier at the more forward c.g.'s and lighter as the airplane is balanced further aft. There is therefore a c.g. at which this gradient vanishes, and this then becomes our final neutral point, termed the stick-free maneuver point, N_m'. Flying any airplane back of the stick-free maneuver point is very hazardous, for when balanced in this condition, the slightest effort on the part of the pilot will start the airplane in a very rapid acceleration divergence during which the airplane may very well fail structurally before corrective action can be taken by the pilot. Fighter airplanes having stick force per g gradients that are too high fatigue the pilot excessively; if the gradient is too low, the airplane will be too sensitive and liable to structural failures as indicated above.

In order to analyze this problem further, the theory underlying the stick forces in accelerated flight will be developed. The general stick force equation may be written as follows:

$$F_s = -GS_e c_e \tfrac{1}{2}\rho V^2 \eta_t (C_{h0} + C_{h_\alpha}\alpha_s + C_{h_{\delta e}}\delta_e + C_{h_{\delta t}}\delta_t) \quad (7\text{-}24)$$

For pull-up maneuvers the following expressions may be substituted into equation (7–24):

$$\alpha_s{}^\circ = \alpha_0{}^\circ + \frac{2n(W/S)}{\rho V^2 a_w}\left(1 - \frac{d\epsilon}{d\alpha}\right) - i_w + i_t + \frac{gl_t}{V^2}(n-1)57.3 \quad (7\text{-}25)$$

$$\delta_e{}^\circ = \delta_{e0}{}^\circ - \frac{2n(W/S)}{\rho V^2 C_{m\delta}}\left(\frac{dC_m}{dC_L}\right)_{\text{Fix}} - \frac{63gl_t}{\tau V^2}(n-1) \quad (7\text{-}26)$$

Substitution of the above into equation (7–24), with some rearranging, gives:

$$F_s = -G\eta_t c_e S_e \tfrac{1}{2}\rho V^2 \Bigg[(C_{h0} + C_{h_\alpha}\alpha_0 + C_{h_\alpha}i_t - C_{h_\alpha}i_w + C_{h\delta}\delta_{e0} + C_{h\delta t}\delta_t)$$

$$- \frac{2n(W/S)C_{h\delta}}{\rho V^2 C_{m\delta}}\left(\frac{dC_m}{dC_L}\right)_{\text{Free}} + \frac{57.3gl_t}{V^2}(n-1)\left(C_{h_\alpha} - \frac{1.1C_{h\delta}}{\tau}\right)\Bigg] \quad (7\text{-}27)$$

Now if the trim tab is set for zero stick force at some speed V_{Trim} at $n = 1$, equation (7–27) can be written as follows:

$$F_s = \frac{-G\eta_t S_e c_e (W/S)C_{h\delta}}{C_{m\delta}}\left(\frac{dC_m}{dC_L}\right)_{\text{Free}}\left(\frac{V^2}{V_{\text{Trim}}{}^2} - n\right)$$

$$- 57.3G\eta_t S_e c_e g l_t \frac{\rho}{2}(n-1)\left(C_{h_\alpha} - \frac{1.1C_{h\delta}}{\tau}\right) \quad (7\text{-}28)$$

When $n = 1$, equation (7–28) reduces to equation (6–51) of Chapter 6, the expression for the variation of stick force with speed in unaccelerated flight.

In order to obtain the gradient of stick force versus g, it is only necessary to differentiate equation (7–28) with respect to n.

$$\left(\frac{dF_s}{dn}\right)_{\text{Pull-up}} = \frac{G\eta_t S_e c_e (W/S) C_{h\delta}}{C_{m\delta}} \left(\frac{dC_m}{dC_L}\right)_{\text{Free}}$$

$$- 57.3 G\eta_t S_e c_e g l_t \frac{\rho}{2} \left(C_{h\alpha} - \frac{1.1 C_{h\delta}}{\tau}\right) \quad (7\text{–}29)$$

Equation (7–29) is the standard expression for stick force per g in pull-ups. The first term is the contribution of the airplane's stick-free stability, while the second term is that due to the airplane's damping.

The stick force per g in steady turns may be developed in a similar fashion:

$$\left(\frac{dF_s}{dn}\right)_{\text{Turns}} = \frac{G\eta_t S_e c_e (W/S) C_{h\delta}}{C_{m\delta}} \left(\frac{dC_m}{dC_L}\right)_{\text{Free}}$$

$$- 57.3 G\eta_t S_e c_e g l_t \frac{\rho}{2} \left(1 + \frac{1}{n^2}\right) \left(C_{h\alpha} - \frac{1.1 C_{h\delta}}{\tau}\right) \quad (7\text{–}30)$$

Besides the stability and damping contributions to the variation of stick force with g, any static weight moment about the elevator hinge will also contribute. If such a weight moment, produced by static unbalance or overbalance of the elevator or the installation of a bob-weight, is represented by the factor K (positive when tending to rotate the elevator downward), the stick force due to this moment will be

$$F_s = -KGn \quad (7\text{–}31)$$

and the stick force per g

$$\frac{dF_s}{dn} = -KG \quad (7\text{–}32)$$

Weights in the elevator system, then, can add to or subtract from the gradient of stick force versus g and are quite often used as a means to finally adjust these gradients to the required limits.

An inspection of equations (7–29) and (7–30) shows that the gradient stick forces per g do not vanish at the stick-free neutral point $[(dC_m/dC_L)_{\text{Free}} = 0]$ but shows that, if the c.g. is moved sufficiently aft of N_0, a position will be reached where $dF_s/dn = 0$. This c.g. position is termed the stick-free maneuver point and is represented by the symbol $N_m{}'$. A typical variation of stick force per g with c.g. position is shown in Figure 7–8.

For this airplane, having zero weight moment about the elevator hinge, the stick force per g is still 3 lb per g at the stick-free neutral point, and therefore is the stick force per g due to the airplane's damping. The stick-free maneuver point is almost always aft of the stick-free neutral point. An analytical expression for finding the stick-free maneuver point comes directly from equation (7–29).

$$N_m' = N_0' + \frac{57.3C_{m\delta}}{(W/S)C_{h\delta}}\left[\frac{\rho}{2}gl_t\left(C_{h\alpha} - \frac{1.1C_{h\delta}}{\tau}\right)\right] \qquad (7\text{–}33)$$

A study of equation (7–29) brings to light many important aspects of this stick force gradient in maneuvering or accelerated flight. In

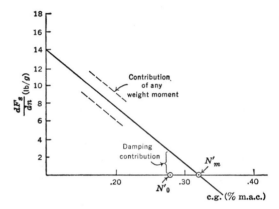

FIGURE 7–8.　Stick force per g versus c.g.

the first place it can be seen that the variation of dF_s/dn with c.g. shift is a function of the hinge moment parameter, $C_{h\delta}$, only. The slope of the curve of dF_s/dn versus c.g. position shown in Figure 7–8 can be expressed as

$$\frac{d(F_s/n)}{d(\% \text{ c.g. shift})} = \frac{(W/S)GS_ec_eC_{h\delta}}{100C_{m\delta}} \qquad (7\text{–}34)$$

This fact is of considerable interest to the flight test engineer, for by determining this variation by flight test it is possible to extract the value of $C_{h\delta}$. It is also apparent that change in stick forces per g with c.g. varies with airplane size and wing loading.

A study of the damping term shows the very important fact that the stick forces in accelerated flight are a function of altitude, falling off as the altitude is increased. This phenomenon has been noticed by fighter pilots, who report a decay in stability with altitude. The importance of keeping the gradient of stick force per g within limits became more and more apparent during the years of World War II.

The Air Force and Navy have set up limits for this gradient for all types of airplanes. It was found that if this gradient was too heavy, the pilot of fighter-type airplanes would tire rapidly in combat, whereas if the gradient was too light, the airplane would be too sensitive and subject to structural failure due to inadvertent excessive acceleration of the airplane. For fighter-type airplanes an upper limit on the stick forces was found to be about 8 lb per g, while the lower limit was placed at 3 lb per g. On larger airplanes such as bombers and cargo planes this upper limit on this gradient is in the neighborhood of 35 lb per g.

The imposition of limits on these gradients places further restrictions on the center of gravity travel. The forward c.g. is limited to that

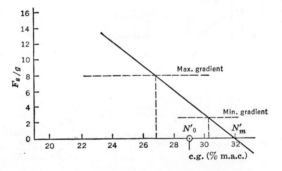

FIGURE 7-9.   Maximum and minimum gradients.

where the maximum gradient is encountered, and the aft c.g. limited to where the minimum gradient is obtained. In many planes, especially fighters, these limits are critical ones and must be reckoned with seriously by the airplane designers. An example of the problem confronting the designer of a typical fighter is pictured in Figure 7-9.

This airplane, a large fighter, in spite of a closely balanced elevator (small negative $C_{h\delta}$), has a rapid increase in stick force per g with forward shift of the neutral point. If the design aft c.g. is taken back to where $dF_s/dn = 3$ lb/g, the c.g. will be aft of the stick-free neutral point, which is undesirable. Therefore to maintain stick-free stability and also not go over the forward limit of $dF_s/dn = 8$ lb/g, the allowable c.g. is severely restricted (27–29 per cent). In order to improve this situation the designer may reduce $C_{h\delta}$ still further. If this is done, the slope of the curve dF_s/dn versus c.g. will flatten, allowing more c.g. range, but this can only be done within limits, for pilots feel that too low values of $C_{h\delta}$ are undesirable. For large fighters the possibility of obtaining correct stick forces in accelerated flight is much more remote than for small fighters, and it requires that the airplane's actual c.g. spread be kept down to an absolute minimum.

The analytical treatment developed for the gradients dF_s/dn has been for steady-state conditions where dynamic equilibrium has been achieved. Although this allows a simple explanation and development of these gradients, it must be realized that this picture may not be quite so simple as represented. In many actual cases the airplane's motion may be a transient where the steady state is never reached. For these cases the simple relations given no longer apply. In fast maneuvers the acceleration reached for a given stick force depends largely on the type of aerodynamic balance and the rapidity at which the stick force is applied. For instance, if the elevator is very closely balanced ($C_{h\delta}$ very small), the proper stick force per g can be obtained by making $C_{h\alpha}$ slightly positive. However, if the stick force is applied rapidly enough, large elevator deflections may be obtained before the airplane's response builds up the stick force through the floating tendency $C_{h\alpha}$. This will give large airplane accelerations for undesirably small stick forces in the transient maneuver, while in the steady-state maneuver, the stick forces may be satisfactory.

It is felt that it is beyond the scope of this book to go into these transient maneuvers, but they are pointed out to introduce the reader to their importance.

7–4 Limits on the Airplane's c.g.

It is felt advisable at this point to gather all the restrictions on the airplane's center of gravity locations together in order to see which are the important ones and which can usually be neglected. In order to do this a part of the mean aerodynamic chord of a typical airplane is shown in Figure 7–10 and the various limitations placed on it.

From Figure 7–10 the critical limitations on the airplane's c.g. are seen to be as follows for a typical case:

1. On the forward c.g.:
    a. The maximum stick force per g gradient.
    b. The elevator required to land at $C_{L\max}$.
2. On the aft c.g.:
    a. Power-on stick-free neutral point.
    b. Minimum gradient, stick force per g.

In order to produce a satisfactory airplane from the point of view of stability and control, it is essential that as much useful c.g. range be provided as possible. There are many airplanes flying today that do not satisfy the stability and control requirements at any c.g. In order to obtain the best results the aerodynamicist must strive to open

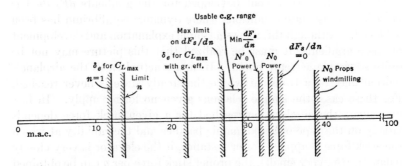

FIGURE 7–10. Summary of stability and control limits on the c.g.

up the useful c.g. range, and the layout men must be conscious of the great importance of decreasing the c.g. range of the airplane to the absolute minimum.

SUGGESTED READING

1. RAE Aero. Report 1740, "Proposal For an Elevator Maneuverability Criterion," by S. B. Gates, 1942.
2. NACA WR L-430, "A Theoretical Investigation of Longitudinal Stability of Airplanes with Free Controls including Effect of Friction in Control System," by Greenberg and Sternfield, 1944. (Static parts.)
3. AAF TR 5167, "Some Theoretical Developments in Airplane Static Longitudinal Stability and Control," by C. D. Perkins, 1944.
4. NACA WR L-322, "Analysis of Wind-tunnel Stability and Control Tests in Terms of Flying Qualities of Full-scale Airplanes," by G. G. Kayten, 1943.
5. AAF TR 5242, "Stability and Control Flight Test Methods," by C. D. Perkins and T. F. Walkowicz, 1945.

PROBLEMS

7–1. Calculate and plot curves of airplane angular velocity in radians per second versus true air speed in miles per hour for the following airplane normal accelerations obtained in pull-ups from level flight: $n = 1.0, 1.5, 2.0, 3.0, 4.0, 6.0, 8.0$. The speeds should range from 100 mph to 400 mph.

7–2. Calculate and plot curves as in Problem 7–1 except that the accelerations are to be obtained in level turns.

7–3. An airplane has the following geometric and aerodynamic characteristics. Calculate and plot curves of elevator angle, δ_e, versus normal acceleration, n, for the following true speeds: $V = 140, 200,$ and 320 mph. Calculate for values of n ranging from one to eight.

$$W/S = 40 \text{ lb per sq ft}$$
$$C_{m\delta} = -.015$$
$$\rho = .002 \text{ slug per cubic foot}$$
$$l_t = 12 \text{ ft}$$
$$\tau = .4$$
$$\left(\frac{dC_m}{dC_L}\right)_{\text{Fix}} = -.05$$
$$\delta_{e_0} = -2°$$
$$\text{c.g.} = .30c$$

7-4. For the airplane of Problem 7-3, calculate and plot the value of the gradient $(d\delta_e/dn)$ in pull-ups, versus c.g. location for c.g. positions from .20c to .40c for true speeds of 100, 200, and 300 mph. What is the stick-fixed maneuver point for this airplane?

7-5. Given an airplane with the following geometric and aerodynamic characteristics:

Wing area = 220 sq ft	$C_{h\alpha} = -.004$
Wing span = 35 ft	$C_{h\delta} = -.009$
$a_0 = .10$ per degree	$C_{h_0} = 0$
m.a.c. = 6.5 ft	
T.R. = 2:1	$S_e = 12 \text{ sq ft}$
Horiz. tail area = 40 sq ft	
Horiz. tail span = 13 ft	$c_e = .9 \text{ ft}$
$a_{0_t} = .10$ per degree	$\eta_t = .9$
Tail length = 16.2 ft	$G = .5 \text{ rad/ft}$
Incidence wing = +2°	
Incidence tail = +1°	$W = 8800 \text{ lb}$
$N_0 = .35c$ props windmilling	$d\epsilon/d\alpha = .48$

Calculate the gradient $(dF_s/dn)_{\text{Pull-up}}$ at sea level for c.g. locations of .20c, .26c, and 32c. Plot these values versus c.g. position, and by graphical interpolations or extrapolations determine the stick-free maneuver point. If it is required that this airplane never have a stick force per g gradient in excess of 8 lb per g and no less than 3 lb per g, what would be the c.g. limits to maintain these gradients with these limitations? Assume the tab is always used to trim out the stick force at $n = 1$.

7-6. For the airplane given in Problem 7-5 calculate and plot the variation of the gradient dF_s/dn with increase in altitude from sea level to 40,000 ft at the stick-free neutral point.

7-7. If the aerodynamic balance of the elevator for the airplane of Problem 6-5 is altered in such a way that $C_{h\alpha} = .002$ and $C_{h\delta} = -.002$, recompute and re-plot the variation of the gradient dF_s/dn with c.g. position and determine the new c.g. limits that will keep the stick force per g gradients within the limits specified in Problem 7-5.

7-8. For the airplane in Problem 7-5 calculate and plot stick force versus normal acceleration for acceleration from $n = 1$ to $n = 8$, for the case of straight pull-ups and then for the case of steady level turns at the same accelerations, both for c.g. locations of .26c.

7-9. If the airplane of Problem 7-5 is equipped with a bobweight in the elevator system tending to push the elevator down with a force equivalent to 10 lb at the stick at one g, calculate the shift in the stick-free maneuver point, N_m'.

7-10. If the airplane of Problem 7-5 is equipped with a down spring in the elevator system tending to pull the elevator down with a force equivalent to 10 lb at the stick, calculate the shift in stick-free maneuver point.

CHAPTER 8

DIRECTIONAL STABILITY AND CONTROL

8-1 Introduction

The problems discussed in Chapters 5 and 7 are those relating to the stability of the airplane when flying in equilibrium on symmetrical flight paths, and those required to provide the pilot with controls that will permit him to establish any flight path desired, within the aerodynamic and structural limits of the airplane. With the assumption of flight path symmetry, the problem reduced to that of providing control over the airplane's angle of attack and thereby its lift coefficient, and insuring static stability at this angle of attack. These problems are all classed under the general heading of static longitudinal stability and control.

It is now necessary to investigate the characteristics of the airplane when its flight path no longer lies in the plane of symmetry. This means that the relative wind will be making some angle to the airplane's center line, and this angle will be referred to as the angle of sideslip, β. The airplane's angle of sideslip is just like the airplane's angle of attack, except that it lies in another plane and its action is considerably different. The airplane's angle of attack determines the airplane's lift coefficient and very closely its speed of flight, whereas sideslip, on the whole, is quite useless. Sideslip can be used to increase the airplane's drag and thereby its flight path angle during an approach for a landing, may also be useful in accomplishing smooth acrobatics such as slow rolls, and finally can help out during flight with asymmetric power. However, in general it can be said that in almost all flight conditions it is advantageous to maintain zero sideslip, and the easier it is for the pilot to hold zero sideslip the better he will like the flying qualities of his airplane.

The problem of directional stability and control is first to insure that the airplane will tend to remain in equilibrium at zero sideslip and second to provide a control to maintain zero sideslip during maneuvers that introduce moments tending to produce sideslip.

In order to discuss these problems at all, it is necessary to develop some terminology that will permit the complete description of the air-

plane during any maneuver that involves sideslip.   This problem comes
under the general heading of lateral stability and control and involves
the three antisymmetric degrees of freedom, (i.e., angular displacement
or velocity about the X and Z axes and translations along the Y axis).
Figure 8–1 shows the general notation.

The angle of sideslip, β, is equal to $\sin^{-1}(v/V)$ or for the small angles
encountered in normal flight $\beta = v/V$.   It should be noted that, for the

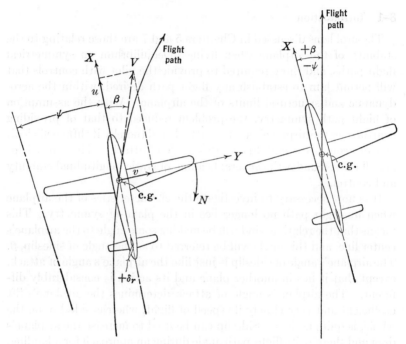

FIGURE 8–1.   Asymmetric flight notation.   FIGURE 8–2.   The straight slip.

curved flight paths shown, the angle of yaw, ψ, is defined as the angular
displacement of the airplane's center line from some azimuth direction
taken as zero at some given instant of time, and ψ does not equal β.
For instance in a 360-degree turn the airplane has yawed 360 degrees,
but may have had no sideslip during the entire maneuver.   For sideslips during which a straight path is maintained, the angle of yaw, ψ,
is equal in magnitude but opposite in sign to the angle of sideslip, β.
(See Figure 8–2.)

Considerable confusion has arisen because of the misunderstanding
of the terms yaw and sideslip.   In most cases the term yaw is used to
define the angle of the relative wind to the plane of symmetry, which

although incorrect in certain phases of the problem is perfectly permissible if the straight flight path conditions are maintained. Yaw, for instance, is almost always used in wind-tunnel work, while sideslip is almost always used in flight test work. In this book yaw will be used as the independent variable as it fits more closely to the majority of present-day data.

8-2   Static Directional Stability Rudder-fixed

Just as in the longitudinal case, where the static stability is defined as the tendency of the airplane to return to a given equilibrium angle of attack or lift coefficient when disturbed, the static directional stability of the airplane is its tendency to develop restoring moments when disturbed from its equilibrium angle of sideslip, usually taken as zero.

The static problem, then, is the study of the airplane's yawing moments developed because of sideslip or yaw, to see if the yawing moment so developed will tend to reduce the sideslip or increase it.

The static directional stability can be developed as was the static longitudinal stability, by summing up the stability contributions of the component parts of the airplane. Each of these components produces yawing moments when flying at angles of sideslip, and the study of the variation of the total yawing moment with angle of sideslip or angle of yaw gives the magnitude of the directional stability.

As has been done in all other phases of aerodynamics the yawing moment coefficient will be discussed instead of the dimensional yawing moment. The coefficient is obtained by dividing the yawing moment in pound feet by the dynamic pressure, q, in pounds per square foot, the wing area, S_w, in square feet, and the wing span, b, in feet.

$$C_n = \frac{N}{qS_wb}$$

The directional stability of the airplane can be assessed if a curve of yawing moment coefficient, C_n, with angle of yaw, ψ, is obtained for any given airplane. A negative slope of this curve is required for static directional stability. A typical stable and unstable directional curve is shown in Figure 8–3. The derivative $dC_n/d\psi$ will be given in the short-hand notation $C_{n\psi}$ and will be given per degree.

The final directional stability curves shown in Figure 8–3 are built up from contributions of many parts of the airplane. In order to obtain the final variation analytically, the magnitude of the contributions from the major components must be developed analytically and then summed up.

The contribution of the wing to the airplane's stability is very small. The factor that affects the contribution the most is its angle of sweepback, Λ. Swept-back wings have slight directional stability, while swept-forward wings are slightly unstable. The stability contribution of straight wings is almost negligible in comparison to the contributions of the other parts of the airplane.

It is interesting to note that the wing's contribution to directional stability is very small, whereas for the longitudinal case it was by far the most important factor. This, of course, is due to the fact that an

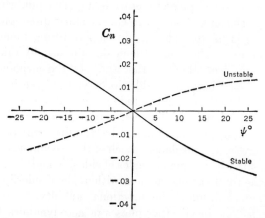

FIGURE 8-3. Typical wind-tunnel curves C_n versus ψ.

angle of sideslip creates only very small cross wind forces on the wing, whereas an angle of attack can produce very large lift forces. In the longitudinal case the large wing forces require that the center of gravity be close by in order to establish equilibrium and make the longitudinal equilibrium and stability very sensitive to slight movements of the center of gravity. In the directional sense, center of gravity movement due to loading changes has a negligible effect on the directional stability.

The wing contribution to directional stability is in many cases negligible. The small contribution of the swept wing can be estimated to a first approximation from the following formula:

$$(C_{n\psi})_{Wing} = -.00006 \ (\Lambda°)^{\frac{1}{2}} \tag{8-1}$$

where Λ is the angle of sweep of the quarter chords in degrees (positive for sweepback), the slope $C_{n\psi}$ also per degree.

The contribution of the fuselage and nacelles to the directional stability of the airplane is usually unstable and certainly one of the major effects. Under the assumption that a fuselage is a slender body of revolution, it was shown in Chapter 5 that the moment created by the potential flow of an ideal fluid at an angle of attack to the body centerline is unstable and is a function of the dynamic pressure, q, and the volume of the fuselage in cubic feet.

$$\frac{dM}{d\alpha} = \frac{(\text{Volume})\, q}{28.7} \qquad (8\text{--}2)$$

This relationship was first established by Munk, who also established the correction factor $(K_2 - K_1)$ given in Figure 5–13. The yawing moment due to an angle of sideslip is exactly the same. Therefore, expressed directionally:

$$\left(\frac{dN}{d\psi}\right)_{\substack{Fus \\ or\ Nac}} = \frac{(\text{Volume})\, q}{28.7} (K_2 - K_1) \qquad (8\text{--}3)$$

The yawing moment of the fuselage at an angle of yaw is a great deal easier to estimate than the pitching moment due to an angle of attack of the fuselage. The reason is that longitudinally the fuselage operates in a very complicated flow regime including upwash ahead of the wing and downwash behind the wing. Directionally it is assumed that the wing does not disturb the sideslip angles and therefore local values of sideslip need not be computed.

If the fuselage contribution to directional stability is expressed in terms of the rate of change of yawing moment coefficient per degree angle of yaw, and the fuselage volume obtained by summing up segments chosen for ease of estimation of the volume:

$$C_{n\psi} = \frac{\pi (K_2 - K_1)}{114.6 S_w b} \int_0^l w_f^2 \, dx \qquad (8\text{--}4)$$

where w_f is taken for each segment such that $(\pi w_f^2/4) \, \Delta x$ gives a close approximation to the segment volume. The fuselage gives a large unstable contribution that for normal airplane configurations varies from .0006 to .0012.

Many empirical approaches to the estimation of the fuselage or nacelle yawing moments have been made during the past few years. One of the best of these is the empirical formula developed by the aerodynamics group of the North American Aviation Company during World War II. This formula is as follows:

$$(C_{n\psi})_{\substack{Fus \\ or\ Nac}} = \frac{.96 K_\beta}{57.3} \left(\frac{S_s}{S_w}\right)\left(\frac{L_f}{b}\right)\left(\frac{h_1}{h_2}\right)^{\frac{1}{2}} \left(\frac{w_2}{w_1}\right)^{\frac{1}{3}} \qquad (8\text{--}5)$$

where S_s is the projected side area in square feet, L_f is the over-all fuselage length, and the dimensions h_1, h_2, w_1, w_2 can be obtained by referring to the sketch in Figure 8-4. The values of the empirical constant K_β are given in Figure 8-4 as functions of fuselage fineness ratio, L_f/h, and location of the c.g. on the body, d/L_f.

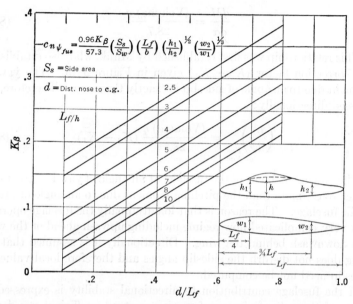

FIGURE 8-4. Fuselage directional stability coefficient. From unpublished data of North American Aviation, Inc.

The directional stability of the combination of wing plus fuselage is usually slightly different from the sum of the two components obtained separately. This is due to the interference flow created at the wing-fuselage juncture. This interference effect is usually slightly stabilizing, but the contribution is never much larger than $\Delta C_{n\psi} = -.0002$ and in the usual case less than $-.0001$. The position of the wing on the fuselage has the major effect, with the interference effect less stabilizing for a low wing. The available experimental data do not allow a very accurate estimation of this interference effect, but the following stability increments are suggested for first estimates.

WING CONFIGURATION	$\Delta_1 C_{n\psi}$
High wing	$-.0002$
Mid wing	$-.0001$
Low wing	0

The running propeller can have large effects on the airplane's directional stability, destabilizing if a tractor and stabilizing if a pusher propeller.   The propeller contribution to directional stability arises from the side force component at the propeller disc created because of yaw as discussed in Chapter 5. (See Figure 8–5.)

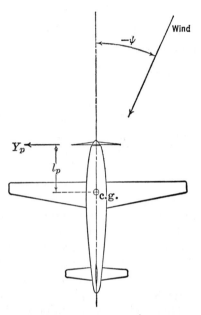

The yawing moment coefficient produced by the side force on the propeller at some angle of yaw is simply:

$$C_n = \frac{Y_p \times l_p}{qS_w b} \qquad (8\text{–}6)$$

The cross wind force, Y_p, can be expressed in terms of the side force coefficient, C_{Y_p}, where $C_{Y_p} = \dfrac{Y_p}{q(\pi D^2/4)}$.   The stability contribution of the running propeller can be obtained by differentiating (8–6) with respect to ψ in degrees.

$$C_{n\psi} = \frac{\pi D^2 l_p \left(\dfrac{dC_{Y_p}}{d\psi}\right) N}{4S_w b} \qquad (8\text{–}7)$$

FIGURE 8–5. Propeller side force moment.

where N is the number of propellers, and l_p the distance from propeller disc to the center of gravity.

The derivative $dC_{Y_p}/d\psi$ is the same as the derivative $(dC_N/d\alpha_p)$ discussed in Chapter 5.   The value of this derivative depends on the characteristics of the propeller, the number of blades, blade angle, etc. This factor can be estimated from propeller data in the literature. For a first estimate of the contribution of the windmilling propeller to the airplane's directional stability the following values of $(dC_{Y_p}/d\psi)_{T_c=0}$ can be used:

	$(dC_{Y_p}/d\psi)_{T_c=0}$
Two-bladed propeller	.00165
Three-bladed propeller	.00235
Four-bladed propeller	.00296
Six-bladed counter rotating	.00510

The directional stability contribution of the propeller with full power is roughly one and a half times that for the windmilling propeller. The directional stability contribution of the propeller at full power can then be roughly estimated as

$$(C_{n\psi})_{p\text{Full power}} = 1.5 \, (C_{n\psi})_{p\text{Props windmilling}} \tag{8-8}$$

The total value of the directional stability, including all the contributions discussed to this point, is as follows:

$$(C_{n\psi})_{Airplane} = (C_{n\psi})_{Wing} + (C_{n\psi})_{\substack{Fus \\ Nac}} + (C_{n\psi})_{Prop} + \Delta_1(C_{n\psi}) \tag{8-9}$$

The summation of the terms in (8-9) usually gives a directionally unstable airplane. In some special cases, however, an airplane with only these components can be made stable. For example, a pure flying wing type airplane with swept-back wings, pusher propellers, and no fuselage will have a slight stability contribution from the wings and pusher propellers and will be directionally stable without any additional stabilizing surfaces. In the normal airplane arrangement with a large fuselage and tractor propellers, the over-all airplane's directional stability as given in (8-9) must necessarily be unstable, and an additional stabilizing surface must be incorporated not only to overcome the instability of the other parts of the airplane, but also to give the desired level of directional stability. This stabilizing surface is the normal vertical tail placed as far aft of the c.g. as practicable.

The stability contribution of the vertical tail surface can be computed as follows. The yawing moment produced by the vertical tail at some angle of yaw can be considered simply as the lift created on the vertical tail, L_v, multiplied by the arm from the aerodynamic center of the vertical tail to the airplane's center of gravity, l_v. See Figure 8-6.

$$N = -L_v \times l_v \tag{8-10}$$

or $$N = - \left(\frac{dC_L}{d\psi}\right)_v \psi q_v S_v l_v \tag{8-11}$$

The yawing moment coefficient, C_n, may be obtained from (8-11) by dividing through by qS_wb, giving:

$$C_n = - \left(\frac{dC_L}{d\psi}\right)_v \psi \frac{S_v}{S_w} \frac{l_v}{b} \frac{q_v}{q} \tag{8-12}$$

The derivative $dC_{Lv}/d\psi$ is not exactly the slope of the lift curve of the vertical tail with angle of attack, a_v, due to the sidewash, σ, created by the wing-fuselage combination at an angle of yaw. The sidewash due to the fuselage arises from the cross wind force created

on the fuselage at some angle of yaw, and the sidewash from the wing
vortex sheet arises as a result of the inboard motion of the air above the
vortex sheet and outboard motion of the air below the vortex sheet. The
sidewash from the fuselage gives a destabilizing flow in the airstream
beside the fuselage, but a stabilizing flow to the air above the fuselage
wake and practically no sidewash to the flow below the wing-fuselage

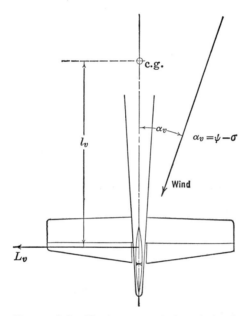

FIGURE 8–6. Yawing moment of vertical tail.

intersection. Almost the entire sidewash comes from the fuselage, as
the sidewash flow due to the wing wake is almost negligible. The
result of this indicates that for low-wing airplanes a stabilizing effect
should be encountered because of the sidewash, whereas a high-wing
design should have very small sidewash influence.

The effective angle of attack of the vertical tail can be given as:

$$\alpha_v = \psi - \sigma \qquad (8\text{--}13)$$

and
$$\frac{d\alpha_v}{d\psi} = 1 - \frac{d\sigma}{d\psi} \qquad (8\text{--}14)$$

Making use of (8–14), equation (8–12) becomes

$$C_n = -a_v \left(1 - \frac{d\sigma}{d\psi}\right) \psi \frac{S_v}{S_w} \frac{l_v}{b} \eta_v \qquad (8\text{--}15)$$

The directional stability contribution of the vertical tail can be obtained from (8–15) by differentiating with respect to ψ:

$$C_{n\psi} = -a_v \left(1 - \frac{d\sigma}{d\psi} \right) \frac{S_v}{S_w} \frac{l_v}{b} \eta_v \qquad (8\text{–}16)$$

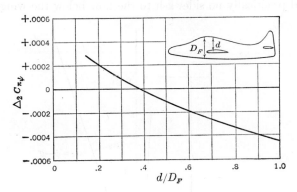

FIGURE 8–7. Interference correction to $C_{n\psi}$.

This expression can be broken down as follows:

$$C_{n\psi} = -a_v \frac{S_v}{S_w} \frac{l_v}{b} \eta_v + a_v \frac{d\sigma}{d\psi} \frac{S_v}{S_w} \frac{l_v}{b} \eta_v \qquad (8\text{–}17)$$

or
$$C_{n\psi} = (C_{n\psi})_v + \Delta_2 C_{n\psi} \qquad (8\text{–}18)$$

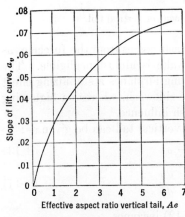

FIGURE 8–8. Slope of lift curve, vertical tail. From NACA TN 775, "Analysis of Wind-tunnel Data on Directional Stability and Control," by H. R. Pass.

The term $\Delta_2 C_{n\psi}$ is that part of the contribution of the vertical tail surface to the airplane's directional stability that arises from the sidewash or interference flow from the wing-fuselage combination. An estimate of this factor can be obtained from the curve given in Figure 8–7.

The slope of the lift curve of the vertical tail can be obtained from Figure 8–8, which is based on the effective aspect ratio of the vertical tail surface.

The effective aspect ratio of the vertical tail surface is greater than the geometrical aspect ratio, defined as $b_v{}^2/S_v$, because of the end plate effect of the horizontal tail on the vertical tail if the vertical tail is mounted over the horizontal

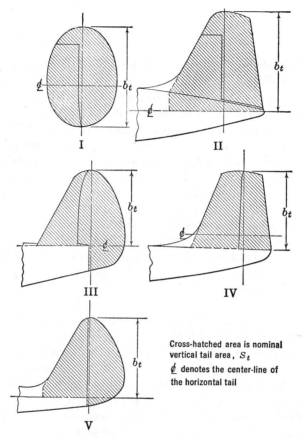

Cross-hatched area is nominal vertical tail area, S_t

$\cancel{}$ denotes the center-line of the horizontal tail

FIGURE 8-9.   Vertical tail definitions.   From NACA TN 775, "Analysis of Wind-tunnel Data on Directional Stability and Control," by H. R. Pass.

tail in the plane of symmetry.   The effective aspect ratio of the vertical tail, A_e, can be estimated as follows:

$$A_e = 1.55 \frac{b_v{}^2}{S_v} \qquad (8\text{--}19)$$

The span and area of the vertical tail surface are usually quite difficult to determine because of the many different types developed during the past years.   The NACA,* however, has laid down some-what arbitrary dimensions for several vertical tail types shown in Figure 8-9.

* NACA TN 775.

The final estimate of the directional stability of the airplane can be obtained from a summation of the various factors discussed heretofore:

$$(C_{n\psi})_{Airplane} = (C_{n\psi})_w + (C_{n\psi})_{Fus \atop Nac} + (C_{n\psi})_{Prop} + (C_{n\psi})_v$$
$$+ \Delta_1 C_{n\psi} + \Delta_2 C_{n\psi} \quad (8\text{-}20)$$

The desirable level of directional stability in terms of the criterion $C_{n\psi}$ is very difficult to express in general terms. There have been practically no cases reported of airplanes having too much directional stability. Usually, the more directional stability built into the airplane, the better it will handle and the more the pilot will like to fly it. In one or two cases there has been a suspicion that some undesirable lateral dynamics could be traced to too much directional stability, but this has never been very well proven. For normal airplane configurations directional stability levels of $C_{n\psi}$ between $-.0015$ and $-.0020$ have been considered good, with $C_{n\psi} = -.0005$ a lower limit. Most fighter-type aircraft operating in World War II had directional stabilities between $-.0008$ and $-.0010$. The use of any upper and lower limit to the derivative $C_{n\psi}$ is somewhat dangerous when the airplane configuration differs from the normal by wide margins. This is due to the yawing moment coefficient being obtained by dividing the yawing moment by the wing area and wing span. The wing, as has been discussed, has only a very small contribution to directional stability, and therefore airplanes with high wing loadings and low aspect ratios would tend to have high slopes $C_{n\psi}$, although their absolute stabilities in terms of actual yawing moments would be the same. It can be argued that the increase in slope $C_{n\psi}$ with decrease in wing area and span, everything else held constant, is justified, as it is the wing that creates the disturbing moments that must be opposed by the directional stability, and therefore the apparent increase in stability is justified. It is felt, however, that some account should be taken of the effect of very high wing loadings and very short wing spans on the desirable level of the derivative $C_{n\psi}$. To this end the following formula is suggested:

$$(C_{n\psi})_{desirable} = -.0005 \left(\frac{W}{b^2}\right)^{\frac{1}{2}} \quad (8\text{-}21)$$

Wind-tunnel curves of yawing moment coefficient, C_n, versus angle of yaw, ψ, usually show this variation to be quite linear, even up to rather large angles of yaw. At very large angles of yaw the stability contribution of the vertical tail starts to fall off because of stalling of this surface; however, this is somewhat offset by the fact that the unstable moments of the fuselage diminish and even change sign to a stable contribution at large angles of yaw. It will be shown later that the tendency of the fuselage to give stable moments at large angles of

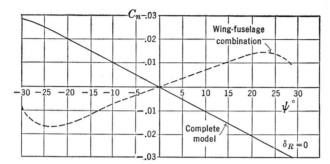

FIGURE 8-10.   Typical wind-tunnel curves.

yaw can be used to advantage, if accentuated by a dorsal or ventral fin, in the reduction of the undesirable phenomenon of rudder lock. Typical wind-tunnel curves of C_n versus ψ are shown in Figure 8-10 for fuselage alone and for the complete airplane.

8-3   Directional Control

Although the airplane, by virtue of its symmetry, will normally be in equilibrium at zero sideslip, there are many flight conditions or maneuvers that introduce yawing moments that must be opposed by some yawing moment control, if the condition of zero sideslip is to be maintained. This yawing moment control is supplied the pilot by means of a rudder, normally a plain flap making up the aft portion of the vertical tail. The magnitude of the rudder power required varies with different types of airplanes and must be analyzed separately for each particular case.

If the airplane has very high directional stability, a tricycle landing gear, and light wing loading, it is quite possible to do away with the rudder altogether, making a very safe non-spinnable airplane. However, on most airplanes the rudder control is a definite requirement and must be designed to provide the yawing moment required for the particular type involved.

The flight condition or maneuvers that introduce yawing moments that must be overcome with the rudder control are as follows:

a. Adverse yaw—When the airplane is rolled into a turn, the rolling control, as will be discussed fully in Chapter 9, together with the in-clination of the lift vectors on the rolling wing, creates yawing moments which will produce sideslip. These yawing moments must be over-come by the rudder. The critical condition is at high lift coefficient with full rolling control. The use of rudder to maintain zero sideslip in turns is called coordinating the turn.

b. Slipstream rotation—The slipstream behind the propeller has a rotational component which changes the angle of attack of the vertical

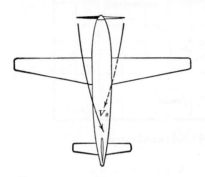

tail and will create sideslip if un-corrected by the rudder. The critical condition for slipstream rotation is for high power at low speed. See Figure 8–11. Counter-rotating propellers, of course, ob-viate this factor.

c. Cross winds during take-off and landings—The rudder should be powerful enough to hold the sideslip required to make cross wind take-offs and landings at least to speeds within 20 per cent of the stalling speed. Below this speed the brakes and/or power can be used as the directional control.

FIGURE 8–11. Slipstream rotation.

d. Spinning—In most cases of high-performance airplanes, the rudder control is the major recovery control. Some light planes can-not be made to spin unless the rudder is hard over and snap right out of the spin as soon as the rudder is released.

e. Antisymmetric Power—In multiengined airplanes, the failure of one engine at low air speeds will create a heavy yawing moment that must be overcome by the rudder in order to maintain flight at zero slip. This is nearly always the design condition for the rudder on multi-engined airplanes.

In the light of the above requirements for the rudder control the following design conditions can be chosen.

For single-engine airplanes having very high directional stability the sideslip due to the adverse yaw can be made so low as to be un-objectionable, placing no requirement on the rudder. If a tricycle gear is employed, the cross wind take-off and landing requirement on the rudder can be neglected. If counter-rotating propellers are used or if the slipstream does not cover the vertical tail, the slipstream rotation requirement is obviated. Finally, if the airplane cannot be made to spin without rudder deflection, it would be foolish to put a rudder on the airplane. If the airplane fits the above requirements, a strong case for a rudderless or two-control airplane can be made.

For fighter or other high-performance airplanes having a single engine with one propeller, the slipstream rotation and spinning requirements design the rudder, whereas if the airplane is equipped with counter-rotating propellers, the spin recovery requirements usually design the rudder.

For all multiengined airplanes, the antisymmetric power requirements design the rudder power.

In all cases high directional stability is desirable, but the rudder power required must be analyzed for each particular case.

The rudder power can be very readily developed theoretically in much the same manner as the elevator power was developed in Chapter 5. The yawing moment due to deflecting the rudder is given as

$$N = -L_v \times l_v \tag{8–22}$$

or in coefficient form:

$$C_n = -\frac{C_{L_v} S_v l_v q_v}{q S_w b} \tag{8–23}$$

The rate of change of yawing moment coefficient per degree change in rudder angle is obtained from (8–23) by differentiating with respect to rudder deflection, δ_r, remembering that the control surface effectiveness factor $d\alpha_v/d\delta_r = \tau$ as given in Chapter 5.

$$\frac{dC_n}{d\delta_r} = -a_v \tau \frac{S_v}{S_w} \frac{l_v}{b} \eta_v \tag{8–24}$$

On normal airplane configurations, the rudder power varies considerably between airplanes, but a good mean value can be taken as $C_{n\delta r} = -.001$. If the directional stability of the airplane is at the mean level of $C_{n\psi} = -.001$, it can be readily seen that 1 degree of rudder will produce 1 degree of yaw. The criterion of 1 degree yaw for 1 degree of rudder deflection has been used as a criterion for rudder design, but as discussed above this is a very poor over-all criterion for all airplanes.

For most rudders, deflections are usually limited to plus or minus 30 degrees, for the effectiveness of the rudder falls off abruptly past this angle. Typical wind-tunnel tests for rudder effectiveness are shown in Figure 8–12, indicating the start of the fall-off in effectiveness as full rudder deflection is reached.

The effectiveness of the rudder varies considerably with speed for full power operation if the vertical tail is in the slipstream. This is due to the effect of the slipstream velocity, V_s, which at low speed gives a high ratio of η_v. Rudders located in the prop wash will be very effective at low speeds and high power, and will lose effectiveness as the speed of the airplane is increased.

The rudder power required to overcome the adverse yaw during rolling maneuvers is usually not very high and very seldom designs the rudder. Adverse yaw is created, as mentioned briefly before, by the normal action of the aileron or lateral control, together with the yawing moment created as a result of the rolling of the wing itself.

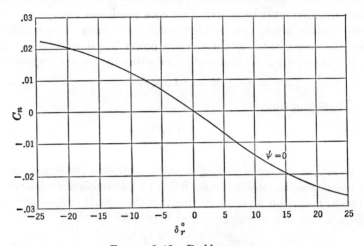

FIGURE 8-12. Rudder power.

These adverse yawing moments are always critical at low speeds, and the rudder must be capable of overcoming these adverse moments at speeds very close to the stall. A typical flight test of the effects of adverse yaw and the magnitude of the rudder deflection to overcome these effects is shown in Figure 8-13. It can be seen in a that when the rudder is held locked and the airplane rolled by abrupt full aileron deflection, the airplane will develop a sideslip angle of 15 degrees because of the adverse yawing moment. From Figure 8-13 it can be seen that a rudder deflection of 20 degrees is required to maintain approximately zero sideslip. As these tests were run at a speed very close to the stalling speed of the airplane, the rudder is quite adequate to meet the condition of adverse yaw. Figure 8-14 shows similar time histories for abrupt rolls at cruising speed. The much smaller rudder deflection for coordination is apparent.

In order to analyze the rudder design, some estimate must be made of the adverse yawing moment anticipated from the lateral controls. If a wind-tunnel model of the airplane is available, the adverse yawing

$V_c = 120$ mph

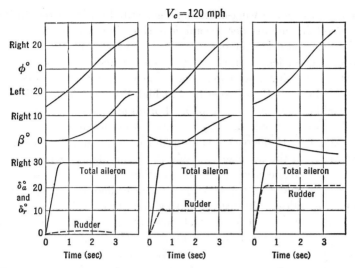

FIGURE 8-13.   Typical time histories of abrupt aileron rolls (low speeds).

$V_c = 240$ mph

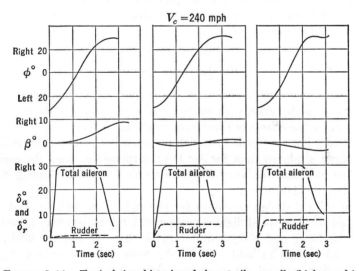

FIGURE 8-14.   Typical time histories of abrupt aileron rolls (high speeds).

moment due to the controls can be measured.   The adverse yawing moment due to the rolling wing can be estimated from the following formula (8-25):

$$C_n = - \frac{C_L}{8} \frac{pb}{2V} \qquad (8\text{-}25)$$

where $pb/2V$ is the wing-helix angle generated during a roll (p is the
rate of roll in radians per second, b is the wing span in feet, and V
is the true speed in feet per second). The factor $pb/2V$ will be discussed at great length in Chapter 9, but a value of $pb/2V = .08$ can
be used for a first approximation. The yawing moment due to the
rolling velocity is therefore critical at high lift coefficients and runs in
the neighborhood of $C_n = \pm.015$. For an airplane whose directional
stability was normal, say $C_{n\psi} = -.001$, the airplane would develop
approximately 15 degrees of sideslip during a severe rolling maneuver
resulting from the adverse yaw due to rolling alone. On top of this
must be added the adverse yaw due to the rolling control itself, which
will usually add $C_n \cong \pm.005$ for standard ailerons and require the
airplane to sideslip 20 degrees during the abrupt rolling maneuvers.
If the rudder is to be made just powerful enough to overcome this
adverse yawing moment, it must produce a $C_n = \pm.020$ for 30 degrees
of rudder throw or $dC_n/d\delta_r = -.00067$. As the average rudder usually
has a rudder power equal to $dC_n/d\delta_r = -.001$, adverse yaw, up to
the present, has not been a design problem on the rudder.

The rudder power required to overcome adverse yaw can be expressed analytically as follows:

$$C_{n\delta r} = \frac{C_n \text{ (roll)} + C_n \text{ (lat. control)}}{\text{Rudder throw (not over 30°)}} \qquad (8\text{-}26)$$

A more severe requirement on the rudder, for single-engine high-
performance airplanes, with single rotation propellers is the need for
overcoming the effects of slipstream rotation. This condition is critical at very low air speeds with high power, being usually more critical
for Navy airplanes than for Air Force airplanes because the carrier
landing technique for Navy airplanes requires full-stall landings with
full flaps at high power.

The critical design condition for the rudder on multiengined airplanes is the low-speed flight condition with full antisymmetric power.
The Air Force and Navy requirement is that the rudder should be
powerful enough to hold zero sideslip with the most critical engine
windmilling and all other engines delivering full power down to
1.2 times the airplane's stalling speed in the take-off configuration.
The yawing moment coefficient due to the antisymmetric thrust can
be given as follows (see Figure 8–15):

$$C_{nT} = \frac{375BHP\ \eta_p l_y}{V_{\text{mph}}qS_w b} \qquad (8\text{-}27)$$

This is a cubic in V_{mph} with the yawing moment coefficient increasing inversely with V^3. A typical curve of C_{nt} versus V_{mph} is shown in Figure 8-16.

The rudder at full throw gives a constant yawing moment coefficient. The intersection of the two curves shown in Figure 8-16 is

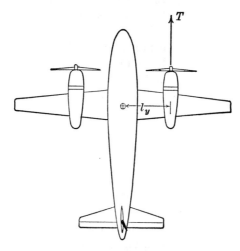

FIGURE 8-15.   Yawing moment due to asymmetric power.

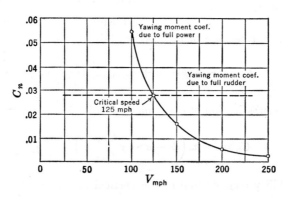

FIGURE 8-16.   Critical speed due to asymmetric power.

the speed below which full rudder will not balance out the moment due to antisymmetric power.

The spinning and cross wing take-off and landing requirements on the rudder are usually not critical, and most rudders are designed only to fulfill the more critical requirements stated above.

8-4 Stick-free Directional Stability

When the rudder is left free to float in response to its hinge moments, it can have large effects on the directional stability of the airplane, in the same manner that freeing the elevator was shown to have large effects on the longitudinal stability.

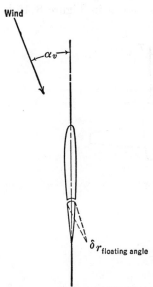

The floating angle of the rudder can be expressed analytically in terms of the two hinge moment coefficient parameters $C_{h_{av}}$ and $C_{h_{\delta r}}$.

$$\delta_{r\text{Floating}} = -\frac{C_{h_{av}}}{C_{h_{\delta r}}}\alpha_v \qquad (8\text{-}28)$$

If the airplane sideslips, the restoring moment due to the tail will be decreased if the rudder floats with the wind and will be increased if the rudder floats against the wind. The floating rudder changes the effective angle of attack of the vertical tail, Figure 8–17.

FIGURE 8–17. The floating rudder.

$$\alpha_{v\text{Effective}} = \psi + \tau\delta_{r\text{Floating}} - \sigma \qquad (8\text{-}29)$$

The restoring yawing moment coefficient developed because of the vertical tail will be:

$$C_n = -a_v(\psi - \sigma + \tau\delta_{r\text{Floating}})\frac{S_v}{S}\frac{l_v}{b}\eta_v \qquad (8\text{-}30)$$

which upon substitution of equation (8–28) for $\delta_{r\text{Floating}}$ gives

$$C_n = -a_v\left(\psi - \sigma - \tau\frac{C_{h_{av}}}{C_{h_{\delta r}}}\psi\right)\frac{S_v}{S}\frac{l_v}{b}\eta_v \qquad (8\text{-}31)$$

The stability contribution of the vertical tail with a free rudder is therefore:

$$(C_{n_\psi})_{v\text{Free rudder}} = -a_v\frac{S_v}{S_w}\frac{l_v}{b}\eta_v\left(1 - \frac{C_{h_{av}}}{C_{h_{\delta r}}}\tau\right) + \Delta_2 C_{n_\psi} \qquad (8\text{-}32)$$

For high-speed airplanes for which close aerodynamic balance of the rudder is required in order to make the rudder pedal forces within the pilot's force limitations, it is essential that the ratio of $C_{h_{av}}/C_{h_{\delta r}}$

be kept very low so as not to lose too much directional stability stick-free.

The stick-free directional stability of the airplane is made manifest to the pilot by the pedal force required to produce sideslip at constant air speed. If this gradient is too low through zero sideslip, it is very difficult for the pilot to hold zero sideslip during maneuvers, and the airplane's directional feel will be objectionable to the pilot. On the other hand, if the gradient of pedal force versus sideslip is high, the airplane will feel stiff directionally to the pilot, and in general he will consider the airplane quite satisfactory from this point of view.

The analytical treatment for the rudder pedal force versus sideslip gradient simulates very closely the development given for longitudinal stability in Chapter 5.

The equation for pedal force, PF, in pounds, in terms of the rudder hinge moments and mechanical gearing, G, in radians per foot, is as follows:

$$PF = G \times HM \qquad (8\text{-}33)$$

If the hinge moment is expressed in terms of the hinge moment parameters, with right pedal force considered positive,

$$PF = G(C_{h_{\alpha v}}\psi + C_{h_{\delta r}}\delta_r + C_{h_{\delta t}}\delta_t)q\eta_v S_r \bar{c}_r \qquad (8\text{-}34)$$

The rudder angle required to produce sideslip can be written simply as

$$\delta_r = \frac{d\delta_r}{d\psi}\psi \qquad (8\text{-}35)$$

where

$$\frac{d\delta_r}{d\psi} = -\frac{(C_{n\psi})_{\text{Rudder fixed}}}{C_{n_{\delta r}}} \qquad (8\text{-}36)$$

Substituting into (8–34), the pedal force equation becomes

$$PF = Gq\eta_v S_r \bar{c}_r \left[C_{h_{\alpha v}}\psi - C_{h_{\delta r}}\frac{(C_{n\psi})_{\text{Fix}}}{C_{n_{\delta r}}}\psi + C_{h_{\delta t}}\delta_t \right] \qquad (8\text{-}37)$$

differentiating with respect to ψ and rearranging gives the required gradient.

$$\frac{dPF}{d\psi} = -\frac{Gq S_r \bar{c}_r \eta_v C_{h_{\delta r}}}{C_{n_{\delta r}}}(C_{n\psi})_{\text{Free}} \qquad (8\text{-}38)$$

The gradient of pedal force versus sideslip varies with the velocity squared for normal aerodynamic balance. A criterion of 5 lb per degree of sideslip at 150 mph indicated has been taken as a minimum level for this gradient.

The above formula holds only for linear hinge moment variation and constant directional stability, and this is true only from approximately $\pm 10°$ of sideslip or rudder deflection. At high angles of sideslip the well-balanced rudder, which floats very slightly with the wind at low angles, will start to lose its aerodynamic balance and float over rapidly with increase in sideslip. This condition becomes accentuated at very high angles if the vertical tail commences to stall, which moves the center of pressure of the vertical tail well aft and causes the rudder to float well over. A typical curve of the floating angle of the rudder versus sideslip is shown in curve a, Figure 8–18, for a closely balanced rudder.

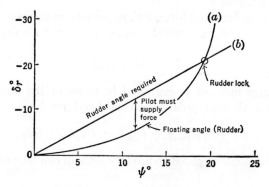

FIGURE 8–18.   Rudder lock.

This tendency of most rudders to float rapidly at high angles of sideslip leads to the phenomenon known as rudder lock. The rudder angle required to produce the sideslip varies somewhat linearly up to rather high angles. The pedal force required of the pilot is a function of the difference between the required rudder angle and the floating angle. At high angles of sideslip the floating angle may catch up with the required angle, at which point the pedal force will be zero. If the sideslip is increased beyond this point, the pedal forces will reverse and the rudder will continue to deflect up to its stops. Considerable force is required of the pilot to break the lock and restore the airplane to zero sideslip.

The difficulty lies in the fact that the requirements on the rudder make it possible to develop large angles of sideslip with full rudder deflection. One way out of the rudder lock problem is to cut down the rudder effectiveness, thereby increasing the rudder deflection required at a given sideslip angle. This artifice can be used on most present-day airplanes, as they usually have more than enough rudder control. For more closely designed rudders on airplanes which encounter rudder lock, the addition of a dorsal fin helps the situation a great deal.

Dorsal fin

FIGURE 8-19. The dorsal fin.

The dorsal fin is an auxiliary fin, as shown in Figure 8-19. The dorsal fin seems to do two things at once. One of these is to increase the fuselage stability at high angles of sideslip, and the second effect is to reduce the tendency of the vertical tail to stall. The increase in directional stability at high sideslip angles will, of course, require more rudder angle for trim and thereby reduce the

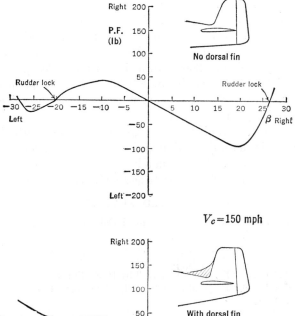

$V_c = 150$ mph

FIGURE 8-20.   Effect of dorsal fin on pedal force versus sideslip.

rudder lock possibilities. A typical example of the effect of a dorsal fin on the pedal force versus sideslip characteristics is shown in Figure 8–20.

The dorsal fin has no effect on the force gradient through zero yaw, its effect being noticed only at large yaw angles.

The final concern of the designer in regard to the directional characteristics of the airplane is pedal force required to handle the directional trim changes with changes of speed. It has been discussed earlier in this chapter how the slipstream rotation changes with changes in airplane speed and power. The rudder angles required to balance out the change in effective angle of attack of the vertical fin due to the change in slipstream rotation can require pedal forces that are extremely large and uncomfortable, and can also be very hard on the pilot while he is trying to maintain accuracy of fire during maneuvers requiring large changes in speed.

For example, a fighter aircraft climbing at full throttle at 180 mph V_i may push over into a dive up to 450 mph V_i. If the airplane is trimmed out directionally in the climb, the pilot may have to exert a pedal force as high as 200 lb to maintain zero sideslip in the dive. It is essential that the aerodynamic balance of the airplane's rudders be very carefully worked out to insure that these pedal force changes with change in speed are maintained very low.

The directional stability of the airplane, together with the design of the rudder, is one of the most important problems affecting the handling of the airplane. Unfortunately in the past, these have been neglected, and our tactical aircraft have suffered through this neglect.

SUGGESTED READING

1. NACA TN 775, "Analysis of Wind-tunnel Data on Directional Stability and Control," by H. R. Pass, 1940.
2. NACA WR L-25, "Notes on the Propeller and Slipstream in Relation to Stability," by H. S. Ribner, 1944.
3. NACA WR L-219, "Propellers in Yaw," by H. S. Ribner, 1943.
4. NACA WR L-336, "Proposal for a Propeller Side Force Derivative," by H. S. Ribner, 1943.
5. NACA WR L-217, "Formulas for Propellers in Yaw and Charts of the Side Force Derivative," by H. S. Ribner, 1943.
6. NACA TN 778, "Notes on the Stalling of the Vertical Tail Surfaces and on Fin Design," by Thompson and Gibruth, 1940.

PROBLEMS

8–1. An airplane is making a steady turn at constant altitude at an angle of bank of 60 degrees. If the air-speed indicator reads 200 mph, the altimeter reads 10,000 ft, and the accelerometer reads $2g$, what is the yawing velocity of the airplane in degrees per second?

8-2. An airplane is placed in a forward slip with the leading right wing held down by the ailerons to allow a straight flight path. If the directional gyro reads a change of 10 degrees, what is the angle of yaw and what is the angle of sideslip?

8-3. A model of a fighter-type jet-propelled airplane is tested in the wind tunnel for its directional stability characteristics, with the vertical tail off. The contributions of the various components to the airplane's directional stability were such that the slope of the yawing moment versus yaw curve was equal to $C_{n\psi} = +.0010$. If the vertical tail is to be positioned at a point on the aft end of the fuselage giving a tail length of 16 ft, with the vertical tail set over the horizontal tail, what vertical tail area is required to give an over-all stability level of $C_{n\psi} = -.0010$, assuming that this area will be proportioned to give a geometric aspect ratio of 2.0? The wing area of this airplane is 200 sq ft, and the wing span is 35 ft. The wing is set in approximately the middle of the fuselage.

8-4. The glider given in Problem 5–5 is to be evaluated for its static directional stability. If the specifications for this glider call for a directional stability level of $C_{n\psi} = -.0015$, will the airplane meet the specification, and if not, how should it be redesigned?

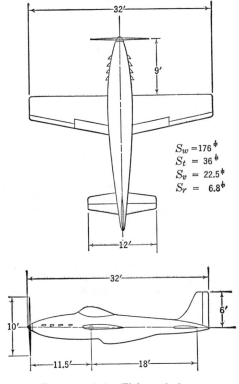

PROBLEM 8–5. Fighter airplane.

8-5. The fighter-type aircraft shown in the figure has the following aerodynamic and geometric characteristics. Estimate the static directional stability of this airplane for the propellers-windmilling condition.

Wing area	176 sq ft
Vertical tail area	22.5 sq ft
Rudder area	6.8 sq ft
Vertical tail airfoil section	NACA 0009
Rudder deflection	±25°
Propeller—four-bladed	Diameter 10 ft
Wing loading	40 lb/sq ft

8-6. What will the directional stability of the airplane in Problem 8–5 be for higher power condition?

8-7. What is the rudder power $(C_{n\delta_r})$ of the fighter airplane given in Problem 8–5? How many degrees of sideslip will be produced by a 1-degree increment in rudder deflection?

8-8. If the fighter airplane of Problem 8–5 is rolled by the application of full right aileron at an indicated airspeed of 150 mph at sea level, it will roll at an angular velocity of 68 degrees per second. The wind-tunnel tests of a model of this airplane indicate that a full deflection of the ailerons to the right will introduce an adverse yawing moment of $C_n = -.005$. How many degrees of rudder must be applied to keep the sideslip zero during this roll?

8-9. Plot the variation of yawing moment coefficient due to full asymmetric power versus airspeed at sea level for a twin-engined transport with the following characteristics:

Brake horsepower = 1200 per engine
Distance airplane centerline to centerline of engine = 14 ft
Wing area = 700 sq ft
Wing span = 70 ft
Assume η_p constant at 75 per cent

8-10. At what speed will the application of full rudder fail to hold the airplane given in Problem 8–9 at zero sideslip in the asymmetric power condition? Full rudder for this airplane creates a yawing moment coefficient of $C_n = \pm.02$.

8-11. What will be the value of the directional stability, stick-free if the rudder hinge moments for the airplane given in Problem 8–5 are as follows,

$$C_{h\alpha v} = -.004$$
$$C_{h\delta_r} = -.009$$

8-12. If the root-mean-square chord of the rudder of the airplane of Problem 8–5 is 1 ft and the gearing between the rudder pedals and the rudder is equal to $G = 1.25$ radians per foot, what will be the gradient of pedal force versus sideslip at 150 mph indicated speed? The hinge moment parameters are as given in Problem 8–11.

CHAPTER 9

DIHEDRAL EFFECT AND LATERAL CONTROL

9-1 Introduction

The discussion of airplane longitudinal and directional stability and control carried out in the past few chapters indicates that one of the determining factors of flight is the angle that the relative wind makes with the airplane, usually broken down into its major components, the angle of attack and angle of sideslip. It has been shown that if the pilot is given control over these angles, and if the airplane is made stable with respect to these angles, the airplane will be safe to fly on any symmetric flight path, within its aerodynamic or structural limits. Throughout this study, the wings of the airplane were considered to be at some arbitrary angle to the vertical (angle of bank, ϕ) usually taken as zero, with no regard to the aerodynamic problem of holding this angle of bank or bringing the airplane into this attitude.

It has been pointed out several times within this volume that a control over the angle of bank is necessary to provide a force to accelerate the flight path in the horizontal plane. It can be argued that with simple rudder control the airplane can be made to sideslip, thereby creating a cross wind or side force that can accelerate the flight path in the horizontal plane. However, this force is a very small one for modern aircraft, and totally inadequate for the rate of turn required.

The problem of holding the wings level or of maintaining some angle of bank is one of control over the rolling moments about the airplane's longitudinal axis. The major control over the rolling moments is the familiar aileron system consisting of flaps on the wing outer panels, which, when deflected asymmetrically, will alter the wings' spanwise lift distribution in such a way that a net rolling moment is created. A secondary control over the rolling moment can be obtained through control over the sideslip angle, as it will be shown that, for certain wing geometry, sideslip will alter the wings' spanwise lift distribution to create a net rolling moment. The phenomenon of rolling moment due to sideslip is termed dihedral effect and is not a static stability in the true sense of the word.

An airplane is said to have stable dihedral effect if a negative rolling moment (left wing down) is created as the result of positive sideslip, β. This definition is somewhat arbitrary but springs from the fact that stable dihedral effect is required for complete dynamic lateral stability, and that stable dihedral effect as defined will require the use of top rudder to pick up a wing that drops because of a gust or any other rolling disturbance.

The rolling moment will be discussed in coefficient form, with the rolling moment coefficient equal to the rolling moment, L, in pound-

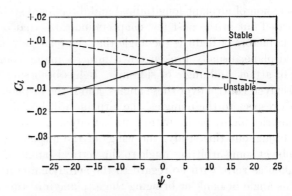

FIGURE 9-1.   Typical wind-tunnel tests for dihedral effect.

feet divided by the dynamic pressure, q, in pounds per square foot, the wing area, S_w, in square feet, and the wing span, b, in feet.

$$C_l = \frac{L}{qS_w b} \qquad (9\text{-}1)$$

The power of the lateral or aileron control will be expressed as the change in rolling moment coefficient per degree deflection of the ailerons, while the dihedral effect will be measured by the change in rolling moment coefficient per degree change in sideslip, β, or in wind-tunnel parlance, with change in yaw, ψ, the negative of β when straight flight paths are considered. The criterion of dihedral effect is the slope of the curve of rolling moment coefficient, C_l, plotted against yaw and is given as the derivative $dC_l/d\psi$ per degree, or in shorthand notation $C_{l\psi}$.

The airplane's dihedral effect will be discussed first, with the aileron control studied in later sections. Typical wind-tunnel curves of the variation of C_l with ψ for a stable and unstable airplane are shown in Figure 9-1.

9-2   Estimation of Airplane Dihedral Effect

The rolling moment due to sideslip (dihedral effect) is mainly created by wing dihedral angle, $\Gamma°$, which is positive for tip chord above the root chord.   In a sideslip the angle of attack of the forward wing will be higher than the angle of attack of the trailing wing.   This will create a lift on the leading wing that will be greater than the lift on the trailing wing, thereby creating a rolling moment about the X axis. See Figure 9-2.

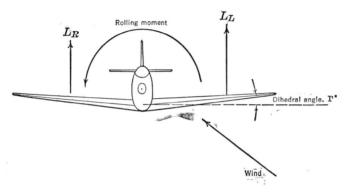

FIGURE 9-2.   Dihedral effect.

The dihedral effect, as mentioned before, is measured by the slope of the curve of rolling moment coefficient versus angle of yaw, C_{l_ψ}. The value of this derivative varies almost directly with wing dihedral angle at the approximate rate $\Delta C_{l_\psi} = .0002\Delta\Gamma°$, and an airplane having C_{l_ψ} equal to .0002 is said to have 1 degree of effective dihedral. The value of the dihedral effect C_{l_ψ} can be stated as follows:

$$C_{l_\psi} = C_{l_{\psi\Gamma=0}} + .0002\Gamma° \qquad (9\text{-}2)$$

The difficult part of estimating the dihedral effect is to estimate the value of this derivative for zero geometric wing dihedral, $C_{l_{\psi\Gamma=0}}$.   It has been found as the result of considerable wind-tunnel experience that this residual dihedral effect varies considerably with the position of the wing on the fuselage.   It has also been found that the dihedral effect will be somewhat invariant with change in wing angle of attack for straight wings, but will change rapidly with angle of attack for swept wings.   The effect of deflected flaps on dihedral effect can be large if the flap hinge line has any sweep, and finally, the effects of power on the dihedral effect can be serious with deflected flaps, especially for airplanes where the slipstream covers the flaps.   To estimate the effective dihedral for an airplane, then, requires considerable experience in allowing for these very complex variables.

In actual practice, the dihedral angle is usually not set from analytical considerations, because of the large errors involved. Most designers set the wing dihedral only after careful analysis of wind-tunnel test data, in which the effects of angle of attack, power, and flap settings are carefully analyzed.

In order to give some idea of the magnitude involved, the estimation of the dihedral effect for an airplane with a straight planform will be

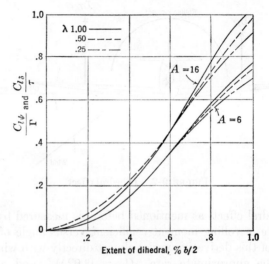

FIGURE 9–3. Chart for estimating C_{l_ψ} due to wing dihedral (Γ). From TR 635, "Theoretical Stability and Control Characteristics of Wings with Various Amounts of Taper and Twist," by Pearson and Jones.

undertaken, while in later sections typical data are given to indicate the changes that might be expected for swept wings at high angles of attack and the changes that can be encountered with deflected flaps and with power.

To calculate the dihedral effect for the wing alone, the curves of Figure 9–3, developed by the NACA,* are presented. These curves are plots of the rolling moment derivative, C_{l_ψ}, per unit dihedral angle as ordinates versus extent of dihedral on wing span as abscissa, for different aspect ratios and taper ratios. For wings whose dihedral extends from the root to the tip, it is merely necessary to interpolate these curves for aspect ratio and taper ratio at extent of dihedral value

* NACA TR 635.

of 1.0. For wings with varying dihedral, each section must be evaluated separately, and the results added. For instance, for the wing shown in Figure 9–4, whose dihedral starts at $.5b/2$, the value of C_{l_ψ}/Γ must be obtained for a dihedral extent of 1.0 and .5; the difference between these values will be the final value of C_{l_ψ}/Γ. The values of C_{l_ψ} and Γ in Figure 9–3 are in radian measure. To convert to C_{l_ψ} per degree, per degree change in dihedral angle, divide by 3290.

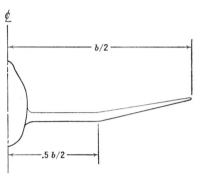

FIGURE 9–4.   Extent of dihedral.

The values given in the curves in Figure 9–3 are for tip shapes for which the maximum ordinate points on the mean lines are in one plane. If the maximum ordinate points on the airfoil upper surface are in one plane, then $\Delta C_{l_\psi} = .0002$ must be added. If the maximum ordinate points on the airfoil lower surface are in one plane, then $\Delta C_{l_\psi} = -.0002$ must be added.

(a) Max. Ord. on Upper Surface in Plane $\Delta C_{l_\psi} = .0002$

(b) Max. Ord. on Mean Lines in Plane $\Delta C_{l_\psi} = 0$

(c) Max. Ord. on Lower Surface in Plane $\Delta C_{l_\psi} = -.0002$

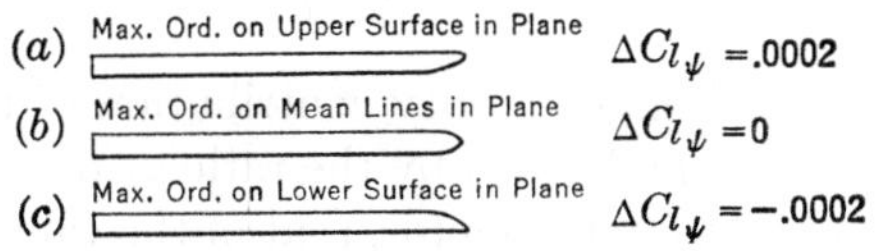

FIGURE 9–5.   Effect of wing tip on C_{l_ψ}.

The increments to the dihedral effect of the wing due to the tips, discussed above, are given in Figure 9–5.

Finally, the dihedral effect of the straight wing alone is given in equation (9–3).

$$(C_{l_\psi})_{Wing} = \frac{C_{l_\psi}}{\Gamma}\Gamma + \Delta C_{l_\psi Tip\ shape} \qquad (9\text{–}3)$$

To complete the analysis of dihedral effect, account must be taken of the interference effects between the wing, the fuselage, and the vertical tail. These effects are very troublesome, as they are most difficult to analyze. From the results of a systematic group of tests made by the NACA, some insight into the magnitude of these interference effects has been established. If the wing is placed high on the fuselage, the interference effects will increase the effective dihedral; the interference effects for a mid wing are negligible, and for a low wing, decrease the effective dihedral. The horizontal position of the wing has only very small effects. These effects can be given as wing-fuselage interference increments as follows:

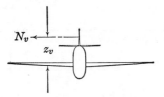

FIGURE 9-6. The rolling moment due to the vertical tail.

High wing $(\Delta C_{l\psi})_1 = +.0006$ (9-4)

Mid wing $(\Delta C_{l\psi})_1 = 0$ (9-5)

Low wing $(\Delta C_{l\psi})_1 = -.0008$ (9-6)

The vertical tail, if located all above the airplane's X axis, can contribute to the dihedral effect which can be computed quite readily from the normal force on the vertical tail created by the sideslip (Figure 9-6). The rolling moment coefficient may be obtained as follows:

$$C_l = a_v \psi \eta_v \frac{S_v}{S_w} \frac{z_v}{b}$$ (9-7)

and the stability contribution

$$(C_{l\psi})_v = a_v \frac{S_v}{S_w} \frac{z_v}{b} \eta_v$$ (9-8)

which is just the directional stability contribution of the vertical tail developed in Chapter 8 multiplied by the ratio z_v/l_v.

Finally, a second interference factor, that of the wing on the vertical tail contribution to $C_{l\psi}$, may be given:

High wing $(\Delta C_{l\psi})_2 = -.00016$ (9-9)

Mid wing $(\Delta C_{l\psi})_2 = 0$ (9-10)

Low wing $(\Delta C_{l\psi})_2 = .00016$ (9-11)

The final equation for the dihedral effect of the complete airplane may be given in the following form:

$$(C_{l\psi})_{Airplane} = (C_{l\psi})_w + (C_{l\psi})_v + (\Delta C_{l\psi})_1 + (\Delta C_{l\psi})_2$$ (9-12)

9-3   Effect of Wing Sweep, Flaps, and Power on Dihedral Effect

The method of estimating the dihedral effect given in the previous section is applicable for an airplane with a straight wing, flaps up with propellers-windmilling. The variation of $C_{l\psi}$ with wing sweep, flap deflection, and high thrust coefficient is a very complex one, and one that is almost impossible to estimate quantitatively by any analytical approach. The dihedral effect for airplanes having swept wing planforms will become a function of the lift coefficient. Airplanes with swept-back wings will have an increasing dihedral effect with lift coefficient, while airplanes with swept-forward wings will have a decreasing dihedral effect with lift coefficient. Typical variations in the dihedral effect parameter, C_{l_ψ}, with airplane lift coefficient are shown in Figure 9-7 for the case of the swept-forward and swept-back wing.

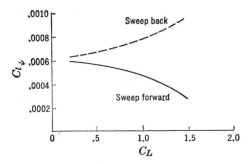

FIGURE 9-7. Typical variation of C_{l_ψ} with C_L for swept wings.

At very low lift coefficient, the sweep of the planform has little effect on C_{l_ψ}, but if the geometric dihedral of the swept planform airplane is set to give a good value of C_{l_ψ} at high speed, then the airplane with swept-back wings will be in danger of having excessive dihedral at low speeds, while the airplane with swept-forward wings will probably encounter negative dihedral effect at low speeds.

The dihedral effect of the airplane can also be seriously affected by flap deflection. Experience indicates that if the hinge line of the flap is unswept, there is little difference in C_{l_ψ} flaps up or down. However, if the hinge line of the flaps is swept back, flap deflection will usually increase C_{l_ψ}, while if the hinge line of the flaps is swept forward, flap deflection will usually decrease C_{l_ψ}. Typical variations in C_{l_ψ} due to flap deflection are shown in Figure 9-8.

The effect of power on dihedral effect is usually serious for only the flap-down condition. This effect usually arises because of the displacement of the slipstream in a sideslip, resulting in one flap being immersed in the slipstream to a greater extent than the other. (See Figure 9-9.) These power effects are at a maximum in full-power, low-speed flight where the ratio of slipstream velocity to free stream velocity is the greatest. The effects of power on C_{l_ψ} with flaps deflected are the greatest for flaps with swept-forward hinge lines, and the least for flaps with swept-back hinge lines, but in almost all cases are destabilizing.

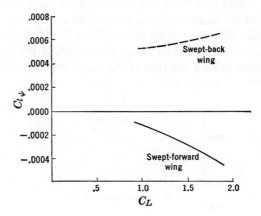

FIGURE 9-8. C_{l_ψ} versus C_L flaps down on swept wings.

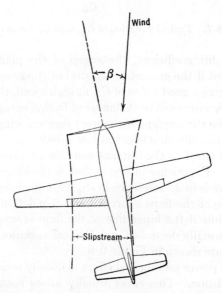

FIGURE 9-9.   Slipstream effect on C_{l_ψ}, flaps down.

If the airplane has a swept-back planform, the stabilizing effect of sweepback at high C_L counteracts somewhat the destabilizing effects of power, while for a swept-forward wing, these effects are additive.

The analytical calculation of the dihedral effect is only approximate, and as mentioned before, should be used only as an indication of the expected magnitude of C_{l_ψ}. The dihedral angle should be adjusted finally only from wind-tunnel or actual flight tests of the airplane.

The desired magnitude of dihedral effect has never been very successfully determined. From the analysis of many stability and control flight tests, it has become apparent that the pilots like to have some dihedral effect, but not too much. The reason for this can be traced to several different sources. If the dihedral is excessive, the airplane becomes very sensitive in roll to the rudder. In other words, the sideslip caused by very slight rudder deflections creates large rolling moments that become very annoying at high speeds, requiring fast lateral control movement to correct. Excessive dihedral effect also is objectionable to the pilot during fast rolling maneuvers. It was pointed out in Chapter 8 that abrupt rolls at high lift create heavy yawing moments, due to the rolling of the wing and the unsymmetric drag of the lateral controls. The sideslip created by these yawing moments will create rolling moments due to the dihedral effect, which, if excessive, will tend to kill the roll. Large dihedral will therefore work against the lateral control and will require very accurate coordination of rudder and aileron by the pilot. Finally, excessive dihedral will be disadvantageous during flight with asymmetric power. If the airplane is allowed to sideslip to relieve the pilot on rudder control, it will introduce large rolling moments that must be corrected by large lateral control deflections and therefore large control forces.

In the problem of airplane dynamics, which will be taken up in Chapter 11, the effect of C_{l_ψ} is shown to be both good and bad. It will be pointed out that heavy dihedral effect will create objectionable lateral dynamics which cannot be tolerated on any successful airplane.

The upper limit on dihedral effect is therefore very important and should not exceed 3 or 4 degrees of effective dihedral. The lower limit on dihedral effect varies with the usage of the airplane. Air Force airplanes are allowed to have negative dihedral for high-power, low-speed, flap-down flight, as long as the dihedral is correct for cruising and high speed. Navy airplanes, on the other hand, which are required to make carrier landings, where close control is required at low speeds with high power and flaps down, are required to maintain dihedral effect right down to the stall. This requirement at times necessitates the airplane having such large dihedral that the high-speed characteristics are compromised.

A criterion for dihedral has been used by the Air Force during the past few years which appears to be conservative and gives the

designer something to aim for.   This criterion is that the dihedral
effect be one-half the magnitude of the directional stability.

$$C_{l_\psi} = -\frac{C_{n_\psi}}{2} \qquad (9\text{–}13)$$

The problem of correct wing dihedral angle setting has never been
well established, but in general it can be said that it is desirable to have
a little stable dihedral effect, but not too much.

9–4   Lateral Control (Introduction)

It was mentioned in Section 9–1 that, although it is possible to
design an airplane to operate with elevator and rudder controls only,
the maneuvering possibilities of an airplane controlled in this manner
are decidedly limited, and in almost all cases
totally inadequate.   It has been found necessary to provide the pilot with a powerful
and definite control over the airplane's angle
of bank in order to satisfy the minimum
maneuvering requirements of the modern airplane.   This lateral control is usually obtained through the use of plain flaps
mounted on the trailing-edge sections of the
outer wing panels which are usually referred
to as ailerons.   The ailerons on each wing deflect asymmetrically, one going up and the
other going down, thereby so altering the
spanwise load distribution that a rolling moment is created about the X axis. A typical
aileron arrangement is shown in Figure 9–10,
indicating the change in the spanwise lift
due to the aileron deflection.

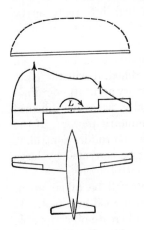

FIGURE 9–10.   Spanwise
loading with deflected
ailerons.

Deflecting the aileron will create a rolling moment that will accelerate
the airplane in roll about the X axis.   As the airplane's rolling velocity
increases, a new lift distribution will be created that is a function of
this rolling velocity and which opposes the rolling moment due to the
aileron deflection.   This moment is known as the damping moment of
the wing.   The alteration of the spanwise lift distribution due to the
rolling velocity comes about as a result of the change in effective angle
of attack at any wing section due to the roll, $\Delta\alpha_s = py/V$, where p is
the rolling velocity in radians per second, y the spanwise location of the
section from the X axis, and V the forward velocity in feet per second.

A typical increment in spanwise lift distribution due to the rolling velocity is given in Figure 9–11.

The damping moments of a wing are very powerful, and therefore the ailerons are required to be very effective indeed if a fast rate of roll is desired. The steady-state rolling velocity is determined when the increment in the rolling moment due to the ailerons is just opposed by the increment in the rolling moment due to the wing damping.

The design of the lateral control for effectiveness and lightness is the most difficult control design problem confronting the aerodynamicist. The reasons for this are that in combat and normal flight maneuvers larger deflections of the ailerons are required than for the other

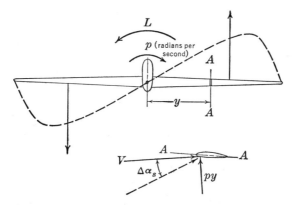

FIGURE 9–11.   Load distribution due to rolling velocity.

controls, and that the lateral control system is geared to a sidewise motion of the stick, a direction in which the pilot can exert only a small effort. In some cases, especially for large airplanes, the ailerons are geared to a wheel, where the mechanical advantage is better and the pilot can apply considerable force. For modern high-speed airplanes extremely close aerodynamic balance is required for direct pilot control, and in some cases this problem has been deemed so difficult that a hydraulic boost system has been incorporated to aid the pilot.

The lateral control must fulfill two basic requirements that determine the size of the control and the amount of aerodynamic balance. The lateral control must be large enough to provide sufficient rolling moment at low speeds to counteract the effects of vertical asymmetric gusts tending to roll the airplane. This requirement is usually one of size of control, because at the low speeds near the stall the pilot will be able to use full control movement and likes to feel that he has plenty of available control to pick up a wing during the landing process.

The second requirement on the lateral control is that it roll the airplane at a sufficiently high rate at high speed for a given stick force. This requirement dictates the amount of aerodynamic balance required.

The function of the lateral control, besides providing the rolling moments to maintain a wings-level attitude during landings and to roll the airplane at high angular velocity at high speed, also must balance the dihedral in sideslips and in flight with asymmetric power. These requirements on the lateral control are very seldom as severe as the two mentioned above, and therefore are considered only casually during the design of the lateral control.

The need for high rates of roll at high speeds for fighter-type aircraft increased tremendously during World War II. This came about because of a basic change in fighter tactics from the low-speed circling combat of World War I to the one-pass fast-breakaway type of maneuver of World War II. In this type of combat, rapid breakaways from a given flight path were the order of the day, and an airplane that could roll up to a vertical bank a few tenths of a second faster than his opponent found himself at an enormous advantage. During the whole war terrific emphasis was placed on rate of roll at high speed, and during this period the design of effective, light ailerons consumed a good proportion of the research facilities of the warring countries.

The design criterion in common use for evaluating lateral control effectiveness is the non-dimensional parameter $pb/2V$, with p the rate of roll in radians per second, b the wing span in feet, and V the true speed in feet per second. It will be shown in subsequent sections that, for geometrically similar airplane and lateral control arrangements, this parameter is a constant, and that for all airplanes the $pb/2V$ that can be produced by full lateral control deflection is a measure of the relative lateral control power available. Flight research over the past years has indicated that the pilot's conception of adequate control power is tied up closer to this factor than to the actual rolling velocity. The term $pb/2V$ is actually the helix angle made by the wing tip during a rolling maneuver. The lateral controls for modern airplanes are designed to give definite values of $pb/2V$ for full aileron deflection. The minimum requirements are as follows:

Cargo and bombardment types: $pb/2V = .07$

Fighter types: $pb/2V = .09$

The aileron size is usually determined by the above requirements, whereas the aerodynamic balance of the aileron is determined from the requirement that the above $pb/2V$ values be maintained up to a given function of the airplane's high speed in level flight.

The aileron control power $(pb/2V)$ per degree of aileron deflection is a constant independent of speed under the assumption of a rigid wing. However, in actual practice the wings' elasticity affects the airplane's response to the lateral control materially, the $pb/2V$ per degree aileron falling off at high speeds because of wing twist.

9-5 Estimation of Lateral Control Power

The rolling performance of any wing-aileron system must be developed from a study of the equation of motion of the airplane in roll. This is usually done by assuming the airplane to be a single-degree of freedom system in roll about the X axis. This is, of course, not strictly accurate, as motion about the X axis couples with the asymmetric degrees of freedom. However, for the purpose at hand the assumption is well justified.

If the wing is considered a rigid structure, the equation of motion in roll can be written very simply, the rolling moments arising only from the aileron deflection, δ_a, and the wing damping due to angular velocity, p.

$$I_x \dot{p} = \frac{\partial L}{\partial p} p + \frac{\partial L}{\partial \delta_a} \delta_a \qquad (9\text{-}14)$$

In this study only the steady-state rolling velocity is sought after, so that the assumption that $\dot{p} = 0$ is made at once, reducing (9-14) to

$$0 = \frac{\partial L}{\partial p} p + \frac{\partial L}{\partial \delta_a} \delta_a \qquad (9\text{-}15)$$

The assumption that only the steady-state rolling velocity is important is well justified by experience, because of the very short duration of the transient motion.

Solving (9-15) for the rolling velocity gives

$$p = - \frac{\partial L/\partial \delta_a}{\partial L/\partial p} (\delta_a) \qquad (9\text{-}16)$$

The partial derivatives $\partial L/\partial \delta_a$ and $\partial L/\partial p$ can be expressed in coefficient form by dividing both numerator and denominator by $qS_w b$.

$$p = - \frac{\partial C_l/\partial \delta_a}{\partial C_l/\partial p} \delta_a \qquad (9\text{-}17)$$

The rolling moment coefficient per degree aileron $(\partial C_l/\partial \delta_a)$ and the rolling moment per degree per second rate of roll can be evaluated by the strip integration method or more accurately from determinations of the spanwise lift distribution. The strip method will be developed first to demonstrate the functional relationship between the variables, while the lift distribution methods will be incorporated later for actual design practice.

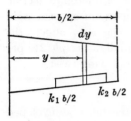

FIGURE 9-12. Strip integration for $C_{l\delta_a}$.

The rolling moment coefficient per degree aileron throw can be developed as follows. Referring to Figure 9-12,

$$dC_l = \frac{cc_l y \, dy}{S_w b} \qquad (9\text{-}18)$$

where c is the wing local chord, and c_l is the section lift coefficient. The section lift coefficient can be expressed as

$$c_l = a_0 \tau \delta_a \qquad (9\text{-}19)$$

where τ is the section flap effectiveness factor.

Substituting (9-19) into (9-18) and integrating between the limits $k_1 b/2$, and $k_2 b/2$ give the rolling moment coefficient for the complete wing.

$$C_l = \frac{2a_w \tau \delta_a}{S_w b} \int_{k_1 b/2}^{k_2 b/2} cy \, dy \qquad (9\text{-}20)$$

where a_w and τ must be for three-dimensional flow. The derivative $\partial C_l/\partial \delta_a$ becomes

$$\frac{\partial C_l}{\partial \delta_a} = \frac{2a_w \tau}{S_w b} \int_{k_1 b/2}^{k_2 b/2} cy \, dy \qquad (9\text{-}21)$$

The rolling moment coefficient due to the angular velocity, p, in radians per second can be developed in a similar fashion. The angle of attack in radians due to the rolling velocity of any section at a distance y from the centerline can be given approximately as:

$$\Delta \alpha = \frac{py}{V} \qquad (9\text{-}22)$$

Therefore the increment in section lift coefficient due to the rolling velocity is

$$c_l = a_0 \frac{py}{V} \qquad (9\text{-}23)$$

Substituting (9–18) into (9–20) and integrating over the span give the rolling moment coefficient for the complete wing.

$$C_l = \frac{2a_w p}{VSb} \int_0^{b/2} cy^2 \, dy \tag{9–24}$$

and the derivative

$$\frac{\partial C_l}{\partial p} = \frac{2a_w}{VSb} \int_0^{b/2} cy^2 \, dy \tag{9–25}$$

Substituting (9–21) and (9–25) into (9–17) gives

$$p = \tau V \frac{\displaystyle\int_{k_1 b/2}^{k_2 b/2} cy \, dy}{\displaystyle\int_0^{b/2} cy^2 \, dy} \cdot \frac{\delta_a}{2} \tag{9–26}$$

For straight tapered wings, the chord is the following function of y, the distance from the centerline:

$$c = c_t \left(TR - \frac{y}{b/2} (TR - 1) \right) \tag{9–27}$$

where c_t is the tip chord, and TR the taper ratio. If (9–27) is substituted into (9–26) and the resulting expression integrated, the following equation is obtained:

$$p = \frac{2\tau V \delta_a{}^\circ {}_{\text{Total}}}{57.3b} \left[\frac{(k_2{}^3 - k_1{}^3)(1 - TR) + 3TR(k_2{}^2 - k_1{}^2)}{TR + 3} \right] \tag{9–28}$$

In equation (9–28) the expression inside the brackets is a function of the extent and location of the aileron along the span, and the wing taper ratio (c_s/c_t). For a given airplane it is therefore a constant. The aileron effectiveness factor, τ, is a constant depending only on the ratio of the aileron chord to the wing chord ahead of it, c_a/c_w. The rate of roll for a given aileron deflection varies directly with the true air speed, V, in feet per second and inversely with the wing span, b, in feet. These are very important basic concepts, for they demonstrate that the rate of roll is lowest for full aileron deflection at low speed, and that for geometrically similar airplanes, the one with the smallest span will be at an immediate advantage.

It can be seen that for a given aileron deflection the ratio $pb/2V$ will be a constant.

$$\frac{pb}{2V} = \frac{\tau \delta_a{}^\circ {}_{\text{Total}}}{57.3} \left[\frac{(k_2{}^3 - k_1{}^3)(1 - TR) + 3TR(k_2{}^2 - k_1{}^2)}{TR + 3} \right] \tag{9–29}$$

The technique used in this development is known as strip integration. It is not a very accurate method, for it assumes an abrupt discontinuity in the spanwise lift distribution at the ends of the ailerons, which does not actually exist. The spanwise lift distribution adjusts itself rapidly but smoothly and continuously at the point of aileron discontinuity. In practice this effect cuts down on the $pb/2V$ calculated by the strip integration method, making it somewhat optimistic. The difference in spanwise load distribution between that assumed in the strip integration method and the actual distribution is shown in Figure 9–13.

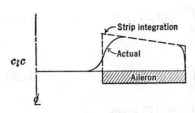

FIGURE 9–13. Comparison of load distributions.

A correction factor of .9 is sometimes applied to equations (9–28) and (9–29) to account for this effect.

$$\frac{pb}{2V} = \frac{.9\tau\delta_a{}^\circ}{57.3}\left[\frac{(k_2{}^3 - k_1{}^3)(1 - TR) + 3TR(k_2{}^2 - k_1{}^2)}{TR + 3}\right] \quad (9\text{--}30)$$

Although the development just carried out is useful in that it demonstrates the relationship between the major variables, it is hardly ever used in practice because of its inherent inaccuracy. The method in nearly universal use at the present time is based on spanwise load distribution data compiled by the NACA.* In this report plots of the ratio $C_{l\delta}/\tau$ are given versus the extent of the aileron on the wing span. These curves were developed from many spanwise lift distribution computations using the Lotz method and are enormously helpful to the airplane designer. These curves are the same as given in Figure 9–3, the ordinates of this graph becoming $C_{l\delta}/\tau$ instead of $C_{l\psi}/\Gamma$.

In the same report, similar calculations gave the damping derivative C_{lp}, defined in this case as:

$$C_{lp} = \frac{dC_l}{d(pb/2V)} \quad (9\text{--}31)$$

The variation of C_{lp} with aspect ratio and taper ratio is given in Figure 9–14.

The steady-state equation of motion in terms of these new derivatives is simply

$$0 = C_{lp}\left(\frac{pb}{2V}\right) - \frac{C_{l\delta}}{\tau}\frac{\tau\delta_a}{2} \quad (9\text{--}32)$$

* NACA TR 635.

or

$$\frac{pb}{2V} = \frac{C_{l\delta}}{\tau}\frac{\tau\delta_a}{114.6C_{l_p}} \qquad (9\text{-}33)$$

In order to determine the value of $C_{l\delta}/\tau$ for a given wing and aileron combination, the values of $C_{l\delta}/\tau$ are found from curves given in Figure 9-3 for values of the location of the outboard and inboard ends of the

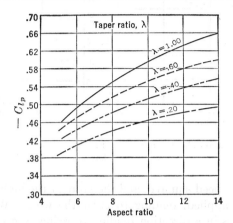

FIGURE 9-14. C_{l_p} versus aspect and taper ratio. From TR 635, "Theoretical Stability and Control Characteristics of Wings with Various Amounts of Taper and Twist," by Pearson and Jones.

aileron in semispans. As the value of $C_{l\delta}/\tau$ is for an aileron deflection of 1 degree on each wing, the total aileron deflection (δ_a) must be divided by 2, as indicated in equation (9-32).

The value of the aileron effectiveness factor, τ, is a function of the aileron chord to wing chord ratio, c_a/c_w. In practice this ratio is never found to be equal to the theoretical value given by thin airfoil theory. From the results of a great many flight tests, experimental values of aileron effectiveness have been obtained and are presented in Figure 9-15. These values are satisfactory for aileron deflections of not more than plus or minus 10 degrees. For deflections beyond 10 degrees, the aileron effectiveness starts to fall off and a correction must be applied to account for this. A multiplying factor K has also been developed* from flight test data to account for high deflections. This correction is given as a curve of K versus $\Delta\delta_a$ in Figure 9-16.

The final equation for aileron performance for a rigid wing follows:

$$\frac{pb}{2V} = \frac{C_{l\delta}}{\tau}\frac{\tau\delta_a{}^\circ K}{114.6C_{l_p}} \qquad (9\text{-}34)$$

* USAF TR 5180.

As the airplane's speed increases, the assumption of a rigid wing becomes less and less accurate. The deflection of the aileron will create a pitching moment tending to twist the wing. This moment

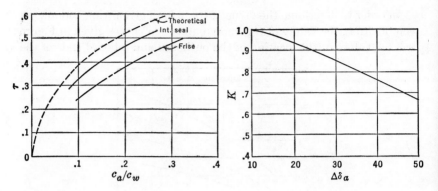

FIGURE 9–15.   Aileron effectiveness.   FIGURE 9–16.   Correction for large aileron deflections. From USAF TR 5180.

varies with the speed squared, and therefore at the high speeds a large torque will be applied, tending to twist the wing. When the wing twists, it rotates in a direction tending to reduce the rolling moment created by the aileron. If the speed is taken high enough, the point

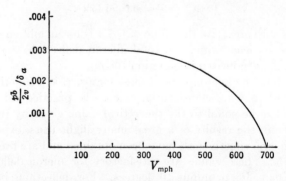

FIGURE 9–17.   Fall-off of $pb/2V$ due to wing twist.

can be reached where the wing twist will just counter the aileron rolling moment and the lateral control will be lost. This speed is known as the aileron reversal speed, V_r, and it is of extreme importance for the designer to insure that the wings are sufficiently rigid in torsion so that the aileron reversal speed is higher than the anticipated high speed of the airplane.

A typical variation of $pb/2V$ per degree aileron is given in Figure

9-17, showing the rapid decay of the lateral control at the higher speeds and the reversal speed at 700 mph.

The calculation of the effects of wing elasticity is beyond the scope of this book. The methods to be used are given in some detail in the literature.

9-6   Aileron Control Forces (Requirements)

Once the size of the lateral control is decided upon, the next step is the investigation of the force that the pilot must apply at the cockpit controls, in order to deflect them. In most cases the hinge moments vary as the velocity squared, and therefore the pilot's force varies essentially in the same manner. It is obvious that for anything

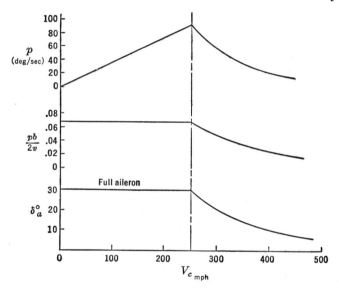

FIGURE 9-18.   Typical airplane rolling characteristics.

except 100 per cent aerodynamic balance, there will always be some speed above which full pilot's force may not be powerful enough to deflect the ailerons fully. For speeds in excess of this limiting speed, the aileron deflection for a given stick force will fall off rapidly. It is the duty of the aerodynamicist to provide some form of balance to insure that adequate lateral control deflections are possible in the usable speed ranges of the airplane. This is the most difficult of all control surface design problems and must be adequately solved before the design can be considered successful.

A typical variation of rolling velocity, $pb/2V$, and aileron deflection is shown in Figure 9-18, plotted against calibrated airspeed, for a maximum pilot's control force of 30 lb applied at the top of the stick. It can be seen that up to a calibrated airspeed of 250 mph the pilot is able to maintain full aileron deflection; the rolling velocity increases linearly with airspeed and $pb/2V$ is constant. Beyond 250 mph V_c, the aileron deflection that the pilot can obtain for 30 lb stick force falls off and so does the $pb/2V$ and rolling velocity.

It is almost impossible to design the aileron to be light enough to give full deflection throughout the speed range, especially for modern high-speed airplanes. The Air Force and Navy require that a $pb/2V$ of .07 for cargo and bombardment types and .09 for fighter types be maintained or exceeded up to 80 per cent of the airplane's maximum indicated level-flight speed, with normal rated power for a pilot's force at the top of the stick of 30 lb or 80 lb applied tangentially to the rim of a wheel control. For very high-speed airplanes the maximum $pb/2V$ is required to be held to only 300 mph V_i for the same control forces. The Air Force and Navy requirements also stipulate that the reduction in rate of roll due to increased hinge moments and wing twist shall not be so great that the airplane's $pb/2V$ is reduced to a value less than .015 at 95 per cent of the limit diving speed.

9-7 Aileron Control Forces

The aileron control forces during rolling maneuvers can be developed along similar lines to those discussed for the elevator and rudder. The aileron control forces are somewhat complicated because of the fact that there are two surfaces involved, one control going up and the other going down, and in some cases moving at different rates. The principles involved, however, are the same, and a little development shows the major problems.

As there are two surfaces involved, it is first necessary to give some notation in order not to become confused when analyzing their combined effects on the stick forces. The term δ_a will refer to the total aileron angle in degrees and will be equal to the deflection of the up, δ_u plus the deflection of the down, δ_d, aileron.

$$\delta_a = \delta_u + \delta_d \qquad (9-35)$$

The terms S_a and c_a are the area and root-mean-square chord of one aileron, respectively. An aileron stick force is considered positive for a force toward the pilot's right, and hinge moments are considered positive for trailing edge down. All aileron deflections are considered positive, as are the control gearings. (See Figure 9-19.)

To develop the aileron control force equation, a roll to the right will be considered, requiring a positive stick force, an upward deflection of the right aileron, and a downward deflection of the left aileron. The gearing of the up aileron will be $d\delta_u/ds$, and the gearing of the down aileron will be $d\delta_d/ds$.

The stick force required at the top of the stick to hold both controls over can be expressed as:

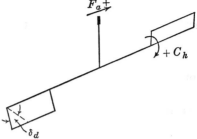

$$F_a = -HM_d\frac{d\delta_d}{ds} + HM_u\frac{d\delta_u}{ds}$$

(9–36)

The hinge moments of the down and up ailerons can be expressed in coefficient form.

FIGURE 9–19.   Aileron force notation.

$$HM_u = C_{h_u}qS_ac_a$$

$$HM_d = C_{h_d}qS_ac_a$$

(9–37)

Equation (9–36) may be rewritten

$$F_a = -qS_ac_a\left(C_{h_d}\frac{d\delta_d}{ds} - C_{h_u}\frac{d\delta_u}{ds}\right)$$ (9–38)

For aileron systems where there is no differential, i.e., the up and down-going ailerons move at the same rate:

$$F_a = -qS_ac_a\left(\frac{d\delta_a}{ds}\right)(C_{h_d} - C_{h_u})$$ (9–39)

If the hinge moment coefficients are considered linear functions of angle of attack and deflection, they may be written as before.

$$C_{h_a} = C_{h_0} + C_{h_\alpha}\alpha + C_{h_\delta}\delta_a$$ (9–40)

The angle of attack distribution across the span is altered because of the rolling velocity, the angles of attack of the wing sections on the down-going wing being increased, while those on the up-going wing are decreased.   If the increment in angle of attack due to the rolling velocity is considered averaged at the spanwise location of the aileron centroid and the distance of this station from the centerline is given as y' feet, the increment in angle of attack at the aileron is:

$$\Delta\alpha° = \frac{py'}{V}\,57.3$$ (9–41)

The rolling velocity, p, can be given in terms of the airplane's $pb/2V$

per degree total aileron:

$$p = \frac{pb/2V}{\delta_a} \delta_a \frac{2V}{b} \qquad (9\text{--}42)$$

or

$$\Delta\alpha° = \frac{pb/2V}{\delta_a} \frac{y'}{b/2} 57.3\delta_a \qquad (9\text{--}43)$$

or

$$\Delta\alpha° = n\delta_a \qquad (9\text{--}44)$$

where

$$n = \frac{pb/2V}{\delta_a} \frac{y'}{b/2} 57.3$$

The hinge moment coefficients of (9-39) can be expressed as follows:

$$C_{h_d} = C_{h0} + C_{h_\alpha}(\alpha - n\delta_a°) + C_{h_\delta}\delta_d \qquad (9\text{--}45)$$

$$C_{h_u} = C_{h0} + C_{h_\alpha}(\alpha + n\delta_a°) + C_{h_\delta}\delta_u \qquad (9\text{--}46)$$

For the case of no differential action

$$\delta_u = \delta_d = \frac{\delta_a}{2} \qquad (9\text{--}47)$$

$$F_a = -qS_a c_a \left(\frac{d\delta_u}{ds}\right)[-2C_{h_\alpha}n\delta_a + C_{h_\delta}\delta_a] \qquad (9\text{--}48)$$

$$F_a = -qS_a c_a G[C_{h_\delta}\delta_a - 2C_{h_\alpha}n\delta_a] \qquad (9\text{--}49)$$

$$F_a = -qS_a c_a G C_{h_\delta}\delta_a \left[1 - 2n\frac{C_{h_\alpha}}{C_{h_\delta}}\right] \qquad (9\text{--}50)$$

Equation (9-50) brings out the very important part played by the floating characteristics of the ailerons. If the ailerons are aerodynamically balanced so as to float up heavily with the wind, the pilot's stick forces will be reduced materially. For a typical case of an airplane with $\frac{pb/2V}{\delta_a} = .0025$ and $\frac{y'}{b/2} = .7$, the term $2n$ will be approximately equal to .2. If the ratio of $C_{h_\alpha}/C_{h_\delta}$ then is made equal to five, the ailerons will float up just enough to overcome the wing damping and the condition of autorotation with the stick free is reached. The term $\left[1 - 2n\frac{C_{h_\alpha}}{C_{h_\delta}}\right]$ is called the aileron response factor and will be referred to by the notation K. Finally,

$$F_a = -qS_a c_a G K C_{h_\delta}\delta_a \qquad (9\text{--}51)$$

The rate of roll for a given stick force can be obtained by solving (9–51) for δ_a and substituting into (9–42):

$$p = -\frac{pb/2V}{\delta_a}\frac{F_a}{qS_ac_aGKC_{h\delta}}\frac{2V}{b} \qquad (9\text{–}52)$$

and

$$\frac{dp}{dF_a} = -\frac{pb/2V}{\delta_a}\frac{4}{\rho V S_a c_a G K C_{h\delta} b} \qquad (9\text{–}53)$$

Now, assuming

$$S_a = K_1 S_w = K_1 bc$$

and

$$c_a = K_2 c_w$$

then

$$S_a c_a = K_3 bc^2 = \frac{K_3 b^3}{A^2} \qquad (9\text{–}54)$$

where A is the wing aspect ratio.

Substituting (9–54) into (9–53) gives

$$\frac{dp}{dF_a} = -\frac{pb/2V}{\delta_a}\frac{4A^2}{\rho V K_3 b^4 G K C_{h\delta}} \qquad (9\text{–}55)$$

For geometrically similar airplanes

$$\frac{dp}{dF_a} = K_4 \frac{1}{\rho V b^4} \qquad (9\text{–}56)$$

which brings out the very important relationship that the rate of roll for a given stick force varies inversely with the density of the air and with the velocity, V, and also inversely with the span to the fourth power. The variation of this response to force factor with span shows the enormous advantage that comes from reducing the size of the airplane if high rate of roll is desired, and indicates why a large fighter cannot be made to compete in this regard with a small fighter.

Equation (9–56) also indicates the rate at which the rate of roll falls off with increasing air speed beyond the speed where full allowable stick force is required (30 lb).

For this case

$$p\rho V = K \qquad (9\text{–}57)$$

The rate of roll will fall off hyperbolically with air speed. A typical case is shown in Figure 9–20.

The calculations just completed are based on the assumption that the aileron hinge moments are linear and that the up- and down-going ailerons move at the same rate (no differential). The hinge moments of ailerons used in actual practice are usually quite non-linear, and differential gearing of the ailerons is often used to give the up-going aileron more deflection than the down-going aileron. For these cases

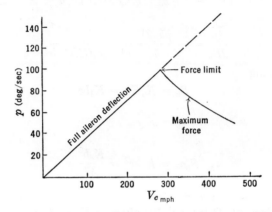

FIGURE 9–20. Typical variation of rate of roll with velocity.

the simplified analysis just completed cannot be used, and it is necessary to return to the more general expression for the stick force required.

$$F_a = -qS_ac_a\left(C_{h_d}\frac{d\delta_d}{ds} - C_{h_u}\frac{d\delta_u}{ds}\right) \qquad (9\text{–}58)$$

The stick force required for a given rate of roll must be computed point by point, making use of curves of the aileron-stick kinematics

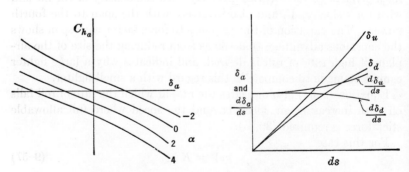

FIGURE 9–21. Typical aileron hinge moments and kinetics.

and the actual hinge moment coefficients plotted against aileron angle for varying angles of attack. Typical curves are shown in Figure 9–21.

Finally the complete picture of the airplane's rolling performance is given in Figure 9-22.

The actual rate of roll versus calibrated air speed falls somewhat short of that predicted by rigid-wing theory, especially at the higher air speeds.   This is due to the effects of wing twist and cable stretch.

The airplane's rate of roll for a given aileron deflection can be affected adversely by secondary moments in yaw developed because of the aileron deflection and the rolling motion of the wing.   These secondary yawing moments are usually grouped under the general heading of adverse yaw.   The adverse yawing moments will cause the airplane to sideslip during the rolling maneuver, and if the dihedral

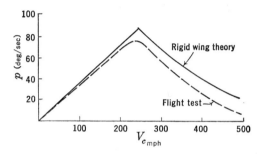

FIGURE 9-22.   Actual and theoretical rate of roll versus velocity.

effect is large, will create rolling moments due to the sideslip that oppose and limit the rolling velocity severely.   The effects of adverse yaw are always the most severe at high lift coefficients.

In order to oppose the adverse yaw, the pilot can deflect the rudder, thereby prohibiting the sideslip.   The aileron and rudder then must move together during rolling maneuvers.   This technique is known as coordinating the roll and becomes most exacting for high lift coefficients and full aileron deflection.   An airplane with high directional stability and low dihedral effect will be the easiest to coordinate and for fixed rudder rolls will lose the least rolling velocity as a result of any adverse yawing moments.   The time histories of two fixed-rudder rolling maneuvers are given in Figure 9-23, showing in a the typical fall-off of rolling velocity for an airplane with low directional stability and high dihedral effect, and in b the more or less constant rate of roll for an airplane with high directional stability and low dihedral effect.

It should be noted in these time histories that the peak values of the rolling velocity are nearly equal.   This is due to the fact that for normal aileron controls the maximum rolling velocity is established rapidly and before the sideslip has time to develop.

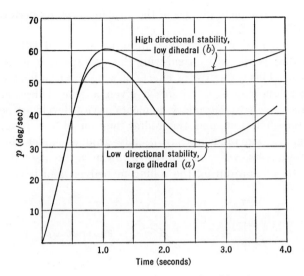

FIGURE 9–23. Typical rolling time histories.

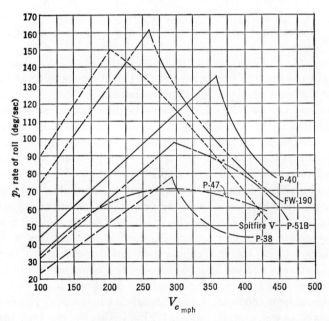

FIGURE 9–24. Comparison of rolling performance for several fighter-type aircraft.

A chart showing the rolling performance of several fighter-type airplanes that engaged each other during World War II is shown in Figure 9-24.

9-8 Balancing the Aileron

As the emphasis during the past few years has been for higher and higher rates of roll at higher and higher speed with ever-increasing airplane sizes, the problem of designing the aileron for lightness has become so difficult that in some instances airplane designers have given up all attempts at aerodynamic balance and gone over to hydraulic boost or assist. When the designer takes this step, he takes the problem out of the hands of the aerodynamicists and places it in the hands of the hydraulic engineer. Although the general trend is towards some sort of boost for the aileron control, it is felt that the major types of lateral control and the methods used for aerodynamic balance should be presented briefly, as they are still used on the majority of present-day aircraft.

The major lateral control is, of course, the aileron, and the development of methods for balancing the aileron took up a large percentage of the aerodynamic testing facilities of this country during World War II. There are about as many methods for balancing the aileron as there are airplane designers and, therefore, any complete summary of the aileron balance picture is beyond the scope of this book. However, the major types of balance will be pointed out with a short discussion of each.

The types of aileron aerodynamic balance can be broken down into two main classes, i.e., nose balance and trailing-edge balance. Aerodynamic balance at the control surface nose consists of variations in nose shape, hinge line set-back, shrouds, gaps, and seals, while the trailing-edge types of aerodynamic balance consist of changes in airfoil contour, balancing tabs, spring tabs, and trailing-edge strips.

One of the most commonly used ailerons in the past and one that is still in use in modified forms is the frise aileron. The pure frise type aileron is characterized by an asymmetrical sharp nose located on the airfoil lower surface so that it will unport as soon as the control is deflected upwards. (See Figure 9-25.) The major purpose of this type of control is to reduce the adverse yaw of the aileron and to provide a means of balancing the aileron for small deflections. The up-going aileron is unstable, as shown in the typical frise aileron hinge moment curve of Figure 9-25, and this unstable up-going aileron helps pull down the stable down-going aileron; if rigged just right, excellent balance is obtained.

An inspection of the hinge moment curves of Figure 9–25 shows that the net balance is a function of the neutral position of the ailerons. If they are both rigged up from neutral, both controls will be overbalanced, giving unstable stick forces for small deflections.   If, on the other hand, the ailerons are rigged down from neutral, both surfaces will be stable and the stick forces will become heavy.   The advantages of the frise aileron are large balance for small hinge line set-back,

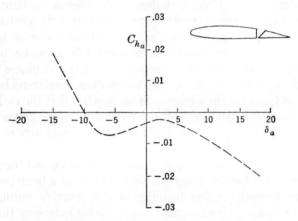

FIGURE 9–25.   Typical frise aileron hinge moments.

simplicity of construction, and reduction of aileron adverse yaw. The disadvantages of the frise aileron are its sensitivity to rigging, the tendency of the air to separate off the lower surface of the up-going aileron, causing aileron buffet and loss of effectiveness, and the tendency of the aileron to overbalance at high speed because of the aileron floating up as a result of control cable stretch.

The disadvantages of the frise ailerons mentioned above are well recognized, and most modern airplanes having ailerons of this type use modified frise ailerons with raised nose shapes which smooth out the hinge moment curves and avoid the high-speed difficulties mentioned above.   Unfortunately at the same time some of the advantages of the pure frise aileron are also eliminated.   In general, however, the characteristics of the pure frise at high speeds and at high deflection are so objectionable that the modified raised nose control is almost mandatory.   There are an infinite number of variations of the frise, and on the whole they have proved to be a very effective and useful control.

The internal seal type of balance is another nose balance that has had widespread use in the past few years (Figure 9–26).   This type of balance has essentially a very sharp nose with a heavily set-back hinge, with curtains covering the balance area, vented close to the hinge line. Typical hinge moment curves for this type of aileron are shown in Figure 9-26 and can be seen to be quite linear. This type of balance is quite popular because of the fact that its hinge moments can be maintained in production, it is not very sensitive to rigging, and it behaves quite well at high speed. Its disadvantage lies in its complicated construction and maintenance problems.

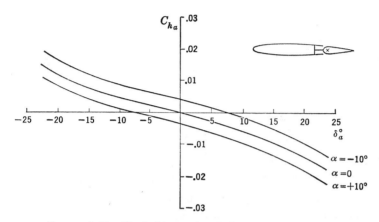

Figure 9-26. Typical internal seal aileron hinge moments.

The types of trailing edge balance include such devices as the beveled trailing edge, the balancing tab, the spring tab, and strips. The beveled trailing-edge balance was developed by the NACA and has been proved to be an effective means of balance, although it had practically no tactical use during World War II. The balancing tab can be used with discretion, but it cuts down the aileron effectiveness and tends to make the airplane laterally unstable stick-free. The English have done a lot of work with trailing-edge chords or strips. These devices are shown to increase both C_{h_α} and C_{h_δ} and are recommended for use with a balancing tab that will lower C_{h_δ} but will leave C_{h_α} high and negative. The English recommend this combination for its very favorable response factor, but it has not been used in this country as yet because of the danger of lateral instability with the stick free with this type of aileron.

The spring tab for ailerons is receiving a lot of attention at the present time. This device is a tab that deflects as a function of the pilot's force only. It has proved to be a very successful device and will be seen more often in the future. Finally, aileron differential coupled with a heavy upfloating tendency due to fixed bent tabs on the aileron trailing edge has been recommended, but as yet has had no actual use in the field. This type of balance makes use of a higher gearing on the up-going aileron than on the down-going aileron. Therefore the increased mechanical advantage of the up-going aileron will help balance out the down-going aileron.

Before leaving the subject of the lateral control, mention must be made of the spoiler-type aileron. This type of lateral control was first investigated by the NACA and used for the first time on a production airplane on the Northrop P-61 night fighter. This type of control creates a rolling moment by spoiling the lift on one wing panel. (See Figure 9–27.) The effectiveness of the spoiler increases as its location on the wing chord moves forward. However, at the same time the lag in the action of the aileron becomes objectionably large, and the location at the present time is limited to a position about 70 per cent of the local wing chord. Spoiler ailerons are useful as they permit more extensive use of flaps and have very low aerodynamic hinge moments.

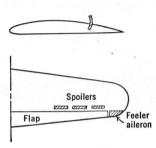

FIGURE 9–27. A spoiler-type aileron installation.

It would take many volumes to do justice to all the types of lateral control and their relative advantages and disadvantages. This short summary is only meant to present the reader with the major design trends of the present time and to refer him to the extensive literature on this subject for further study.

SUGGESTED READING

1. NACA TR 715, "Lateral Control Required for Satisfactory Flying Qualities Based on Flight Tests of Numerous Airplanes," by Gilruth and Turner, 1941.
2. NACA TR 799, "Charts for the Determination of Wing Torsional Stiffness Required for Specified Rolling Characteristics or Aileron Reversal Speed," Pearson and Aiken, 1944.
3. USAF TR 5180, "Prediction of Aileron Effectiveness," by J. D. Bitner, 1945.
4. NACA TR 635, "Theoretical Stability and Control Characteristics of Wings with Various Amounts of Taper and Twist," by Pearson and Jones, 1938.
5. NACA TN 825, "Wind-tunnel Investigation of Effect of Yaw on Lateral Stability Characteristics. III. Symmetrically Tapered Wing at Various Positions on Circular Fuselage with and without Vertical Tail," by Recant and Wallace, 1941. (Also others in this series.)
6. NACA WR L-419, "Collection of Balanced Aileron Data," by F. M. Rogallo, 1944.

PROBLEMS   371

7. NACA RB, "Résumé of Data for Internally Balanced Ailerons," by Rogallo and Lowry, 1943.

8. NACA WR L-169, "Résumé of Hinge Moment Data for Unshielded Horn-Balance Control Surfaces," by J. G. Lowry, 1943.

9. NACA TR 548, "Effect of Tip Shape and Dihedral on Lateral Stability Characteristics," by J. A. Shortal.

10. USAF MR ENG-M-51/VF 18 Add 1, "The Computation of the Critical Speeds of Aileron Reversal, Wing Torsional Divergence, and Wing Aileron Divergence," by L. N. Shornick, 1942.

11. NACA TN 1245, "Summary of Lateral Control Research," by Langley Research Department, compiled by T. A. Toll, March 1947.

PROBLEMS

9–1. A fighter airplane with a 40-ft wing span and ailerons extending from .90 to .50 semispan gives a $pb/2V$ for full aileron deflection of .09. What would be the estimated rate of roll in degrees per second for a geometrically similar airplane at 200 mph indicated speed at 10,000 ft for full aileron deflection?

9–2. An airplane having a straight tapered wing ($TR = 2$ and $A = 10$) has .20c internal seal ailerons extending from .90 semispan to .55 semispan. If the ailerons deflect up 20° and down 10° at full deflection, estimate the airplane's $pb/2V$. If the wing span of this airplane is 50 ft, plot the rate of roll in degrees per second versus indicated airspeed in miles per hour for an altitude of 10,000 ft.

9–3. Compare the value of $pb/2V$ obtained by the spanwise loading method for the example given in Problem 9–2 with the $pb/2V$ obtained from the strip integration method discussed in this chapter.

9–4. The ailerons of the airplane given in Problem 9–2 have hinge moment characteristics such that 30 lb stick force will just give full aileron deflection at 200 mph indicated speed at 10,000 ft.

a. Plot the variation of $pb/2V$ per degree aileron throw for a speed range of 0–500 mph V_i.

b. Plot the variation of total aileron deflection throughout the same speed range.

c. Plot the variation of rate of roll in degrees per second throughout this speed range.

9–5. The internal seal type of ailerons on a typical fighter-type aircraft extend from the tip to .60 semispan. The aileron chord is a constant 20 per cent of the wing chord. If the wing of this airplane has an aspect ratio $A = 6$ and a 2 : 1 taper ratio, what will be the $pb/2V$ per degree aileron throw? The area of each aileron is 8 sq ft, the root-mean-square chord $c_a = .8$ ft, and the centroid of the aileron area is at .75 $b/2$. If these ailerons have no differential and go up and down 15 degrees maximum with a gearing $d\delta_a/ds = .35$ radian per foot, what will be the indicated air speed beyond which full deflection cannot be obtained with a 30-lb force applied at the top of the stick? The balance characteristics of these ailerons are as follows:

$$C_{h\alpha} = -.005$$
$$C_{h\delta} = -.010$$

9–6. If the floating tendency ($C_{h\alpha}$) in Problem 9–5 is reduced to zero and $C_{h\delta}$ is left unchanged, what would the limiting speed become?

9–7. With the floating tendency, $C_{h\alpha}$, of the airplane in Problem 9–5 held equal to zero, what would be the value of $C_{h\delta}$ required to permit full deflection of the ailerons up to 300 mph V_i?

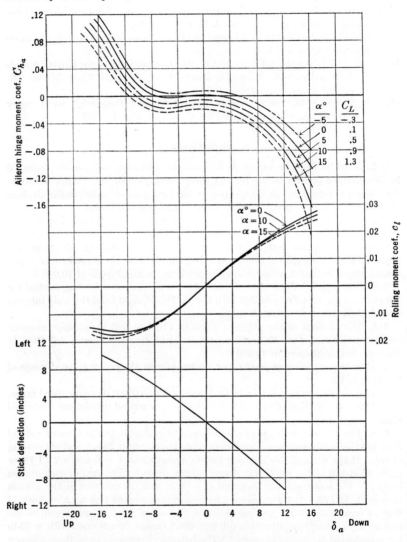

PROBLEM 9-8.   Aerodynamic moments due to left aileron deflection.

9-8. A fighter-type aircraft is equipped with frise ailerons, whose aerodynamic characteristics are given in the accompanying figure.   This airplane weighs 10,000 lb and has a wing span of 40 ft and a wing area of 250 sq ft.   Its taper ratio is 2 : 1.   If the aileron centroid is at .75 $b/2$ and the stick aileron kinematics are as given in the figure, calculate and plot the maximum rate of roll available for a 30-lb maximum stick force in degrees per second versus true airspeed in miles per hour for a speed range from the stall $C_{L_{max}} = 1.3.$ to 500 mph. Estimate for an altitude of 10,000 ft. The area of each aileron is 10 sq ft, and the root-mean-square chord is .70 ft.

9–9. For the airplane given in Problem 9–8, plot the variation of stick force versus aileron deflection for an indicated airspeed of 200 mph.

9–10. If the ailerons of the airplane in Problem 9–8 are rigged up 3°, plot the variation of stick force versus aileron deflection for an indicated air speed of 200 mph. Repeat with the ailerons rigged down 3°.

CHAPTER 10

LONGITUDINAL DYNAMICS

10–1 Introduction

The stability and control characteristics of the airplane discussed in the previous chapters have included only the steady-state phenomena of equilibrium and static stability. In order to understand the requirements for static stability and control it is necessary to study the dynamic characteristics of the airplane, investigating the types of motion that characterize the response of the airplane to a disturbance from some equilibrium flight condition and the nature of the transient motions of the airplane in response to the movement of its controls.

The characteristic modes of motion of the airplane must be known in order to determine the nature of the control required to fly the airplane. For example, if the motion of the airplane in response to some disturbance was a very slow divergence, the control requirements would be completely different from those needed if the divergence was extremely rapid. The ability of the human pilot to react and apply the controls is a factor which must be kept well in mind during all studies of airplane dynamics. In the previous chapters the problem of designing the airplane controls for gross power has been discussed; however, it is still necessary to determine whether or not the pilot can apply these controls in time to control the airplane's motion. The solution to this problem requires some knowledge of the transient response of the airplane to a disturbance or to these controls.

Dynamic systems in general have four different modes of motion when responding to a disturbance from an equilibrium position. These modes are oscillatory or aperiodic, damped or undamped. The four major modes are shown simply in Figure 10–1 as the response of an angle θ versus time, t.

Some types of dynamic systems have only one characteristic mode of motion, whereas other, more complicated systems have two or more progressing at the same time. The airplane, being a somewhat complicated dynamic system, will move in several different modes at the same time, and it is essential for the aerodynamicist to understand the nature of these modes and to study their importance in relation to the handling qualities of the airplane from the pilot's point of view.

In order to determine the characteristic modes of motion for the airplane, it is necessary to set up and solve the airplane's equations of motion. The equations of motion are developed by application of Newton's laws for each of the airplane's degrees of freedom in turn, and the characteristics of the airplane's modes of motion are obtained by solution of the resulting simultaneous differential equations.

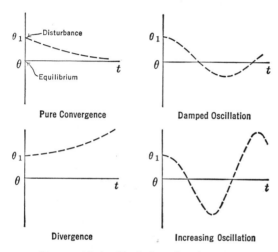

FIGURE 10–1. Typical modes of motion.

The airplane, considered as a rigid body in space, is a dynamic system in six degrees of freedom. Its motion in space must be defined by six components of velocity along and about the airplane axis system defined in Figure 10–2.

The airplane axis system is a right-hand system of Cartesian co-ordinates, with the X and Z axes in the airplane plane of symmetry and the Y axis perpendicular to the plane of symmetry out the right wing. The origin of the airplane axis system is taken at the airplane's center of gravity, and the six velocity components are the linear velocities u, v, and w along these axes and the angular velocities p, q, and r about these axes. The airplane axes move with the airplane.

The mathematical treatment of airplane dynamics is based on methods introduced many years ago by men such as Bryant, Lanchester, and Glauert. The basic theory through which the dynamic characteristics of the airplane are studied is based on the assumption that the disturbed motion of the airplane is one of small oscillations about some steady-state flight condition, and also on the assumption that the changes in the external forces and moments acting on the airplane, because of a small departure from the steady-state motion, depend entirely on the displacement and disturbance velocities along

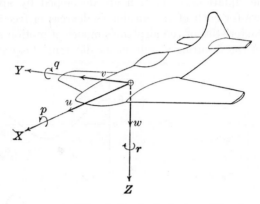

FIGURE 10–2. The airplane's six degrees of freedom.

and about the airplane axes and do not depend on the accelerations involved. It has been proved that these assumptions are well justified for the case of the airplane, and the results obtained correspond with sufficient accuracy to the actual results obtained through flight testing the airplane.

10–2　Development of Equations of Motion (Controls Locked)

The equations of motion for the airplane with controls locked can be written down in accordance with the Newtonian laws of motion which state that the summation of all external forces in any direction must equal the time rate of change of momentum, and the summation of all the moments of the external forces must equal the time rate of change of moment of momentum, all measured with respect to axes fixed in space.

By these laws the six equations of motion of the airplane with locked controls can be written down as follows:

$$\Sigma F_x = ma_x$$

$$\Sigma F_y = ma_y$$

$$\Sigma F_z = ma_z \qquad (10\text{–}1)$$

$$\Sigma L = \frac{dH_x}{dt}$$

$$\Sigma \, M = \frac{dH_y}{dt}$$

$$\Sigma \, N = \frac{dH_z}{dt}$$

where F_x, F_y, and F_z are the summation of the external forces; L, M, and N the moments of the external forces; and H_x, H_y, and H_z the moments of momentum along and about the fixed axes X, Y, and Z, respectively. As the acceleration and rates of change of moments of momentum must all be expressed along axes fixed in space, and the

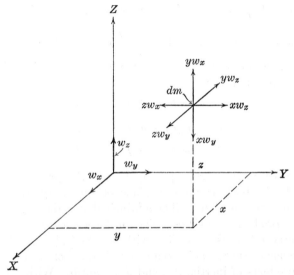

FIGURE 10-3. Moments of momentum of dm.

axes chosen to represent the airplane (Figure 10-2) are moving axes, the acceleration and rates of change of moment of momentum along and about the airplane axes must be referred back to the axes fixed in space.

In order to determine the moment of momentum of a body about the X, Y, and Z axes, consider a body having an angular velocity, w, with components w_x, w_y, and w_z about OX, OY, and OZ, respectively.

The moment of momentum of dm about OX, OY, and OZ is as follows (Figure 10-3):

$$dh_x = w_x(y^2 + z^2) \, dm - w_y xy \, dm - w_z zx \, dm$$
$$dh_y = w_y(z^2 + x^2) \, dm - w_z yz \, dm - w_x xy \, dm \qquad (10\text{-}2)$$
$$dh_z = w_z(x^2 + y^2) \, dm - w_x zx \, dm - w_y yz \, dm$$

For the whole body, the moments of momentum about the three axes are integrals of equations (10–2):

$$h_x = w_x \int (y^2 + z^2)\, dm - w_y \int xy\, dm - w_z \int zx\, dm$$

$$h_y = w_y \int (z^2 + x^2)\, dm - w_z \int yz\, dm - w_x \int xy\, dm \qquad (10\text{–}3)$$

$$h_z = w_z \int (x^2 + y^2)\, dm - w_x \int zx\, dm - w_y \int yz\, dm$$

The integral $\int (y^2 + z^2)\, dm$ is the moment of inertia of the body about the X axis, I_x, and the integral $\int xy\, dm$ is the product of inertia, J_{xy}. Equations (10–3) in terms of moments of inertia and products of inertia are as follows:

$$h_x = w_x I_x - w_y J_{xy} - w_z J_{xz}$$

$$h_y = w_y I_y - w_z J_{yz} - w_x J_{xy} \qquad (10\text{–}4)$$

$$h_z = w_z I_z - w_x J_{xz} - w_y J_{yz}$$

If the axes are chosen as principal axes, the products of inertia vanish. If the body contains a plane of symmetry, then the axis perpendicular to this plane will be a principal axis, thereby eliminating two of the products of inertia. The airplane is a body with a plane of symmetry taken as coinciding with the X-Z plane. The Y axis, perpendicular to this plane of symmetry, is therefore a principal axis, and the products of inertia, J_{xy} and J_{yz}, vanish. With the airplane motion terminology that the angular velocities w_x, w_y, and w_z are simply p, q, and r, and taking the Y axis as a principal axis, the moments of momentum of the airplane with respect to the X, Y, Z axes are as follows:

$$h_x = pI_x - rJ_{xz}$$

$$h_y = qI_y \qquad (10\text{–}5)$$

$$h_z = rI_z - pJ_{xz}$$

The equations of motion given in (10–1) relate to axes fixed in space. If the motion of the airplane is given relative to axes fixed in space, the problem becomes very unwieldy, as the moments and products of inertia would vary from instant to instant. To overcome this difficulty use is made of moving or Eulerian axes which coincide in some particular manner from instant to instant with a definite set of axes fixed in the airplane.

The use of moving axes introduces the complication that accelerations measured with respect to them are not the true accelerations in fixed space to which the Newtonian laws apply. Expressions for the true acceleration may be readily developed as follows.

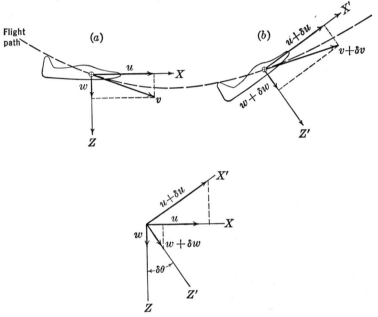

FIGURE 10-4. Acceleration along moving axes referred to fixed space.

Consider the airplane in Figure 10-4 moving at the velocity V as shown at point a. A short interval of time, δt, later the airplane is moving with the velocity $V + \delta V$ as shown at point b. During this interval of time the airplane axes have rotated through the angle $\delta \theta$. From this figure it can be seen that the time acceleration in the direction of the original X axis is:

$$a_x = \frac{(u + \delta u) \cos \delta \theta + (w + \delta w) \sin \delta \theta - u}{\delta t} \tag{10-6}$$

Letting $\cos \delta \theta = 1$ and $\sin \delta \theta = \delta \theta$ and neglecting higher-order terms, equation (10-6) reduces to (10-7) as δt approaches zero:

$$a_x = \frac{du}{dt} + w \frac{d\theta}{dt} = \dot{u} + wq \tag{10-7}$$

In the general case with the airplane axes rotating with the angular velocities p, q, and r, the accelerations relative to fixed space a_x, a_y, and a_z can be developed from Figure 10–5 as follows:

$$a_x = \dot{u} - vr + wq$$
$$a_y = \dot{v} - wp + ur \qquad (10\text{–}8)$$
$$a_z = \dot{w} - uq + vp$$

The rate of change of moment of momentum relative to fixed space can be developed in a similar way from Figure 10–6.

$$\frac{dH_x}{dt} = \frac{dh_x}{dt} - h_y r + h_z q$$
$$\frac{dH_y}{dt} = \frac{dh_y}{dt} - h_z p + h_x r \qquad (10\text{–}9)$$
$$\frac{dH_z}{dt} = \frac{dh_z}{dt} - h_x q + h_y p$$

In these expressions, a_x, a_y, a_z, dH_x/dt, dH_y/dt, dH_z/dt are all measured relative to fixed axes, whereas u, v, w, h_x, h_y, and h_z are measured relative to the moving axis. Equations (10–1) therefore may be written as follows, using equations (10–8) and (10–9):

$$\Sigma\, F_x = m(\dot{u} - vr + wq)$$
$$\Sigma\, F_y = m(\dot{v} - wp + ur)$$
$$\Sigma\, F_z = m(\dot{w} - uq + vp)$$
$$\Sigma\, L = (\dot{h}_x - h_y r + h_z q) \qquad (10\text{–}10)$$
$$\Sigma\, M = (\dot{h}_y - h_z p + h_x r)$$
$$\Sigma\, N = (\dot{h}_z - h_x q + h_y p)$$

Making use of equations (10–5), the equations of motion of the airplane relative to the moving or Eulerian axes become:

$$\Sigma\, F_x = m(\dot{u} - vr + wq)$$
$$\Sigma\, F_y = m(\dot{v} - wp + ur)$$
$$\Sigma\, F_z = m(\dot{w} - uq + vp)$$
$$\Sigma\, L = \dot{p}I_x - \dot{r}J_{xz} + (I_z - I_y)qr - pqJ_{xz} \qquad (10\text{–}11)$$
$$\Sigma\, M = \dot{q}I_y + rp(I_x - I_z) + (p^2 - r^2)J_{xz}$$
$$\Sigma\, N = \dot{r}I_z - \dot{p}J_{xz} + (I_y - I_x)pq + J_{xz}qr$$

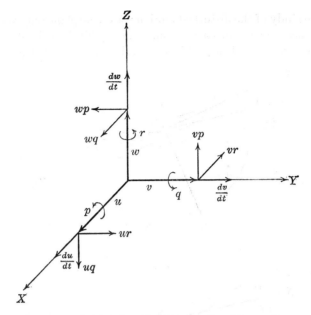

FIGURE 10-5. Acceleration components referred to moving axes.

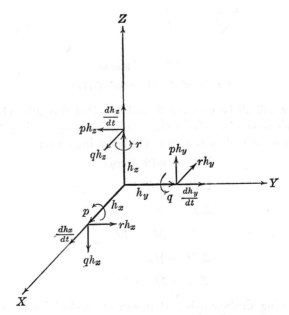

FIGURE 10-6. Moments of momentum referred to moving axes.

In the study of the disturbed motions of the airplane only very small displacements or disturbances from some equilibrium flight condition will be considered. Under this assumption the disturbance velocities

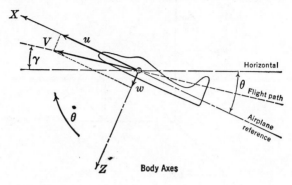

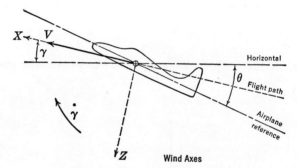

FIGURE 10-7. Body and wind axes.

p, q, and r will all be very small, and it is therefore allowable to disregard their products. Under the condition of very small disturbances the equations of motion reduce to the following form:

$$\Sigma F_x = m(\dot{u} + wq)$$

$$\Sigma F_y = m(\dot{v} + ur - wp)$$

$$\Sigma F_z = m(\dot{w} - uq)$$

$$\Sigma L = \dot{p}I_x - \dot{r}J_{xz} \qquad (10\text{-}12)$$

$$\Sigma M = \dot{q}I_y$$

$$\Sigma N = \dot{r}I_z - \dot{p}J_{xz}$$

The moving airplane axis system can be fixed with reference to the airplane in two different ways. One of these ways is to consider the axes fixed to the airplane under all conditions. These axes are termed body axes, with the X axis along the thrust line or fuselage center line. The other possibility is to consider the X axis always pointing in the direction of the motion or into the relative wind. These axes are usually referred to as wind axes. The X and Z axes therefore rotate slightly with respect to the airplane as the motion proceeds. The wind axes are usually convenient as the component of airplane motion along the X axis will be the airplane's forward velocity $u = V$. The lift and drag forces will therefore always lie along these axes and the components of velocity along the Z axis will be zero ($w = 0$). The angular velocity about the Y axis will be $d\gamma/dt$, where γ is the angle of inclination of the flight path. These axes are shown in Figure 10-7.

If wind axes are chosen as the airplane axes, and the assumption is made that the changes of moments of inertia and products of inertia are negligible for small disturbances, the equations of motion with respect to the wind axes become:

$$\Sigma F_x = m\dot{V}$$

$$\Sigma F_y = m(\dot{v} + Vr)$$

$$\Sigma F_z = -mV\dot{\gamma}$$

$$\Sigma L = \dot{p}I_x - \dot{r}J_{xz} \qquad (10\text{-}13)$$

$$\Sigma M = \dot{q}I_y$$

$$\Sigma N = \dot{r}I_z - \dot{p}J_{xz}$$

Because of the airplane's plane of symmetry small symmetric disturbances will not introduce changes in the external forces and moments along or about the axes outside the plane of symmetry. Conversely small disturbances in roll, yaw, and sideslip will not introduce changes in the symmetric forces and moments. The symmetric degrees of freedom therefore do not couple with the asymmetric degrees of freedom, and it is possible to break the problem of the airplane dynamics down into two separate ones. The first of these involves motion along the X and Z axes and about the Y axis and is termed the airplane's longitudinal or symmetric motion. The second involves motion along the Y axis and about the X and Z axes and is called the lateral motion of the airplane. The problem of longitudinal dynamics will be discussed first.

10-3　Equations of Longitudinal Motion

The equations of longitudinal motion for the airplane with free controls involve the three symmetric equations of (10-13) plus the equation of motion of the elevator control about its own hinge. These equations are as follows:

$$\Sigma F_x = m\dot{V}$$

$$\Sigma F_z = -mV\dot{\gamma}$$

$$\Sigma M = \dot{q}I_y$$

$$\Sigma HM_e = \ddot{\delta}_e I_e$$

$$(10\text{-}14)$$

where HM_e is the total elevator hinge moment in pound-feet, and I_e is the inertia of the elevator about its hinge line.

The major variables to be considered are the change in forward speed, ΔV, the change in angle of attack, $\Delta\alpha$, the change in attitude angle, $\Delta\theta$, and the change in elevator angle, $\Delta\delta_e$. It is assumed that the total change in the forces and moments is made up of the partial derivatives of the forces and moments taken with respect to each of the major variables and with the rate of change of these variables with time.

For example, with the four major variables taken as ΔV, $\Delta\alpha$, $\Delta\theta$, and $\Delta\delta_e$, the total change in the forces along the X axis from some equilibrium can be written as follows:

$$dF_x = \frac{\partial F_x}{\partial V}dV + \frac{\partial F_x}{\partial \alpha}d\alpha + \frac{\partial F_x}{\partial \theta}d\theta + \frac{\partial F_x}{\partial \delta_e}\delta_e + \frac{\partial F_x}{\partial \dot{V}}d\dot{V}$$

$$+ \frac{\partial F_x}{\partial \dot{\alpha}}d\dot{\alpha} + \frac{\partial F_x}{\partial \dot{\theta}}d\dot{\theta} + \frac{\partial F_x}{\partial \dot{\delta}_e}d\dot{\delta}_e \quad (10\text{-}15)$$

If it is assumed that the displacements of each of the variables from the equilibrium condition are small, the partial derivations in (10-15) can be considered linear and the differentials replaced by the actual increments:

$$\Delta F_x = \frac{\partial F_x}{\partial V}\Delta V + \frac{\partial F_x}{\partial \theta}\Delta\theta + \frac{\partial F_x}{\partial \alpha}\Delta\alpha + \frac{\partial F_x}{\partial \delta_e}\Delta\delta_e + \frac{\partial F_x}{\partial \dot{V}}\dot{V}$$

$$+ \frac{\partial F_x}{\partial \dot{\alpha}}\dot{\alpha} + \frac{\partial F_x}{\partial \dot{\theta}}\dot{\theta} + \frac{\partial F_x}{\partial \dot{\delta}_e}\dot{\delta}_e \quad (10\text{-}16)$$

The partial derivatives of (10-16) are usually referred to as stability derivatives and under the assumption of linearity are taken as constants for a given airplane. In many cases several of the partial derivatives are zero, or in other words, a change in some of the variables does not introduce forces along the X axis. For example, a small change in elevator angle, $\Delta \delta_e$, will introduce only a negligible change in forces along the X axis and can therefore be neglected ($\partial F_x/\partial \delta_e = 0$). In like manner it will be shown later on that the only partial derivatives in (10–16) that do exist are $\partial F_x/\partial V$, $\partial F_x/\partial \alpha$, and $\partial F_x/\partial \theta$. In other words, the change in force along the X axis is almost entirely determined by the change in forward velocity, ΔV, the change in angle of attack, $\Delta \alpha$, and the change in the pitch angle, $\Delta \theta$. This, of course, is not at all surprising and as a matter of fact is what might be expected. Equation (10–16) then reduces to the following simple form:

$$\Delta F_x = \frac{\partial F_x}{\partial V} \Delta V + \frac{\partial F_x}{\partial \alpha} \Delta \alpha + \frac{\partial F_x}{\partial \theta} \Delta \theta \qquad (10\text{--}17)$$

The equation of motion for the degree of freedom along the X axis can be written using equations (10–14) and (10–17).

$$\frac{\partial F_x}{\partial V} \Delta V + \frac{\partial F_x}{\partial \alpha} \Delta \alpha + \frac{\partial F_x}{\partial \theta} \Delta \theta = m\dot{V} \qquad (10\text{--}18)$$

The equation of motion for the degree of freedom along the Z axis can be developed in the same manner with $\dot{\gamma} = \dot{\theta} - \dot{\alpha}$,

$$\frac{\partial F_z}{\partial V} \Delta V + \frac{\partial F_z}{\partial \alpha} \Delta \alpha + \frac{\partial F_z}{\partial \theta} \Delta \theta = -mV(\dot{\theta} - \dot{\alpha}) \qquad (10\text{--}19)$$

The equation of motion for the degree of freedom in pitch is developed as follows. In this case the only variables that do not affect the airplane's pitching moment are the pitch angle, θ, and the airplane's acceleration, $\dot{V}$.

$$\frac{\partial M}{\partial V} \Delta V + \frac{\partial M}{\partial \alpha} \Delta \alpha + \frac{\partial M}{\partial \delta_e} \Delta \delta_e + \frac{\partial M}{\partial \dot{\theta}} \dot{\theta} + \frac{\partial M}{\partial \dot{\alpha}} \dot{\alpha} + \frac{\partial M}{\partial \dot{\delta}_e} \dot{\delta}_e$$

$$= mk_y^2 \ddot{\theta} \qquad (10\text{--}20)$$

where k_y is the airplane's radius of gyration about the Y axis.

The equation of motion of the elevator about its hinge line is slightly more complicated because of the mass effects of the elevator. If the stability derivatives are considered as arising from purely aerodynamic or pressure forces, the only negligible variables will be the pitch angle, θ, and the acceleration, $\dot{V}$. The elevator statically balanced with its c.g. off the hinge line will develop hinge moments due to the airplane's acceleration components. These can be seen readily from Figure

10-8. m_e is elevator mass, and x_e is distance of elevator c.g. behind the hinge line.

The hinge moments developed because of the acceleration components on the airplane can be given as follows:

$$HM_e = -\,I_e\ddot{\theta} - m_e x_e l_t \ddot{\theta} + m_e x_e V(\dot{\theta} - \dot{\alpha}) \qquad (10\text{-}21)$$

or by letting $m_e x_e = H_e$, the mass moment of the elevator

$$HM_e = -I_e\ddot{\theta} - H_e(V\dot{\alpha} - V\dot{\theta} + l_t\ddot{\theta}) \qquad (10\text{-}22)$$

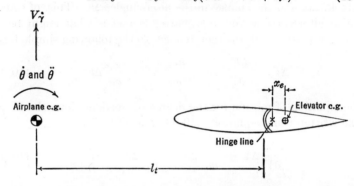

FIGURE 10-8.   Elevator kinematics.

The equation of motion of the elevator about its hinge line becomes:

$$\frac{\partial HM}{\partial V}\Delta V + \frac{\partial HM}{\partial \alpha}\Delta\alpha + \frac{\partial HM}{\partial \delta_e}\Delta\delta_e + \frac{\partial HM}{\partial \dot{\theta}}\dot{\theta} + \frac{\partial HM}{\partial \dot{\alpha}}\dot{\alpha}$$

$$+\,\frac{\partial HM}{\partial \dot{\delta}_e}\dot{\delta}_e = I_e(\ddot{\theta} + \ddot{\delta}_e) + H_e(V\dot{\alpha} - V\dot{\theta} + l_t\ddot{\theta}) \qquad (10\text{-}23)$$

The final equations of motion for the airplane in its longitudinal degrees of freedom are the four equations (10-18), (10-19), (10-20), and (10-23) taken simultaneously.

$$\frac{\partial F_x}{\partial V}\Delta V + \frac{\partial F_x}{\partial \alpha}\Delta\alpha + \frac{\partial F_x}{\partial \theta}\Delta\theta = m\dot{V} \qquad (10\text{-}24a)$$

$$\frac{\partial F_z}{\partial V}\Delta V + \frac{\partial F_z}{\partial \alpha}\Delta\alpha + \frac{\partial F_z}{\partial \theta}\Delta\theta = -mV(\dot{\theta} - \dot{\alpha}) \qquad (10\text{-}24b)$$

$$\frac{\partial M}{\partial V}\Delta V + \frac{\partial M}{\partial \alpha}\Delta\alpha + \frac{\partial M}{\partial \delta_e}\Delta\delta_e + \frac{\partial M}{\partial \theta}\dot{\theta}$$

$$+\,\frac{\partial M}{\partial \dot{\alpha}}\dot{\alpha} + \frac{\partial M}{\partial \dot{\delta}_e}\dot{\delta}_e = m k_y{}^2\ddot{\theta} \qquad (10\text{-}24c)$$

$$\frac{\partial HM}{\partial V}\,\Delta V + \frac{\partial HM}{\partial \alpha}\,\Delta \alpha + \frac{\partial HM}{\partial \delta_e}\,\Delta \delta_e + \frac{\partial HM}{\partial \dot{\alpha}}\,\dot{\alpha}$$

$$+ \frac{\partial HM}{\partial \dot{\theta}}\,\dot{\theta} + \frac{\partial HM}{\partial \dot{\delta}_e}\,\dot{\delta}_e = I_e(\ddot{\theta} + \ddot{\delta}_e)$$

$$+ H_e(V\dot{\alpha} - V\dot{\theta} + l_t\ddot{\theta}) \quad (10\text{-}24d)$$

In order to solve these simultaneous equations for the longitudinal modes of motion, it is necessary to evaluate the partial derivatives, the constants of (10–24), and to bring the equation into simpler form for ease of solution.

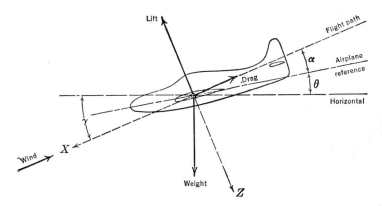

FIGURE 10–9. Airplane forces in plane of symmetry, power off.

If equation (10–24a) is taken first, the summation of forces along the X axis in a power-off ($T_c = 0$) glide is given in (10–25) as developed from Figure 10–9.

$$F_x = -C_D \tfrac{1}{2}\rho V^2 S - W(\theta - \alpha) \quad (10\text{-}25)$$

The partial derivatives of F_x with respect to the four variables $(V, \alpha, \theta, \delta_e)$ are as follows:

$$\frac{\partial F_x}{\partial V} = -C_D \rho S V \quad (10\text{-}26)$$

$$\frac{\partial F_x}{\partial \alpha} = -\frac{dC_D}{d\alpha}\frac{1}{2}\rho V^2 S + W \quad (10\text{-}27)$$

$$\frac{\partial F_x}{\partial \theta} = -W \quad (10\text{-}28)$$

$$\frac{\partial F_x}{\partial \delta_e} = 0 \quad (10\text{-}29)$$

Equation (10–24a) may therefore be rewritten

$$-C_{D\rho}SV\,\Delta V - \frac{dC_D}{d\alpha}\frac{1}{2}\rho V^2 S\,\Delta\alpha + W\,\Delta\alpha - W\,\Delta\theta = m\dot{V} \quad (10\text{–}30)$$

Dividing equation (10–30) by ρSV^2,

$$-C_D\frac{\Delta V}{V} - \frac{1}{2}\frac{dC_D}{d\alpha}\Delta\alpha + \frac{W}{\rho SV^2}\Delta\alpha - \frac{W}{\rho SV^2}\Delta\theta = \frac{m}{\rho SV^2}\dot{V} \quad (10\text{–}31)$$

If the substitution $C_L = \dfrac{2(W/S)}{\rho V^2}$ is made and if the speed ratio

$\Delta V/V$ is called simply u and $\dfrac{\dot{V}}{V} = \dot{u}$, equation (10–31) becomes

$$-C_D u - \left(\frac{1}{2}\frac{dC_D}{d\alpha} - \frac{C_L}{2}\right)\Delta\alpha - \frac{C_L}{2}\Delta\theta = \frac{m}{\rho SV}\dot{u} \quad (10\text{–}32)$$

The left-hand side of (10–32) is non-dimensional. The derivative $\dot{u}$ has the dimension of one over time $(1/T)$, while the factor $m/\rho SV$ has the dimensions of time. If the symbol τ is assigned to represent the factor $m/\rho SV$, and if time is counted in terms of the time ratio t/τ, the right-hand side of (10–22) can be expressed as $\dfrac{du}{d(t/\tau)}$. If use is made of the operator $\mathrm{d} = \dfrac{d}{d(t/\tau)}$, the right-hand side of (10–32) becomes simply $\mathrm{d}u$. Finally, if all terms of (10–32) are collected on the left-hand side and the short-hand notation used $(C_{D\alpha} = dC_D/d\alpha)$, it follows that:

$$(C_D + \mathrm{d})u + \left(\frac{C_{D\alpha}}{2} - \frac{C_L}{2}\right)\Delta\alpha + \frac{C_L}{2}\Delta\theta = 0 \quad (10\text{–}33)$$

This is then the final non-dimensional form of (10–14a) expressed in familiar aerodynamic terms.

The development of equation (10–24b) involving the summation of forces along the Z axis proceeds in a similar manner. If Figure 10–9 is again referred to, the forces acting along the Z axis become:

$$F_z = -C_L\tfrac{1}{2}\rho V^2 S + W \quad (10\text{–}34)$$

The partial derivatives of F_z with respect to the four variables are as follows:

$$\frac{\partial F_z}{\partial V} = -C_L\rho SV \quad (10\text{–}35)$$

$$\frac{\partial F_z}{\partial \alpha} = - \frac{dC_L}{d\alpha} \frac{1}{2} \rho V^2 S \qquad (10\text{--}36)$$

$$\frac{\partial Z}{\partial \theta} = 0 \qquad (10\text{--}37)$$

$$\frac{\partial Z}{\partial \delta_e} = 0 \qquad (10\text{--}38)$$

Substituting (10–35) and (36) into (10–24b) and non-dimensionalizing in the same way described before give the final developed equation for the Z axis:

$$C_L u + \left(\frac{C_{L\alpha}}{2} + d\right) \Delta\alpha - d\, \Delta\theta = 0 \qquad (10\text{--}39)$$

The partial derivatives for the equation of motion in pitch (10–24c) can be put in terms of the pitching moment coefficients as follows:

$$M = C_m \tfrac{1}{2}\rho V^2 Sc \qquad (10\text{--}40)$$

and the partial derivatives become

$$\frac{\partial M}{\partial V} = \frac{\partial C_m}{\partial V} \frac{1}{2} \rho V^2 Sc \qquad (10\text{--}41)$$

$$\frac{\partial M}{\partial \alpha} = \frac{\partial C_m}{\partial \alpha} \frac{1}{2} \rho V^2 Sc \qquad (10\text{--}42)$$

$$\frac{\partial M}{\partial \delta_e} = \frac{\partial C_m}{\partial \delta_e} \frac{1}{2} \rho V^2 Sc \qquad (10\text{--}43)$$

$$\frac{\partial M}{\partial \dot\theta} = \frac{\partial C_m}{\partial \dot\theta} \frac{1}{2} \rho V^2 Sc \qquad (10\text{--}44)$$

$$\frac{\partial M}{\partial \dot\alpha} = \frac{\partial C_m}{\partial \dot\alpha} \frac{1}{2} \rho V^2 Sc \qquad (10\text{--}45)$$

$$\frac{\partial M}{\partial \dot\delta_e} = \frac{\partial C_m}{\partial \dot\delta_e} \frac{1}{2} \rho V^2 Sc \qquad (10\text{--}46)$$

If (10–41) to (46) are substituted into (10–24c), and the resulting equation divided through by $\tfrac{1}{2}\rho V^2 Sc$:

$$\frac{\partial C_m}{\partial V} \Delta V + \frac{\partial C_m}{\partial \alpha} \Delta\alpha + \frac{\partial C_m}{\partial \delta_e} \Delta\delta_e + \frac{\partial C_m}{\partial \dot\alpha} \dot\alpha$$

$$+ \frac{\partial C_m}{\partial \dot\theta} \dot\theta + \frac{\partial C_m}{\partial \dot\delta_e} \dot\delta_e = \frac{2mk_y^2}{\rho SV^2 c} \ddot\theta \qquad (10\text{--}47)$$

The first term of (10–47) can be made non-dimensional by multiplying and dividing by V, giving $(\partial C_m/\partial u)u$. Also the time derivatives, for example $(\partial C_m/\partial \dot{\alpha})\dot{\alpha}$, can be multiplied and divided by the time parameter, τ, giving $\dfrac{\partial C_m}{\partial\left(\dfrac{d\alpha}{d(t/\tau)}\right)} \dfrac{d\alpha}{d(t/\tau)}$, or in operator form $C_{md\alpha}d\alpha$;

likewise $\dfrac{\partial C_m}{\partial \dot{\theta}}\,\dot{\theta}$ becomes $C_{md\theta}d\theta$, etc. With these substitutions and with the usual derivative short-hand, (10–47) becomes

$$C_{m_u}u + C_{m_\alpha}\Delta\alpha + C_{m\delta_e}\Delta\delta_e + C_{md\alpha}d\alpha + C_{md\theta}d\theta$$
$$+ C_{md\delta_e}d\delta_e = \frac{2mk_y^2}{\rho SV^2c}\ddot{\theta} \quad (10\text{–}48)$$

The right-hand side of (10–48) can be placed in terms of non-dimensional time as before, giving

$$\frac{2mk_y^2\ddot{\theta}}{\rho SV^2c} = 2\left(\frac{k_y}{c}\right)^2 \frac{\rho Sc}{m}\, d^2\theta \quad (10\text{–}49)$$

The factor $m/\rho Sc$ is dimensionless and for convenience is assigned the symbol μ. It is usually referred to in the literature as the airplane's relative density factor. If use is made of this symbol, (10–49) simply becomes $\dfrac{2k_y^2}{\mu c^2}\, d^2\theta$, and equation (10–48) reduces to:

$$C_{m_u}u + (C_{m_\alpha} + C_{md\alpha}d)\alpha + \left(C_{md\theta}d - \frac{2k_y^2}{\mu c^2}\, d^2\right)\theta$$
$$+ (C_{m\delta} + C_{md\delta}d)\delta_e = 0 \quad (10\text{–}50)$$

In this equation the variables $\Delta\alpha$ and $\Delta\theta$ are written simply α and θ; the Δ is implied.

This is the final non-dimensional form of equation (10–24c). The evaluation of the derivatives (C_{m_u}, C_{m_α}, $C_{md\alpha}$, etc.) will be discussed in the next section.

The partial derivatives for the equation of motion of the elevator about its hinge (10–24d) can be put in terms of the hinge moment coefficient as follows:

$$HM = C_h \tfrac{1}{2}\rho V^2 S_e c_e \quad (10\text{–}51)$$

The partial derivatives of H with respect to the variables become

$$\frac{\partial HM}{\partial V} = \frac{\partial C_h}{\partial V}\frac{1}{2}\rho V^2 S_e c_e \quad (10\text{–}52)$$

$$\frac{\partial H}{\partial \alpha} = \frac{\partial C_h}{\partial \alpha} \frac{1}{2} \rho V^2 S_e c_e \qquad (10\text{-}53)$$

$$\frac{\partial H}{\partial \delta_e} = \frac{\partial C_h}{\partial \delta_e} \frac{1}{2} \rho V^2 S_e c_e \qquad (10\text{-}54)$$

$$\frac{\partial H}{\partial \dot{\alpha}} = \frac{\partial C_h}{\partial \dot{\alpha}} \frac{1}{2} \rho V^2 S_e c_e \qquad (10\text{-}55)$$

$$\frac{\partial H}{\partial \dot{\theta}} = \frac{\partial C_h}{\partial \dot{\theta}} \frac{1}{2} \rho V^2 S_e c_e \qquad (10\text{-}56)$$

$$\frac{\partial H}{\partial \dot{\delta}_e} = \frac{\partial C_h}{\partial \dot{\delta}_e} \frac{1}{2} \rho V^2 S_e c_e \qquad (10\text{-}57)$$

Substituting these partial derivatives into (10–24d) and dividing by $\frac{1}{2}\rho V^2 S_e c_e$:

$$C_{h_u} u + C_{h_\alpha} \Delta\alpha + C_{h_{\delta e}} \Delta\delta_e + C_{h_{\dot\alpha}} \dot\alpha + C_{h_{\dot\theta}} \dot\theta$$

$$+ C_{h_{\dot\delta_e}} \dot\delta_e = \frac{I_e(\ddot\theta + \ddot\delta_e)}{\frac{1}{2}\rho V^2 S_e c_e} + \frac{H_e(V\dot\alpha - V\dot\theta + l_t\ddot\theta)}{\frac{1}{2}\rho V^2 S_e c_e} \qquad (10\text{-}58)$$

Making use of the airplane's relative density factor, μ, and defining an elevator density parameter, $\mu_e = m_e/\rho S_e c_e$, the two terms on the right-hand side of (10–58) become

$$\frac{I_e(\ddot\theta + \ddot\delta_e)}{\frac{1}{2}\rho V^2 S_e c_e} = \frac{2\mu_e k_e^2}{\mu^2 c^2} d^2(\theta + \delta_e) \qquad (10\text{-}59)$$

and

$$\frac{H_e}{\frac{1}{2}\rho V^2 S_e c_e}(V\dot\alpha - V\dot\theta + l_t\ddot\theta) = \frac{2\mu_e x_e}{\mu c} d(\alpha - \theta) + \frac{2\mu_e x_e l_t}{\mu^2 c^2} d^2\theta \qquad (10\text{-}60)$$

The following parameters are defined for convenience:

$$\frac{2k_y^2}{\mu c^2} = h \qquad (10\text{-}61)$$

$$\frac{2\mu_e x_e}{c\mu} = h_1 \qquad (10\text{-}62)$$

$$\frac{2\mu_e k_e^2}{\mu^2 c^2} = h_2 \qquad (10\text{-}63)$$

$$\frac{2\mu_e x_e l_t}{\mu^2 c^2} = l_1 \qquad (10\text{-}64)$$

Making use of (10–59) to (64) in equation (10–58), the final non-dimensional hinge moment equation of motion is obtained:

$$C_{h_u}u + (C_{h_\alpha} - h_1\mathrm{d})\Delta\alpha + (C_{h_{d\theta}}\mathrm{d} - h_2\mathrm{d}^2 + h_1\mathrm{d} - l_1\mathrm{d}^2)\Delta\theta$$
$$+ (C_{h\delta} + C_{h_{d\delta}}\mathrm{d} - h_2\mathrm{d}^2)\Delta\delta_e = 0 \qquad (10\text{–}65)$$

Collecting equations (10–33), (10–39), (10–50), and (10-64), the four non-dimensional simultaneous equations are obtained:

$$(C_D + \mathrm{d})u + \frac{1}{2}(C_{D_\alpha} - C_L)\alpha + \frac{C_L}{2}\theta = 0 \qquad (10\text{–}66a)$$

$$C_L u + (\tfrac{1}{2}C_{L_\alpha} + \mathrm{d})\alpha - \mathrm{d}\theta = 0 \qquad (10\text{–}66b)$$

$$C_{m_u}u + (C_{m_\alpha} + C_{m_{d\alpha}}\mathrm{d})\alpha + (C_{m_{d\theta}}\mathrm{d} - h\mathrm{d}^2)\theta$$
$$+ (C_{m\delta} + C_{m_{d\delta}}\mathrm{d})\delta_e = 0 \qquad (10\text{–}66c)$$

$$C_{h_u}u + (C_{h_\alpha} - h_1\mathrm{d})\alpha + (C_{h_{d\theta}}\mathrm{d} - h_2\mathrm{d}^2 + h_1\mathrm{d} - l_1\mathrm{d}^2)\theta$$
$$+ (C_{h\delta} + C_{h_{d\delta}}\mathrm{d} - h_2\mathrm{d}^2)\delta_e = 0 \qquad (10\text{–}66d)$$

10–4 Evaluation of Stability Derivatives

The four equations (10–66a–d) are simultaneous homogeneous differential equations with constant coefficients. The constant coefficients of these equations are made up of the airplane mass and inertia parameters and the so-called stability derivatives. In this section the evaluation of these derivatives will be studied in order that solution of the simultaneous equations can proceed.

For convenience all the angles in the sections on dynamics will be taken in radian measure.

1. C_{L_α}, the slope of the lift curve, a function of the airfoil section lift characteristics, a_0, and the wing aspect ratio. This derivative can be estimated from the curves given in Figure 5–5.

2. C_{D_α}, the rate of change of the drag coefficient with angle of attack. Obtained from the basic drag formula

$$C_D = C_{D_f} + \frac{C_L{}^2}{\pi e A}$$
$$(10\text{–}67)$$
$$C_{D_\alpha} = \frac{2C_L}{\pi e A}C_{L_\alpha}$$

3. C_{m_α}, the static longitudinal stability criterion $dC_m/d\alpha$

$$\frac{dC_m}{d\alpha} = \frac{dC_m}{dC_L}\cdot\frac{dC_L}{d\alpha} \qquad (10\text{–}68)$$

or $C_{m_\alpha} = C_{L_\alpha}(x_{cg} - N_0)$, where N_0 is the stick-fixed neutral point.

4. $C_{m_{d\alpha}}$, the rate of change of pitching moment coefficient with rate of change of angle of attack with respect to t/τ. This derivative arises because of the lag in the wing downwash getting from wing to tail. For an airplane whose angle of attack is increasing at the rate $d\alpha/dt$, the tail angle of attack, at any instant of time, corresponding to the wing angle of attack, α_w, will be:

$$\alpha_t = \alpha_w - \epsilon - i_w + i_t \qquad (10\text{-}69)$$

where

$$\epsilon = \frac{d\epsilon}{d\alpha}\left(\alpha_w - \frac{d\alpha}{dt}\,\Delta t\right) \qquad (10\text{-}70)$$

where Δt is the time it takes an air particle to go from the wing to the tail, $\Delta t = l_t/V$.

The pitching moment coefficient due to the tail becomes

$$C_{m_t} = -a_t\bar{V}\eta_t\left[\alpha_w - \frac{d\epsilon}{d\alpha}\left(\alpha_w - \frac{d\alpha}{dt}\frac{l_t}{V}\right) - i_w + i_t\right] \qquad (10\text{-}71)$$

therefore

$$\frac{dC_m}{d(d\alpha/dt)} = -a_t\bar{V}\eta_t\frac{l_t}{V}\frac{d\epsilon}{d\alpha} \qquad (10\text{-}72)$$

Dividing each side of (10-72) by the time parameter, τ, and making use of the relative airplane density factor, μ, we obtain:

$$\frac{dC_m}{d\left(\dfrac{d\alpha}{d(t/\tau)}\right)} = -a_t\bar{V}\eta_t\frac{l_t}{V}\frac{d\epsilon}{d\alpha}\frac{1}{\tau} \qquad (10\text{-}73)$$

or

$$C_{m_{d\alpha}} = -a_t\bar{V}\eta_t\frac{1}{\mu}\frac{l_t}{c}\frac{d\epsilon}{d\alpha} \qquad (10\text{-}74)$$

5. $C_{m_{d\theta}}$, the airplane's damping in pitch. There are several contributions to the longitudinal damping of the airplane, but the largest of these is the contribution of the horizontal tail. The wing, fuselage, and propeller all add to this derivative, but as in most normal airplane designs the horizontal tail is by far the largest factor; it is normal practice to evaluate the damping contribution of the horizontal tail and then lump all other contributions as a multiplying factor on the tail damping. This factor is usually taken equal to 1.10.

The damping contribution of the horizontal tail can be estimated by determining the change in tail angle of attack due to the airplane's angular velocity, $\dot{\theta}$.

$$\Delta\alpha_t = \dot{\theta}\frac{l_t}{V} \qquad (10\text{-}75)$$

The tail pitching moment produced by this tail angle of attack becomes

$$C_{mt} = -a_t \bar{V} \eta_t \frac{d\theta}{d_t} \frac{l_t}{V} \tag{10-76}$$

$$\frac{dC_{mt}}{d\left(\dfrac{d\theta}{dt}\right)} = -a_t \bar{V} \eta_t \frac{l_t}{V} \tag{10-77}$$

Dividing each side by τ,

$$\frac{dC_{mt}}{d\left(\dfrac{d\theta}{d(t/\tau)}\right)} = -a_t \bar{V} \eta_t \frac{l_t}{V\tau} \tag{10-78}$$

or

$$C_{md\theta} = -a_t \bar{V} \eta_t \frac{l_t}{c} \frac{1}{\mu} \tag{10-79}$$

to account for the rest of the airplane multiply by 1.10, giving finally:

$$C_{md\theta} = -1.1 a_t \bar{V} \eta_t \frac{l_t}{c} \frac{1}{\mu} \tag{10-80}$$

6. $C_{m\delta}$, the elevator power term discussed in Chapter 5,

$$C_{m\delta} = -a_t \bar{V} \eta_t \tau \tag{10-81}$$

7. The pitching moment due to the rate of elevator deflection with t/τ can be computed from the following formula:

$$C_{md\delta} = -\frac{1}{2\mu} [A + a_t B] \bar{V} \frac{c_t}{c_w} \tag{10-82}$$

where A and B are constants given in Figure 10–10, with c_t equal to the mean chord of the horizontal tail.

8. $C_{h\alpha}$, the rate of change of elevator hinge moment coefficient with wing angle of attack. The variation of the elevator hinge moment with tail angle of attack, $C_{h\alpha t}$, was discussed in Chapter 6. The derivative $C_{h\alpha}$ can be developed in terms of $C_{h\alpha t}$ by accounting for the wing downwash.

$$C_{h\alpha} = C_{h\alpha t}\left(1 - \frac{d\epsilon}{d\alpha}\right) \tag{10-83}$$

9. $C_{h\delta}$, the rate of change of elevator hinge moment with control deflection. Also treated in detail in Chapter 6.

10. $C_{hd\delta}$, the elevator damping derivative, can be developed from the following formula:

$$C_{h_{d\delta}} = -\frac{1}{2\mu}[C + Da_t]\frac{c_t}{c_w} \qquad (10\text{-}84)$$

where C and D are constants given in Figure 10-10, also taken from NACA TR 709.

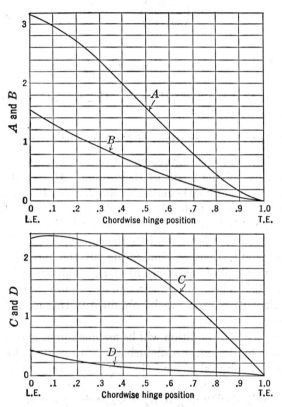

FIGURE 10-10.   Flap rate derivatives.   From NACA TR 709, "An Analysis of the Stability of the Airplane with Free Controls," by Jones and Cohen.

11. $C_{h_{d\theta}}$, the rate of change of elevator hinge moment with rate of change of airplane pitch angle with t/τ. This derivative arises as a result of the change in angle of attack of the horizontal tail with airplane pitching velocity and the floating tendency of the elevator.

$$C_h = C_{h_{\alpha_t}}\theta\frac{l_t}{V} \qquad (10\text{-}85)$$

$$\frac{dC_h}{d\left(\dfrac{d\theta}{dt}\right)} = C_{h_{\alpha t}} \frac{l_t}{V} \qquad (10\text{--}86)$$

Dividing both sides of (10–86) by τ,

$$\frac{dC_h}{d\left(\dfrac{d\theta}{d(t/\tau)}\right)} = C_{h_{\alpha t}} \frac{l_t}{V\tau} \qquad (10\text{--}87)$$

or

$$C_{h_{d\theta}} = C_{h_{\alpha t}} \frac{l_t}{\mu c} \qquad (10\text{--}88)$$

It is assumed that the derivatives C_{m_μ}, C_{h_μ}, $C_{h_{d\alpha}}$, $C_{h_{\mathrm{d}^2\alpha}}$, and $C_{m_{\mathrm{d}^2\alpha}}$ are either zero or negligible.

10–5 Solution of the Equations of Motion (Stick-fixed Case)

The solutions to the simultaneous differential equations (10–66) can be obtained by solving them as they stand. However, it has been found convenient and mathematically simpler to solve these equations in two steps. The first step is to consider the elevator fixed, which eliminates the $\Delta\delta_e$ terms in (10–66c) and all of equation (10–66d), while the second step is to consider the elevator free, with the speed assumed constant, thereby eliminating all the terms containing u as a variable and all of equation (10–66a). This breakdown is allowable because it has been shown through experience that the important modes of airplane motion can be obtained in this manner with only small losses in accuracy.

The elevator-fixed condition will be assumed first, and the nature of the airplane's motion after a disturbance with controls locked will be investigated by solution of the following reduced equations of motion:

$$(C_D + \mathrm{d})u + \frac{1}{2}(C_{D_\alpha} - C_L)\alpha + \frac{C_L}{2}\theta = 0 \qquad (10\text{--}89a)$$

$$C_L u + (\tfrac{1}{2}C_{L_\alpha} + \mathrm{d})\alpha - \mathrm{d}\theta = 0 \qquad (10\text{--}89b)$$

$$(C_{m_\alpha} + C_{m_{d\alpha}}\mathrm{d})\alpha + (C_{m_{d\theta}}\mathrm{d} - h\mathrm{d}^2)\theta = 0 \qquad (10\text{--}89c)$$

The solutions to these equations of motion are obtained by assuming the solution in the form:

$$u = u_1 e^{\lambda\, t/\tau} \qquad \alpha = \alpha_1 e^{\lambda\, t/\tau} \qquad \theta = \theta_1 e^{\lambda\, t/\tau} \qquad (10\text{--}90)$$

where λ is a real or complex constant of equal value for each variation, and where u_1, α_1, and θ_1 are also real or complex constants.

If the above values for the variables are substituted into equation (10–89), together with their first and second derivatives (i.e., $du = u_1\lambda e^{\lambda\,t/\tau}$ and $d^2u = u_1\lambda^2 e^{\lambda\,t/\tau}$, etc.) and the common term $e^{\lambda\,t/\tau}$ divided out, equation (10–89) reduces to three algebraic equations in the unknown λ, and the new variables u_1, α_1, and θ_1.

$$(C_D + \lambda)u_1 + \frac{1}{2}(C_{D\alpha} - C_L)\alpha_1 + \frac{C_L}{2}\theta_1 = 0$$

$$C_L u_1 + (\tfrac{1}{2}C_{L\alpha} + \lambda)\alpha_1 - \lambda\theta_1 = 0 \qquad (10\text{–}91)$$

$$(C_{m\alpha} + C_{md\alpha}\lambda)\alpha_1 + (C_{md\theta}\lambda - h\lambda^2)\theta_1 = 0$$

These equations are now homogeneous algebraic equations, and as such are subject to the requirement that, for them to be consistent, the determinant of their coefficients must vanish.

$$\begin{vmatrix} C_D + \lambda & \tfrac{1}{2}(C_{D\alpha} - C_L) & \dfrac{C_L}{2} \\[2mm] C_L & (\tfrac{1}{2}C_{L\alpha} + \lambda) & -\lambda \\[2mm] 0 & (C_{m\alpha} + C_{md\alpha}\lambda) & (C_{md\theta}\lambda - h\lambda^2) \end{vmatrix} = 0 \quad (10\text{–}92)$$

By expanding this determinant a quartic equation in λ is obtained, the roots of which are the four values of λ that determine the final solution.

$$A\lambda^4 + B\lambda^3 + C\lambda^2 + D\lambda + E \qquad (10\text{–}93)$$

$$u = u_1 e^{\lambda_1\,t/\tau} + u_2 e^{\lambda_2\,t/\tau} + u_3 e^{\lambda_3\,t/\tau} + u_4 e^{\lambda_4\,t/\tau} \text{ etc.} \qquad (10\text{–}94)$$

In order to study the characteristic motions of any dynamic system, it is not necessary to establish the values of the coefficients u_1, u_2, u_3, and u_4, as these will vary only as a function of the initial disturbance when $t/\tau = 0$. The thing that is important is to determine the character of the motion, if oscillatory the period and damping, and if aperiodic the rate of convergence or divergence. This can be accomplished readily by investigating the four values of λ obtained from the quartic (10–93).

If all the λ's come out as real numbers, the motion is aperiodic, convergent if negative, divergent if positive. If any of the λ's form a complex pair, the motion is oscillatory, damped if the real part is negative, undamped if the real part is positive. This is demonstrated in Figure 10–11.

The study, then, of the character of the motion of the airplane after a disturbance from equilibrium is a study of the roots of the quartic

(a) All λ's real: 2 negative, 2 positive

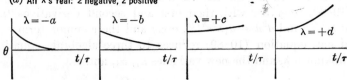

(b) One complex pair, 2 real

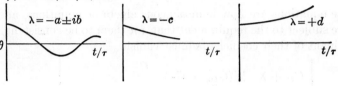

(c) Two complex pairs, 1 damped, 1 undamped

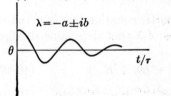
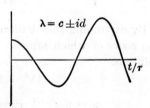

FIGURE 10–11. Typical motions, four roots.

equation in λ (10–93). This equation is often referred to as the stability quartic.

The coefficients of the stability quartic can be obtained by expansion of the determinant in λ (10–93) and collection of like powers of λ. The result of this expansion yields the following:

$$A = 1$$

$$B = \frac{1}{2} C_{L\alpha} + C_D - \frac{1}{h} C_{md\theta} - \frac{1}{h} C_{md\alpha}$$

$$C = \frac{1}{2} C_D C_{L\alpha} + \frac{C_L{}^2}{2} - \frac{C_{md\theta}}{2h} C_{L\alpha} - \frac{C_D}{h} C_{md\theta} - \frac{C_L}{2} C_{D\alpha}$$

$$- \frac{1}{h} C_{m\alpha} - \frac{C_{md\alpha}}{h} C_D \quad (10\text{–}95)$$

$$D = \frac{C_L}{2h} C_{D\alpha} C_{m\mathrm{d}\theta} - \frac{C_D}{2h} C_{m\mathrm{d}\theta} C_{L\alpha} - \frac{C_L{}^2}{2h} C_{m\mathrm{d}\alpha} - \frac{C_L{}^2}{2h} C_{m\mathrm{d}\theta}$$
$$- \frac{C_D}{h} C_{m\alpha}$$

$$E = - \frac{C_L{}^2}{2h} C_{m\alpha}$$

Once the coefficients A–E above, of the quartic in λ, are determined for any given airplane and flight conditions, inspection of the quartic will yield valuable information concerning the motion before the roots are obtained.   If all the coefficients are positive, there can be no positive real root and there is no possibility of a pure divergence.   If a combination of the coefficients known as Routh's discriminant $(BCD - AD^2 - B^2E)$, is positive, then there is no possibility of the real part of any complex pair being positive and there will be no undamped oscillation.   If Routh's discriminant is equal to zero, there will be a neutrally damped oscillation; if negative, there will be one complex pair with a positive real part, implying an undamped oscillation.   If the term $E = 0$, then, one of the roots is zero and one of the modes can continue unchanged indefinitely; while if one of the coefficients is negative, there can be either an increasing oscillation or a pure divergence in one of the modes.

Several methods are available for extracting the roots from quartic equations.   One of these is the normal analytical treatment given in any book on algebra.   Another method,* which at times is much faster, especially for the case where all roots are complex, is a graphical method and is the one described herein.   The graphical method first assumes the quartic equation broken down into two quadratic equations.

$$(\lambda^2 + a_1\lambda + b_1)(\lambda^2 + a_2\lambda + b_2) = 0 \qquad (10\text{–}96)$$

Expanding these quadratics and equating the like powers of λ yield the following:

$$A = 1$$
$$B = a_1 + a_2$$
$$C = b_1 + a_1 a_2 + b_2 \qquad (10\text{–}97)$$
$$D = a_2 b_1 + a_1 b_2$$
$$E = b_1 b_2$$

* NACA TR 589.

As these equations are symmetric in a and b, it is possible by elimination to form the following two general equations:

$$a = \frac{B}{2} \pm \sqrt{\left(\frac{B}{2}\right)^2 - C + b + \frac{E}{b}}, \quad \text{and} \quad a = \frac{b^2 B - bD}{b^2 - E} \quad (10\text{--}98)$$

If these equations are plotted on rectangular coordinates a versus b, there will be two intersections, (a_1, b_1) and (a_2, b_2). These values are placed in the quadratics (10–96) which can readily be solved for the roots. After experience is gained in this method, the roots can be obtained quite rapidly and with any degree of accuracy required.

If the aerodynamic characteristics of a typical airplane are substituted into equations (10–95), the coefficients $(A-E)$ of the stability quartic in λ are obtained. The roots of this quartic for a statically stable airplane in almost all cases combine into two complex pairs.

$$\lambda_{1,2} = \xi_1 \pm i\eta_1$$
$$\lambda_{3,4} = \xi_2 \pm i\eta_2$$

$(10\text{--}99)$

The combining of the roots into two complex pairs indicates that the airplane's longitudinal motion, after a disturbance with the elevator locked, has two oscillatory modes. The period and damping of these modes can be obtained by making use of the relations given below:

$$\text{Period} = \frac{2\pi}{\eta} \tau \text{ seconds}$$

$$\text{Time to damp to } \tfrac{1}{2} \text{ amplitude} = \frac{.693}{\xi} \tau \text{ seconds}$$

Typical values for the roots are as follows:

$$\lambda_{1,2} = -.02 \pm i\,0.30$$
$$\lambda_{3,4} = -2.0 \pm i\,2.5$$

These indicate the following oscillations for an airplane with a time characteristic $\tau = 1.5$.

$\lambda_{1,2}$—Period 31.5 seconds, time to damp to half amplitude 52 seconds

$\lambda_{3,4}$—Period 3.77 seconds, time to damp to half amplitude .52 second

The characteristic modes of stick-fixed longitudinal motion for nearly all airplanes are two oscillations, one of long period with poor damping and the other of short period with heavy damping. The first of these oscillations is usually referred to as the phugoid mode or long period mode, while the second is referred to merely as the short period mode or second mode.

The second or short period mode is always so heavily damped that it is of very little consequence in itself. The pilot of the airplane is hardly ever cognizant of its existence. It is felt by the pilot only as a bump when a gust is encountered or in the response of the airplane to an abrupt control movement. The long period or phugoid mode has very weak damping, and many airplane designs today have negatively damped phugoid modes during part of the lift coefficient range through which they fly. The period of the phugoid mode is so long and the response of the airplane so slow that negative damping of this oscillation

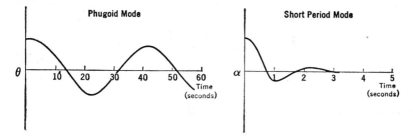

FIGURE 10–12. Typical longitudinal modes, controls locked.

has little bearing on the pilot's opinion of the flying qualities of the airplane. It is for this reason that neither the Air Force nor Navy has any requirements on the damping of the phugoid mode, and it is unimportant therefore to go into great detail on methods for designing the airplane to insure a damped phugoid.

The important things to understand about the two oscillatory modes of longitudinal motion are, first, their existence; second, a simple physical interpretation of them; and finally, a generalized picture of what the effects of the major airplane variables are on these modes.

The first step, that of their existence, has been developed theoretically and is well proved from flight test. Typical examples are given in Figure 10–12.

The second phase, that of obtaining a more simple physical explanation of this oscillation, is possible from flight test experience coupled with a knowledge of the dynamic system involved. The phugoid oscillation is one in which there is a large amplitude variation of air speed, pitch, and altitude, but during which the variation of angle of attack is very small and can be considered nearly constant. The motion is so slow that the effects of inertia forces and damping forces are very low. The whole phugoid oscillation can be thought of as a slow interchange of kinetic and potential energy about some equilibrium energy level, or as the attempt of the airplane to re-establish the equilibrium $C_L V^2 = K$, from which it has been disturbed.

Under the assumption of no change in angle of attack, no damping, no inertia, and no airplane drag, equations (10–89) reduce to the following simple form:

$$du + \frac{C_L}{2} \theta = 0$$

$$C_L u - d\theta = 0 \tag{10-100}$$

Assumption of the solution in the form $u = u_1 e^{\lambda t/\tau}$ and substitution into (10–100) above give the following coefficient determinant:

$$\begin{vmatrix} \lambda & \dfrac{C_L}{2} \\ C_L & -\lambda \end{vmatrix} = 0 \tag{10-101}$$

Expanding $$\lambda^2 + \frac{C_L{}^2}{2} = 0$$

solving $$\lambda = \pm i \sqrt{\frac{C_L{}^2}{2}}$$

and finally, $$u = u_1 e^{\pm i \sqrt{C_L{}^2/2}\, t/\tau}$$

This is an oscillation whose period is $2\pi/\sqrt{C_L{}^2/2} \ \tau$ seconds.

Substituting $C_L = 2(W/S)/\rho V^2$ and $\tau = (W/S)/\rho g V$, the period equals $.138V$ seconds, indicating a very slow oscillation at normal flight speeds.

The solution of the complete equations and experience from flight tests indicate that the linear variation of period with speed is verified, although the constant .138 is somewhat higher, nearly .178.

This approximation to the phugoid oscillation indicates no damping, with the oscillation continuing forever. If the assumption of zero drag is removed, equation (10–89) will reduce as follows:

$$(C_D + d)u + \frac{C_L}{2} \theta = 0$$

$$C_L u - d\theta = 0 \tag{10-102}$$

Again assuming the solution in the form $u = u_1 e^{\lambda t/\tau}$, etc., and expanding the determinant of the algebraic equation in λ give:

$$\lambda^2 + C_D \lambda + \frac{C_L{}^2}{2} = 0 \tag{10-103}$$

or

$$\lambda = -\frac{C_D}{2} \pm \sqrt{\left(\frac{C_D}{2}\right)^2 - \frac{C_L{}^2}{2}} \qquad (10\text{--}104)$$

The solution of equation (10–102) then gives a damped oscillation which, as $C_L \gg C_D$, has a period, as before, equal to $.138V$ seconds. The time to damp the oscillation to $\frac{1}{2}$ amplitude is $1.386/C_D$ τ seconds. The damping of the phugoid is therefore a direct function of the airplane drag coefficient. Although the results of the complete equations show other factors influencing the damping somewhat, the drag is nevertheless a large factor and places the airplane designer in somewhat of a dilemma. The cleaner he makes the airplane, the more difficult it will be for him to insure damping of the phugoid mode.

Finally, the phugoid can be summed up as a long-period, slow oscillation of the airplane's velocity as it attempts to re-establish the equilibrium condition $C_L V^2 = K$, from which it has been disturbed. The lift coefficient is held essentially constant during the oscillation and the motion is damped, mainly because of the effects of the airplane's drag. The more drag, the better is the damping.

Flight test experience and analytical study of the short period mode indicate that this oscillation proceeds essentially at constant speed. This is due to the fact that the motion of the airplane in the short period mode proceeds so rapidly that the motion is completely damped out by the time the speed has time to change. Under the assumption of no change in speed ($u = 0$), equations (10–89) reduce as follows:

$$\begin{vmatrix} \frac{1}{2}C_{L\alpha} + \lambda & -\lambda \\ C_{m\alpha} + C_{md\alpha}\lambda & C_{md\theta}\lambda - h\lambda^2 \end{vmatrix} = 0 \qquad (10\text{--}105)$$

Expanding

$$\lambda^3 - \left(\frac{1}{h}C_{md\theta} - \frac{1}{2}C_{L\alpha} + \frac{1}{h}C_{md\alpha}\right)\lambda^2$$
$$- \left(\frac{1}{h}C_{m\alpha} + \frac{C_{L\alpha}}{2h}C_{md\theta}\right)\lambda = 0 \qquad (10\text{--}106)$$

This equation may be divided through by λ, giving one zero root and the resulting quadratic:

$$\lambda^2 - \left(\frac{1}{h}C_{md\theta} - \frac{1}{2}C_{L\alpha} + \frac{1}{h}C_{md\alpha}\right)\lambda$$
$$- \left(\frac{1}{h}C_{m\alpha} + \frac{C_{L\alpha}}{2h}C_{md\theta}\right) = 0 \qquad (10\text{--}107)$$

This quadratic in general yields a pair of complex roots defining a heavily damped, short-period oscillation for all statically stable airplanes. The coefficient of λ to the first power in equation (10-107) is the damping term, and upon inspection it can be seen that for any airplane all terms will add to the damping, making the damping very heavy. This oscillation can be thought of as the response of the air-

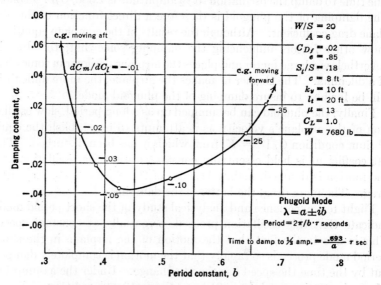

FIGURE 10-13. Effect of c.g. movement on phugoid mode.

plane to a disturbance from an equilibrium flight condition, in which the very heavy stability moments coupled with the heavy damping create a short period oscillation of the airplane's angle of attack at nearly constant speed.

The third phase of the discussion of the stick-fixed modes of longitudinal motion relates to the effects of the major airplane variables on these modes as deduced from the solutions of the complete equations of motion. These effects can be studied in two different ways. One of these is to change one of the variables slowly, holding all other variables constant and solving the equations of motion for the period and damping of the longitudinal modes. The results can be plotted as shown in Figure 10-13 for a particular airplane configuration.

The second way of presenting the variation of the airplane's dynamics with change in the airplane stability derivatives is to determine the boundaries between stability and instability by use of Routh's discriminant. Routh's discriminant, as mentioned previously, is a grouping of the coefficients of the stability quartic which if equal to zero will indicate a boundary between damped and undamped oscillations. It is usual to plot these boundaries in a plane of two of the major variables. For example, in Figure 10–14 the boundary between

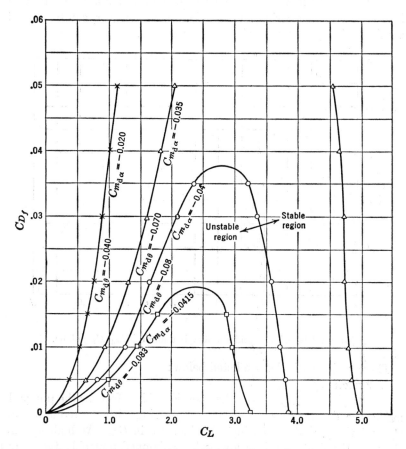

FIGURE 10–14. Phugoid stability boundaries, C_{D_f}-C_L plane.

stability and instability is plotted in the C_L, C_{D_f} plane for given values of the damping derivative, $C_{m_{d\theta}}$.

The stability boundaries for the same airplane as given in Figure 10–14 are shown in Figure 10–15 plotted in the plane of airplane damping parameter, $C_{m_{d\theta}}$, and the static stability parameter, C_{m_α}, for various values of the lift coefficient.

The results of systematic studies such as shown above are given in great detail in the literature and indicate the following characteristics of the stick-fixed longitudinal modes of motion.

In general it can be assumed that wide variations in airplane stability parameters will have little effect on the short period mode, which therefore constitutes no design problem and will not be dealt with further. The long period, or phugoid, mode, however, can be altered considerably by change in airplane parameters.

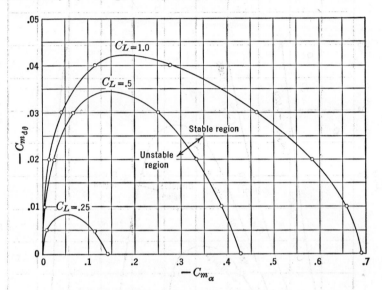

FIGURE 10–15. Phugoid stability boundaries, $C_{m_{d\theta}}$-C_{m_α} plane.

These major variables are listed below:

a. Variation with C_L

The airplane lift coefficient has considerable effect on the phugoid mode, the damping of the airplane improving with the reduction in C_L and the period decreasing with increase in C_L. At high speeds (low C_L) the airplane nearly always has a damped phugoid; therefore when flight testing for this mode, the test should always be run at high C_L.

b. Variation with C_{D_f}

Increasing the parasite drag of the airplane will improve the damping of the long period mode and have small effect on the period.

c. Variation with Airplane Damping

The damping of the phugoid improves with increase in the damping parameter $C_{m_{d\theta}}$, but the period is only slightly affected.

d. Variation with Inertia, k_y

Increasing the inertia tends to reduce the damping of the long period mode and has little effect on the period.

e. Variation with Static Stability

There is usually some c.g. position for which the damping of the phugoid is poorest. Forward or aft shift of the c.g. from this critical position will improve the damping. Moving the c.g. aft tends to lengthen the period.

It is repeated that the Air Force and Navy do not require that the long period mode be damped, as a great many flight tests of the flying qualities of the airplane indicate that there is no correlation between the damping of this mode and the pilot's opinion of the handling qualities of the airplane. It is usual practice for airplane designers, of military aircraft at least, to concentrate on designing airplanes to meet other more important requirements than to attempt to insure damping of the phugoid oscillation throughout the flight range.

10–6   Solution of the Equations of Motion (Stick-free Case)

The equations of motion studied in the preceding section were for the elevator-fixed case. If the limitation of the fixed elevator is removed, the reduced equations (10–89) no longer represent the physical facts and the complete equations of motion (10–66) must be re-examined. It is possible to solve these equations as they stand, but the process is very laborious. The complete solutions of these equations indicate three oscillatory modes, one of which is of long period and weak damping and very similar in all respects to the stick-fixed phugoid mode as defined by the solution of the reduced equations of motion (10–89). The other two oscillatory modes are of short period; one of these is similar to the stick-fixed short period mode except that it has a possibility of weak or even negative damping, while the other is of even shorter period, but with very heavy damping. These three modes are usually referred to as the stick-free phugoid and the second and third stick-free modes.

As the phugoid is only slightly affected by freeing the elevators and as it is of only small importance at best, the problem will be simplified by assuming that the motion proceeds at constant speed, which eliminates the phugoid at once and allows a simpler mathematical attack on the second and third modes with only slight loss in accuracy. Under this assumption, equation (10–66a) is completely eliminated, as is the first term of each of the remaining three equations.

Equations (10–66) reduce to the following three equations:

$$(\tfrac{1}{2}C_{L_\alpha} + \mathrm{d})\alpha - \mathrm{d}\theta = 0$$

$$(C_{m_\alpha} + C_{m_{\mathrm{d}\alpha}}\mathrm{d})\alpha + (C_{m_{\mathrm{d}\theta}} - h\mathrm{d})\mathrm{d}\theta$$
$$+ (C_{m_\delta} + C_{m_{\mathrm{d}\delta}}\mathrm{d})\delta_e = 0 \quad (10\text{–}108)$$

$$(C_{h_\alpha} - h_1\mathrm{d})\alpha + (C_{h_{\mathrm{d}\theta}} + h_1 - h_2\mathrm{d} - l_1\mathrm{d})\mathrm{d}\theta$$
$$+ (C_{h_\delta} + C_{h_{\mathrm{d}\delta}}\mathrm{d} - h_2\mathrm{d}^2)\delta_e = 0$$

If the solutions to equations (10–108) are assumed in the form $\alpha = \alpha_1 e^{\lambda\, t/\tau}$, etc., as before, substitution into (10–108) will yield the determinant of the coefficients as follows:

$$\begin{vmatrix} \tfrac{1}{2}C_{L_\alpha} + \lambda & -1 & 0 \\ (C_{m_\alpha} + C_{m_{\mathrm{d}\alpha}}\lambda) & (C_{m_{\mathrm{d}\theta}} - h\lambda) & (C_{m_\delta} + C_{m_{\mathrm{d}\delta}}\lambda) \\ (C_{h_\alpha} - h_1\lambda) & (C_{h_{\mathrm{d}\theta}} + h_1 - h_2\lambda - l_1\lambda) & (C_{h_\delta} + C_{h_{\mathrm{d}\delta}}\lambda - h_2\lambda^2) \end{vmatrix} = 0$$
$$(10\text{–}109)$$

Expanding this determinant, a quartic in λ is obtained:

$$A\lambda^4 + B\lambda^3 + C\lambda^2 + D\lambda + E = 0 \quad (10\text{–}110)$$

where

$$A = 1$$

$$B = -\frac{C_{h_{\mathrm{d}\delta}}}{h_2} - \frac{C_{m_{\mathrm{d}\theta}}}{h} + \frac{C_{L_\alpha}}{2} - \frac{C_{m_{\mathrm{d}\alpha}}}{h} + \left(\frac{h_2 + l_1}{hh_2}\right)C_{m_{\mathrm{d}\delta}}$$

$$C = -\frac{C_{h_\delta}}{h_2} + \frac{C_{h_{\mathrm{d}\delta}}}{hh_2}\left(C_{m_{\mathrm{d}\theta}} - \frac{hC_{L_\alpha}}{2} + C_{m_{\mathrm{d}\alpha}}\right)$$
$$- \frac{C_{m_{\mathrm{d}\delta}}}{hh_2}\left[C_{h_{\mathrm{d}\theta}} - \left(\frac{h_2 + l_1}{2}\right)C_{L_\alpha}\right] - \frac{C_{L_\alpha}}{2h}C_{m_{\mathrm{d}\theta}}$$
$$- \frac{C_{m_\alpha}}{h} + \frac{h_2 + l_1}{hh_2}C_{m_\delta}$$

$$D = \frac{C_{m_{\mathrm{d}\theta}}}{hh_2}\left(C_{h_\delta} + \frac{C_{L_\alpha}}{2}C_{h_{\mathrm{d}\delta}}\right) \quad (10\text{–}111)$$
$$- \frac{C_{m_{\mathrm{d}\delta}}}{hh_2}\left(C_{h_\alpha} + \frac{C_{L_\alpha}}{2}C_{h_{\mathrm{d}\theta}} + \frac{h_1 C_{L_\alpha}}{2}\right)$$
$$- \frac{C_{h_{\mathrm{d}\delta}}C_{m_\delta}}{hh_2} + \frac{C_{h_{\mathrm{d}\delta}}C_{m_\alpha}}{hh_2} + \frac{C_{m_{\mathrm{d}\alpha}}C_{h_\delta}}{hh_2}$$

$$+ \frac{(h_2 + l_1)}{2hh_2}(C_{m\delta}C_{L\alpha}) - \frac{C_{L\alpha}C_{h\delta}}{2h_2}$$

$$E = \frac{C_{h\delta}}{hh_2}\left(\frac{C_{L\alpha}C_{m\mathrm{d}\theta}}{2} + C_{m\alpha}\right)$$

$$- \frac{C_{m\delta}}{hh_2}\left(C_{h\alpha} + \frac{C_{L\alpha}C_{h\mathrm{d}\theta}}{2} + \frac{h_1}{2}C_{L\alpha}\right)$$

The stability derivatives in (10–111) were discussed in section (10–3) and formulae developed for evaluating them. The procedure used in solving for the stick-free modes is to substitute the values of the stability derivatives in (10–111) in order to determine the coefficients of the stability quartic (10–110). When these have been obtained, the roots of the quartic are extracted, and the period and damping of the two oscillatory modes obtained.

Typical values of the roots of the quartic for present-day airplanes are as follows:

$$\lambda_2 = -.2 \pm i\,3.9$$

$$\lambda_3 = -5.5 \pm i\,7.8$$

which indicate two oscillations: the first with a period of 1.61τ seconds, damping to one-half amplitude in 2.14 cycles, and the second with a period of $.807\tau$ seconds, damping to one-half amplitude in .091 cycle. For an airplane whose time constant $\tau = 1.5$ seconds, the periods of these oscillations will be 2.42 and 1.21 seconds, respectively. This example shows the two characteristic modes of stick-free motion at constant air speed. The second of these is of quite short period and heavy damping and can be thought of as the characteristic flapping motion of the elevator about its hinge line. The first of these modes has a period in the neighborhood of 2 seconds and for the example shown has marginal damping. In actual practice wide variations in airplane parameters have little effect on the damping of the very short period mode, which therefore will not be dealt with further. The damping of the longer period mode can, however, become very weak and under certain design conditions can become neutrally damped and even unstable. This mode when neutrally damped is usually called "porpoising," and has been encountered in flight in a mild form on several slow-speed airplanes, manifesting itself to the pilot as a pumping of the control column and mild oscillation of the normal acceleration. Neutrally damped or unstable porpoising has also been encountered on high-speed airplanes, in which the oscillation of normal acceleration became so severe that the pilot was injured seriously and the airplane damaged before he could stop the oscillation by slowing the airplane down.

Of all the oscillatory modes considered, the porpoising mode is the only one that the airplane designer should consider. The phugoid modes are so slightly unstable at the worst and the period of the oscillation so long that the pilot has no difficulty whatsoever in controlling the motion and, as a matter of fact, is usually unaware of its existence. The short period stick-fixed mode and the very short period stick-free mode are both so heavily damped that they are rarely considered in the airplane's design stages. The porpoising mode, however, has a period just short enough to be critical from the point of view of the pilot controlling the oscillation, and the damping can possibly become so weak that control over this mode must be kept in mind at all times. Instability in this porpoising mode has become more and more possible as the airplane has become heavier and faster, mainly because of the need for closer aerodynamic balancing required for the elevator, which, as will be seen later, has adverse effects on the damping.

The major airplane variables governing the damping of the porpoising mode are the elevator hinge moment parameters, C_{h_α} and C_{h_δ}, the elevator static unbalance factor, $\mu_e x_e/c$, the airplane's relative density, μ, and the elevator's damping derivative, $C_{h_{d\delta}}$. Other airplane characteristics have small influence on the porpoising mode, and, as they are small, they will not be discussed herein.

If the stability boundary for the porpoising mode is obtained by setting Routh's discriminant equal to zero and solving in terms of any two of the above variables, the general effects of the variables can be traced. As most modern airplanes have mass balanced elevators, the term $\mu_e x_e/c$ is usually close to zero. The stability boundary for the porpoising mode is usually plotted in a plane of C_{h_α}, C_{h_δ}. An example of this boundary for a typical modern airplane is shown in Figure 10–16; boundaries are shown for several values of airplane relative density factor, μ.

The data given in Figure 10–16 indicate that close aerodynamic balance gives a possibility of instability in the porpoising mode. In order to obtain instability, it is necessary for $C_{h_{\alpha t}}$ to be overbalanced or to be positive, and for low airplane density C_{h_δ} must also be positive, a very unlikely state of affairs and one that accounts for the small amount of trouble encountered with this mode before the advent of the heavily loaded high-speed airplane. As the airplane's density increases, the possibility of instability exists even with negative C_{h_δ}. The trend in elevator balancing is in the direction of making C_{h_δ} as small as possible but still negative, while C_{h_α} is permitted to go positive in order to improve the stick-free stability and to lower the change in stick force per g with c.g. movement. This design trend is running the airplane directly into the unstable region, and difficulty with porpoising becomes a distinct possibility. The trend of hinge moment parameters is also shown in Figure 10-16.

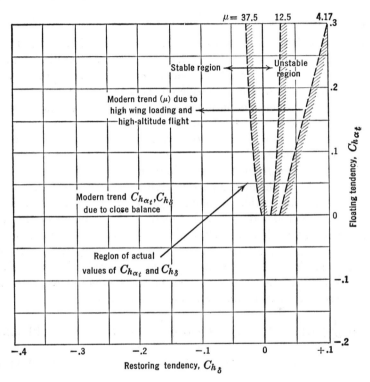

FIGURE 10-16. Typical stick-free oscillatory boundaries.

The effect of static unbalance of the elevator is adverse if this unbalance tends to make the elevator hang trailing edge down. If any unbalance exists, the instability regions of Figure 10-16 are increased and porpoising becomes even more of a possibility. For modern aircraft the trend is toward static and dynamic balance of the elevator in order to insure against control surface flutter.

One control surface characteristic that has a great influence on the porpoising mode is solid control surface friction. If solid friction exists, it will tend to establish steady oscillations with an amplitude proportional to the amount of friction. A positive floating tendency is required of the elevator to support these oscillations, but again the trend of modern elevator control balance is toward a positive C_{h_α}.

An exhaustive treatment of the "porpoising" mode of airplane motion will not be undertaken herein, as it would be beyond the scope of this book. The literature contains some excellent papers* on this subject, and the reader is referred to them for a more complete study.

10–7   The Response of the Airplane to the Longitudinal Control

The studies just completed have given an indication of the types of longitudinal motion that the airplane experiences after disturbances from some equilibrium flight condition.   In these studies the nature of the disturbance was not considered, and the entire investigation focused on the frequency equation in order to determine the characteristic modes of the longitudinal motion. To complete the picture of longitudinal dynamics it is worth while to investigate the actual response of the airplane to the disturbance created because of a deflection of the elevator from its equilibrium or trim position, as this will shed some light on the airplane's sensitivity and resulting feel.

If the disturbing action is taken as an instantaneous deflection of the elevator, considered as a step function, the response of the airplane's velocity, angle of attack, and angle of pitch can be computed from the general equations of longitudinal motion (10–66).   If the elevator is considered locked at its new angle, and neglecting the derivative $C_{m_{\mathrm{d}\delta}}$, the equation of motion can be written as follows, with the forcing term $C_{m_{\delta_e}}\delta_e$ taken on the right-hand side.

$$(C_D + \mathrm{d})u + \tfrac{1}{2}(C_{D_\alpha} - C_L)\alpha + \frac{C_L}{2}\theta = 0 \qquad (10\text{–}112a)$$

$$C_L u + (\tfrac{1}{2}C_{L_\alpha} + \mathrm{d})\alpha - \mathrm{d}\theta = 0 \qquad (10\text{–}112b)$$

$$(C_{m_\alpha} + C_{m_{\mathrm{d}\alpha}}\mathrm{d})\alpha + (C_{m_{\mathrm{d}\theta}}\mathrm{d} - h\mathrm{d}^2)\theta = -C_{m_{\delta_e}}\delta_e(t) \qquad (10\text{–}112c)$$

The solution of these equations for V, α, θ as functions of time for a typical airplane responding to a push-down elevator step-function is given in Figure 10–17.

The very slow oscillation of the airplane's velocity, V, and pitch angle, θ, about their new steady-state values is apparent and is, of course, the manifestation of the phugoid mode.   The large amplitude changes in velocity and angle of pitch can be seen in comparison to the much smaller amplitude change in the angle of attack.   The short period mode can be seen mainly to involve the change in angle of

* NACA WR L–430.

attack, and its influence vanishes after about 3 seconds. During the first seconds, when the change in angle of attack is large, the change in velocity is very small, and for any study of this initial response in angle of attack the assumption of constant speed is well justified.

It was determined in Chapter 5 that for every elevator angle there is an equilibrium lift coefficient and, therefore, speed. However, if the elevator deflection is changed fast enough, the airplane's speed

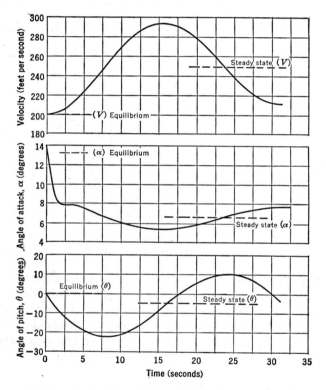

FIGURE 10–17. Typical time history after an abrupt elevator deflection.

will not be able to follow the rapid change in angle of attack, and accelerated flight conditions will result. As a matter of fact, if it is desired to move the elevator so that the speed can be changed while maintaining 1 g, the motion would necessarily be very slow.

When the airplane is responding to an abrupt or very fast elevator forcing action, at essentially constant speed, the normal acceleration becomes proportional to and in phase with the change in airplane angle of attack. The loads on the airplane then respond very rapidly to elevator action and introduce the requirement that the short period characteristics of the airplane never become so marginal that small pilot action can introduce breaking loads before the pilot has time to take corrective action.

A study of the response of the airplane's angle of attack to abrupt elevator action at constant speed can be obtained from equations (10–112).

$$(\tfrac{1}{2}C_{L\alpha} + \mathrm{d})\alpha - \mathrm{d}\theta = 0 \qquad (10\text{–}113a)$$

$$(C_{m\alpha} + C_{m\mathrm{d}\alpha}\mathrm{d})\alpha + (C_{m\mathrm{d}\theta}\mathrm{d} - h\mathrm{d}^2)\theta = -C_{m\delta e}\delta_e(t) \quad (10\text{–}113b)$$

If equations (10–113) are solved for α, they reduce to an equivalent one-degree-of-freedom system.

$$\mathrm{d}^2\alpha + \left(\frac{C_{L\alpha}}{2} - \frac{C_{m\mathrm{d}\alpha}}{h} - \frac{C_{m\mathrm{d}\theta}}{h}\right)\mathrm{d}\alpha$$
$$- \left(\frac{C_{m\alpha}}{h} + \frac{C_{L\alpha}C_{m\mathrm{d}\theta}}{2h}\right)\alpha = \frac{C_{m\delta e}}{h}\delta_e(t) \quad (10\text{–}114)$$

If the terms contained in the bracket operated upon by $\mathrm{d}\alpha$ are thought of as effective damping terms and called simply b, and if the terms in the bracket operated on by α are thought of as effective spring constant terms and called k, and if the ratio $C_{m\delta e}/h$ is called m_0, equation (10–114) can be rewritten as

$$\mathrm{d}^2\alpha + b\,\mathrm{d}\alpha + k\alpha = m_0\delta_e(t) \qquad (10\text{–}115)$$

This equation is, of course, a simple second order differential equation with constant coefficients, and can be solved by obtaining the complementary or transient solution from the reduced equation with the right-hand zero, and the particular integral or steady-state solution from the nature of the variation of the elevator angle with time. For the step-function elevator input the solution becomes

$$\Delta\alpha = \frac{m_0\delta_e}{k}\left[1 - \frac{\sqrt{k}}{\sqrt{k - (b/2)^2}}e^{-b/2\,t/\tau}\right.$$
$$\left. \sin\left(\sqrt{k - \left(\frac{b}{2}\right)^2}\frac{t}{\tau} + \phi\right)\right] \quad (10\text{–}116)$$

where

$$\phi = \tan^{-1}\frac{\sqrt{k - (b/2)^2}}{b/2}$$

The variation of angle of attack, $\Delta\alpha$, with time parameter, t/τ, is given in Figure 10–18 for a typical airplane with varying degrees of static longitudinal stability, C_{m_α}. It should be noted that for stable values of C_{m_α} the response is oscillatory but heavily damped, and the motion will not become divergent at the stick-fixed neutral point. The motion does become divergent, however, if the c.g. is moved far enough aft of the stick-fixed neutral point. The c.g. for divergence of the short period motion is the stick-fixed maneuver point discussed previously.

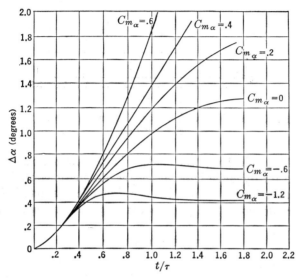

Figure 10-18. Effect of static stability on airplane response to elevator step-function.

The effect of varying the longitudinal damping is shown in Figure 10-19. The damping parameter, $C_{m_{d\theta}}$, taken for the airplane of the example was arbitrarily divided by ten, which gives a damping derivative very close to that for a tailless airplane. The response of the tailless airplane's angle of attack to an abrupt elevator is much more rapid and will tend to become divergent very close to the stick-fixed neutral point. This effect makes it more dangerous to fly a statically unstable tailless airplane than airplanes with normal configurations, and for similar wing loadings and static margins makes the tailless airplane's "feel" more delicate.

In the years before World War II, the dynamics of the airplane were studied by the aerodynamicists but seldom used by the airplane designer because of the fact that, when the static requirements of the airplane were satisfied, the dynamics usually took care of themselves. This situation was changed somewhat during World War II, when dynamic instabilities with controls free were encountered, as a result of the trend towards higher airplane density and close aerodynamic balance of the control surface. Since the war, however, great advances in the field

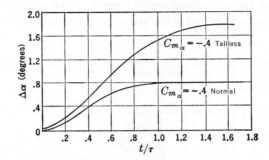

FIGURE 10-19. Effect of longitudinal damping on airplane response to elevator step-function.

of airplane dynamics have been made. These have come about because of the need for dynamic information for automatic controls and new flight test techniques in which the aerodynamic characteristics of the airplane are deduced from its response to given forcing actions.

SUGGESTED READING

1. NACA TR 521, "An Analysis of Longitudinal Stability in Power-off Flight with Charts for Use in Design," by Charles H. Zimmerman.
2. NACA TR 578, "Flight Measurements of the Dynamic Longitudinal Stability of Several Airplanes and a Correlation of the Measurements with Pilots' Observations of Handling Characteristics," by H. A. Soulé.
3. NACA TR 709, "An Analysis of the Stability of an Airplane with Free Controls," Jones and Cohen, 1941.
4. NACA WR L-444, "A Flight Investigation of Short Period Longitudinal Oscillations of an Airplane with Free Elevator," by W. H. Phillips, 1942.
5. Durand, Volume V, "Dynamics of the Airplane," by B. M. Jones, 1935.
6. NACA TN 828, "Methods of Analyzing Wind-tunnel Data for Dynamic Flight Conditions," by Donlan and Recant, 1941.
7. NACA WR L-430, "A Theoretical Investigation of Longitudinal Stability of Airplanes with Free Controls Including Effect of Friction," by Greenberg and Sternfield, 1944.

PROBLEMS

10-1. A fighter-type airplane has the aerodynamic, geometric, mass, and inertia characteristics given below. If this airplane is placed in a glide at 10,000 ft with power off at $C_L = 1.0$, calculate the stability derivatives $C_{L\alpha}$, $C_{D\alpha}$, $C_{m\alpha}$, $C_{m\dot{\alpha}}$, and $C_{m\dot{\theta}}$.

Inertia about Y axis $= 15,000$ slug ft^2	Horizontal tail span $= 13$ ft
Weight $= 10,000$ lb	Stick-fixed neutral point $(T_c = 0) = .38c$
Wing area $= 250$ sq ft	Center of gravity location $= .33c$
Wing span $= 40$ ft	Rate of change of downwash, $d\epsilon/d\alpha = .5$
m.a.c. $= 8$ ft	$a_o = .10$
Taper ratio $= 2 : 1$	$\eta_t = .90$
Horizontal tail area $= 50$ sq ft	$e = .85$
Distance c.g. to tail a.c. $= 20$ ft	$C_{D_f} = .02$

10–2. The response of an airplane in pitch, θ, to a disturbance is given in the following form:

$$\theta = \theta_1 e^{\lambda_1 t} + \theta_2 e^{\lambda_2 t} + \theta_3 e^{\lambda_3 t} + \theta_4 e^{\lambda_4 t}$$

The values of λ in this equation can be determined from the following quartic:

$$\lambda^4 + 4\lambda^3 + 10\lambda^2 + 1.0\lambda + 3.8 = 0$$

Extract the roots from this quartic and describe the characteristic modes involved. If any of the modes are oscillatory, determine the period of the oscillation and the time for the oscillation either to damp to half amplitude, or to double its amplitude. Repeat this calculation, but for the case where the quartic in λ is as follows:

$$\lambda^4 + 10\lambda^3 + 100\lambda^2 + 600\lambda + 2000 = 0$$

10–3. If the coefficient of the quartic equation in λ $(A\lambda^4 + B\lambda^3 + C\lambda^2 + D\lambda + E)$ when arranged $(BCD - AD^2 - B^2 E)$ equals zero, then a condition of zero damping is defined. This grouping of the coefficients is known as Routh's discriminant. Prove this relationship. (*Hint:* the complex values of λ combine to define an oscillation $\lambda = a \pm ib$, where the value of a gives the magnitude of the damping. For the condition of zero damping, $\lambda = \pm ib$ with a equal to zero.)

If the motion of the airplane is defined by the roots of a cubic $(A\lambda^3 + B\lambda^2 + C\lambda + D = 0)$, what is Routh's discriminant for this cubic?

10–4. Determine the characteristic modes of longitudinal motion stick-fixed for the airplane given in Problem 10–1. What is the period and what is the damping of the oscillatory modes? Determine the periods in seconds and the time to damp to $\frac{1}{2}$ amplitude in both seconds and number of cycles.

10–5. Plot the boundary between stability and instability of the phugoid mode in the $C_{m\alpha} - h$ plane for values of $C_L = .5$, 1.0, and 1.5 for the airplane given in Problem 10–1.

10–6. If the airplane given in Problem 10–1 is equipped with an elevator with the following characteristics, determine the stability derivatives, $C_{m\delta}$, $C_{m\dot\delta}$, $C_{h\dot\theta}$, $C_{h\dot\delta}$.

Elevator area, $S_e = 15$ sq ft
Mean chord, $c_e = 1.5$ ft
Hinge line at 70 per cent of horizontal tail chord
Mean chord horizontal tail, $c_t = 4$ ft
$C_{h\alpha t} = -.20$
$C_{h\delta} = -.50$

10–7. If the elevator of the airplane given in Problem 10–1 is completely mass balanced, and weighs 20 lb with a radius of gyration k_e equal to .6 ft, determine the nature of the control-free modes of longitudinal motion under the assumption of constant speed. Repeat with $C_{h\alpha t} = .6$, $C_{h\delta} = 0$. Repeat both under assumption that $I_e = 0$.

10–8. If all the characteristics of the airplane discussed in Problems 10–1, etc., are held constant, but the balance of the elevator is changed in order to vary the hinge moment parameters, $C_{h\alpha}$ and $C_{h\delta}$, plot the boundary between stability and instability of the proposing mode in the $C_{h\alpha}$, $C_{h\delta}$ plane. Neglect elevator inertia.

10–9. The same airplane as discussed in Problem 10–8 is flying in level flight at an indicated air speed of 200 mph at 10,000 ft. The pilot pulls the elevator up ½ degree from trim so rapidly that the forcing action can be considered a step-function. Plot the response of the airplane's angle of pitch, θ, angle of attack, α, in degrees, and velocity in miles per hour true, versus time in seconds for at least 2 cycles of the phugoid mode.

10–10. Assuming constant speed, calculate and plot the variation of airplane angle of attack, α, in degrees and normal acceleration, n, in units of g, versus time in seconds for the same step-function as given in Problem 10–9. Plot for at least 2 cycles of the short period mode. Repeat for a c.g. location at .38c, and for a c.g. location at .43c.

CHAPTER 11

LATERAL DYNAMICS

11-1 Introduction

In the preceding chapter on longitudinal dynamics, the motions of the airplane in response to a disturbance from some equilibrium flight condition and to the longitudinal controls were investigated under the assumption that the airplane's plane of symmetry prevented any coupling between the motions in the plane and those motions outside the plane of symmetry. For example, it was assumed that pitching motions would not introduce rolling moments about the X axis and so on. Under the same assumptions, the characteristic asymmetric or lateral motions of the airplane will be developed, as will the response of the airplane to its lateral controls.

It was pointed out in Chapter 10 that, although there are only four degrees of freedom for the longitudinal case, there are five degrees of freedom for the lateral case because of the fact that there are two lateral controls, the rudder and the aileron.

The five lateral degrees of freedom are as follows, and the general notation is given in Figure 11-1.

1. Velocity along the Y axis.
2. Rotation about the X axis.
3. Rotation about the Z axis.
4. Rotation of rudder about its hinge.
5. Rotation of aileron about its hinge.

The equations of motion of the airplane in the lateral degrees of freedom, using the results of (10–13), become simply:

$$\Sigma F_y = m(\dot{v} + Vr)$$

$$\Sigma L = \dot{p}I_x - \dot{r}J_{xz}$$

$$\Sigma N = \dot{r}I_z - \dot{p}J_{xz} \qquad (11\text{-}1)$$

$$\Sigma H_a = I_a\ddot{\delta}_a$$

$$\Sigma H_r = I_r\ddot{\delta}_r$$

where ΣF_y is the summation of all the forces along the Y axis, ΣL and ΣN the summation of all rolling and yawing moments, respectively, and ΣH_a and ΣH_r the summation of all hinge moments about the aileron and rudder hinge axes, respectively.

In the past it has been safely assumed that the X axis is essentially a principal one and therefore the products of inertia J_{xz} would be zero.

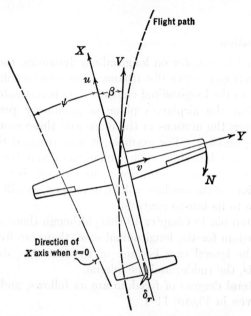

FIGURE 11-1. Axis system for lateral dynamics.

In the past few years, however, it has been demonstrated that under certain conditions of high airplane density the products of inertia can be of considerable importance. In order to keep the developments simple, however, the X axis will be assumed a principal axis and therefore J_{xz} vanishes. The acceleration along the Y axis, a_y, can be expressed in terms of the sideslip velocity, v, and yawing velocity by letting $\beta = v/V$:

$$a_y = V(\dot{\beta} + \dot{\psi}) \qquad (11\text{-}2)$$

With these assumptions equations (11-1) reduce to the following:

$$\Sigma F_y = mV(\dot{\beta} + \dot{\psi})$$
$$\Sigma L = \dot{p}I_x$$
$$\Sigma N = \dot{r}I_z \qquad (11\text{-}3)$$
$$\Sigma H_a = I_a\ddot{\delta}_a$$
$$\Sigma H_r = I_r\ddot{\delta}_r$$

In developing the equations of motion, the five independent variables will be β, ψ, ϕ, δ_a, and δ_r. It is assumed that the total change in the forces or moments from some equilibrium can be expressed as the sum of the partial derivatives of these forces or moments taken with respect to each of these variables, with the rate of change of these variables, and with the acceleration of these variables. In many cases these partial derivatives do not exist, but in general for small changes in the variable it can be assumed that the change in Y force from equilibrium will be

$$dF_y = \frac{\partial F_y}{\partial \beta}\, d\beta + \frac{\partial F_y}{\partial \psi}\, d\psi + \frac{\partial F_y}{\partial \phi}\, d\phi$$

$$+ \cdots \frac{\partial F_y}{\partial \dot{\beta}}\, d\dot{\beta} + \cdots \frac{\partial F_y}{\partial \ddot{\beta}}\, d\ddot{\beta},\ \text{etc.} \quad (11\text{--}4)$$

If each of the above partial derivatives is examined, it will be found that only two of them are important. These can be developed from a look at the forces along the airplane's Y axis or its side force. It is known that any airplane at some angle of sideslip, β, will develop

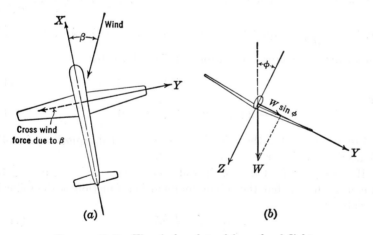

FIGURE 11-2. The airplane lateral force, level flight.

a cross wind or side force; the magnitude of this side force varies with fuselage depth and flap position and is obtainable directly from the wind tunnel (Figure 11-2a).

Another component of side force is developed as a result of the airplane's angle of bank, ϕ, which introduces a component of the airplane's weight along the Y axis (Figure 11-2b). None of the other variables (ψ, $\dot{\psi}$, $\dot{\beta}$, $\dot{\phi}$, etc.) introduces forces along the Y axis, so it can be assumed for small deflections for which the derivatives are linear and the sine of the angle equal to the angle in radians that:

$$\Delta F_y = \frac{\partial F_y}{\partial \beta} \Delta \beta + \frac{\partial F_y}{\partial \phi} \Delta \phi \qquad (11\text{--}5)$$

In coefficient form with $F_y = C_y q S$, and β, ϕ, and $\psi = 0$ at $t = 0$:

$$C_y = \frac{\partial C_y}{\partial \beta} \beta + \frac{\partial C_y}{\partial \phi} \phi \qquad (11\text{--}6)$$

The equation of motion along the Y axis for level flight will therefore become

$$C_{y\beta}\beta + C_L\phi = \frac{mV}{qS_w} (\dot{\beta} + \dot{\psi}) \qquad (11\text{--}7)$$

where

$$C_{y\beta} = \frac{\partial C_y}{\partial \beta}$$

In order to non-dimensionalize equation (11–7), the time parameter $\tau = m/\rho S_w V$ is introduced as before. Then $\tau\dot{\beta}$ will become $\dfrac{d\beta}{d(t/\tau)}$, or, letting $\dfrac{d}{d(t/\tau)}$ equal the operator d, becomes simply dβ. Making use of the parameter τ, equation (11–7) reduces to:

$$C_{y\beta}\beta + C_L\phi = 2\text{d}(\beta + \psi) \qquad (11\text{--}8)$$

or

$$(C_{y\beta} - 2\text{d})\beta - 2\text{d}\psi + C_L\phi = 0 \qquad (11\text{--}9)$$

Equation (11–9), then, is the non-dimensional equation of motion along the Y axis for level flight ($\gamma = 0$).

If the equation of motion in roll about the X axis is investigated, it can be shown that the rolling moment is a function of the following variables:

$$L = f(\beta, \dot{\psi}, \dot{\phi}, \delta_a, \dot{\delta}_a) \qquad (11\text{--}10)$$

and that either the rolling moment is not a function of the remaining variables or its variation is negligible with respect to them. With this information in hand, the change in rolling moment due to a small disturbance from some equilibrium can be given as:

$$\Delta L = \frac{\partial L}{\partial \beta} \beta + \frac{\partial L}{\partial \dot{\psi}} \dot{\psi} + \frac{\partial L}{\partial \dot{\phi}} \dot{\phi} + \frac{\partial L}{\partial \delta_a} \delta_a + \frac{\partial L}{\partial \dot{\delta}_a} \dot{\delta}_a \qquad (11\text{--}11)$$

and the equation of motion in roll in terms of rolling moment coefficient, C_l, becomes:

$$C_{l_\beta}\beta + C_{l_{\dot\psi}}\dot\psi + C_{l_{\dot\phi}}\dot\phi + C_{l_{\delta_a}}\delta_a + C_{l_{\dot\delta_a}}\dot\delta_a = \frac{I_x}{qSb}\ddot\phi \quad (11\text{-}12)$$

It was shown in Chapter 9 that C_{l_p} is conveniently expressed non-dimensionally as $\dfrac{dC_l}{d(pb/2V)}$, where $pb/2V$ is the non-dimensional rolling parameter. In order to make use of this convenient form of the damping derivative in roll, the following is presented:

$$C_{l_{\dot\phi}}\dot\phi = \frac{dC_l}{d(pb/2V)} \cdot \frac{b}{2V} \cdot \dot\phi \quad (11\text{-}13)$$

By letting $C_{l_p} = \dfrac{dC_l}{d(pb/2V)}$ and making use of the airplane density parameter $\mu = m/\rho Sb$, the following is obtained:

$$C_{l_{\dot\phi}}\dot\phi = \frac{C_{l_p}}{2}\frac{\tau}{\mu}\dot\phi = \frac{C_{l_p}}{2\mu}\,d\phi \quad (11\text{-}14)$$

Similarly, $C_{l_{\dot\psi}}$ can be expressed in terms of the non-dimensional yawing parameter, $rb/2V$, where $C_{l_r} = \dfrac{dC_l}{d(rb/2V)}$, giving

$$C_{l_{\dot\psi}}\dot\psi = \frac{C_{l_r}}{2\mu}\,d\psi \quad (11\text{-}15)$$

Finally, if $C_{l_{\dot\delta_a}}\dot\delta_a$ is multiplied and divided by the time parameter, τ, then

$$C_{l_{\dot\delta_a}}\dot\delta_a = C_{l_{d\delta_a}}d\delta_a \quad (11\text{-}16)$$

Equation (11-12) becomes:

$$C_{l_\beta}\beta + \frac{C_{l_r}}{2\mu}\,d\psi + \frac{C_{l_p}}{2\mu}\,d\phi + C_{l_{\delta_a}}\delta_a + C_{l_{d\delta_a}}d\delta_a = \frac{I_x}{qSb}\ddot\phi \quad (11\text{-}17)$$

If the airplane's inertia, I_x, is given as $mk_x{}^2$, where k_x is the radius of gyration about the X axis, and making use of the parameters μ and τ, then,

$$\frac{I_x}{qSb}\ddot\phi = \frac{2}{\mu}\left(\frac{k_x}{b}\right)^2 d^2\phi \quad (11\text{-}18)$$

The final equation of motion in roll becomes:

$$\mu C_{l_\beta}\beta + \frac{C_{l_r}}{2}\,d\psi + \left[\frac{C_{l_p}}{2}\,d - 2\left(\frac{k_x}{b}\right)^2 d^2\right]\phi$$
$$+ \mu(C_{l_{\delta_a}} + C_{l_{d\delta_a}}d)\delta_a = 0 \quad (11\text{-}19)$$

The equation of motion about the Z axis is studied in the same manner. The yawing moment, N, is a function of the following variables:

$$N = f(\beta, \dot{\psi}, \dot{\phi}, \delta_r, \text{ and } \dot{\delta}_r) \tag{11-20}$$

The change in yawing moment for small disturbances from equilibrium will be

$$\Delta N = \frac{\partial N}{\partial \beta} \beta + \frac{\partial N}{\partial \dot{\psi}} \dot{\psi} + \frac{\partial N}{\partial \dot{\phi}} \dot{\phi} + \frac{\partial N}{\partial \delta_r} \delta_r + \frac{\partial N}{\partial \dot{\delta}_r} \dot{\delta}_r \tag{11-21}$$

The equation of motion in yaw in terms of the yawing moment coefficient, C_n, becomes:

$$C_{n\beta}\beta + C_{n\dot{\psi}}\dot{\psi} + C_{n\dot{\phi}}\dot{\phi} + C_{n\delta_r}\delta_r + C_{n\dot{\delta}_r}\dot{\delta}_r = \frac{I_z}{qS_wb}\ddot{\psi} \tag{11-22}$$

Placing $C_{n\dot{\psi}}\dot{\psi} = \dfrac{C_{n_r}}{2\mu}\,d\psi$, etc., as before, the equation is finally developed as:

$$\mu C_{n\beta}\beta + \left[\frac{C_{n_r}}{2} - 2\left(\frac{k_z}{b}\right)^2 d\right]d\psi + \frac{C_{n_p}}{2}\,d\phi$$
$$+ \mu(C_{n\delta_r} + C_{n\dot{\delta}_r}d)\delta_r = 0 \tag{11-23}$$

The equation of motion of the aileron controls about their hinge axes can be developed under the assumption that the aileron hinge moment is a function of the following variables:

$$H_a = f(\beta, \dot{\phi}, \delta_a, \dot{\delta}_a, \ddot{\phi}) \tag{11-24}$$

The change in aileron hinge moment for a small disturbance from equilibrium will be therefore:

$$\Delta H_a = \frac{\partial H_a}{\partial \beta}\beta + \frac{\partial H_a}{\partial \dot{\phi}}\dot{\phi} + \frac{\partial H_a}{\partial \delta_a}\delta_a + \frac{\partial H_a}{\partial \dot{\delta}_a}\dot{\delta}_a + \frac{\partial H_a}{\partial \ddot{\phi}}\ddot{\phi} \tag{11-25}$$

The equation of motion, then, in terms of the aileron hinge moment coefficient, $C_{h_a} = H_a/qS_a\bar{c}_a$, becomes:

$$C_{h_a\beta}\beta + C_{h_a\dot{\phi}}\dot{\phi} + C_{h\delta_a}\delta_a + C_{h_a\dot{\delta}_a}\dot{\delta}_a + C_{h_a\ddot{\phi}}\ddot{\phi} = \frac{I_a}{qS_a\bar{c}_a}\ddot{\delta}_a \tag{11-26}$$

where I_a is the inertia of the aileron control system. If the inertia I_a is expressed as $m_a k_a{}^2$, with k_a equal to the radius of gyration of the aileron control system, and a new aileron density parameter is established, $\mu_a = m_a/\rho S_a \bar{c}_a$, the right-hand side of (11–26) becomes:

$$\frac{I_a}{qS_a c_a}\ddot{\delta}_a = \frac{2\mu_a}{\mu^2}\left(\frac{k_a}{b}\right)^2 d^2\delta_a \tag{11-27}$$

Letting $2\mu_a/\mu^2 (k_a/b)^2 = h_a$, equation (11–26) becomes:

$$C_{h_a\beta}\beta + (C_{h_{ad\phi}}\mathrm{d} + C_{h_{ad^2\phi}}\mathrm{d}^2)\phi$$

$$+ (C_{h_{a\delta a}} + C_{h_{ad\delta a}}\mathrm{d} - h_a\mathrm{d}^2)\delta_a = 0 \quad (11–28)$$

The equation of motion of the rudder about its hinge axis can be developed under the assumption that the rudder hinge moments are a function of the following variables:

$$H_r = f(\beta,\dot\psi,\delta_r,\dot\delta_r,\ddot\psi,\ddot\phi) \quad (11–29)$$

The equation of motion in terms of rudder hinge moment coefficient $C_{h_r} = H_r/qS_r\bar c_r$ becomes:

$$C_{h_r\beta}\beta + C_{h_r\dot\psi}\dot\psi + C_{h_r\delta_r}\delta_r + C_{h_r\dot\delta_r}\dot\delta_r + C_{h_r\ddot\phi}\ddot\phi + C_{h_r\ddot\psi}\ddot\psi$$

$$= \frac{I_r}{qS_r\bar c_r}\ddot\delta_r \quad (11–30)$$

Letting $I_r = m_r k_r^2$ and introducing the rudder density parameter, $\mu_r = m_r/\rho S_r\bar c_r$, equation (11–31) can be given the form:

$$C_{h_r\beta}\beta + (C_{h_{rd\psi}} + C_{h_{rd^2\psi}}\mathrm{d})\mathrm{d}\psi + (C_{h_{rd^2\phi}}\mathrm{d}^2)\phi$$

$$+ (C_{h\delta r} + C_{h_{rd\delta r}} - h_r\mathrm{d}^2)\delta_r = 0 \quad (11–31)$$

where

$$h_r = \frac{2\mu_r}{\mu^2}\left(\frac{k_r}{b}\right)^2 \quad (11–32)$$

The five simultaneous equations of motion in their finally developed form (equations 11–9, 11–19, 11–23, 11–28, and 11–31) are given below:

$$(C_{y\beta} - 2\mathrm{d})\beta - 2\mathrm{d}\psi + C_L\phi = 0 \quad (11–33a)$$

$$\mu C_{l\beta}\beta + \frac{C_{l_r}}{2}\mathrm{d}\psi + \left[\frac{C_{l_p}}{2}\mathrm{d} - 2\left(\frac{k_x}{b}\right)^2\mathrm{d}^2\right]\phi$$

$$+ \mu(C_{l\delta a} + C_{l d\delta a}\mathrm{d})\delta_a = 0 \quad (11–33b)$$

$$\mu C_{n\beta}\beta + \left[\frac{C_{n_r}}{2} - 2\left(\frac{k_z}{b}\right)^2\mathrm{d}\right]\mathrm{d}\psi + \frac{C_{n_p}}{2}\mathrm{d}\phi$$

$$+ \mu(C_{n\delta r} + C_{n d\delta r}\mathrm{d})\delta_r = 0 \quad (11–33c)$$

$$C_{h_a\beta}\beta + (C_{h_{ad\phi}}\mathrm{d} + C_{h_{ad^2\phi}}\mathrm{d}^2)\phi$$

$$+ (C_{h_{a\delta a}} + C_{h_{ad\delta a}}\mathrm{d} - h_a\mathrm{d}^2)\delta_a = 0 \quad (11–33d)$$

$$C_{h_r\beta}\beta + (C_{h_{rd\psi}} + C_{h_{rd^2\psi}}\mathrm{d})\mathrm{d}\psi + C_{h_{rd^2\phi}}\mathrm{d}^2\phi$$

$$+ (C_{h\delta r} + C_{h_{rd\delta r}}\mathrm{d} - h_r\mathrm{d}^2)\delta_r = 0 \quad (11–33e)$$

It should be noted that in these equations there are no partial derivatives with respect to ψ, the lowest order being $d\psi$. This means that the airplane has no directional sense, being satisfied to fly north as well as south. For this reason $d\psi$ will be taken as the variable instead of ψ.

The five simultaneous equations (11–33) can be solved for the five variables β, $d\psi$, ϕ, δ_a, and δ_r, but the labor in so doing is enormous. In order to keep the mathematics within the scope of this book, the problem will be investigated piecemeal. First, the motion of the airplane with fixed controls will be studied, and finally the motions of the airplane with free aileron and then with free rudder will be discussed.

11–2 Characteristic Motions of the Airplane with Controls Locked

If the controls of the airplane are assumed locked, all the partial derivatives with respect to control deflection and rate of deflection will vanish, as will the two hinge moment equations (11–33d and e). The equations for level flight become:

$$(C_{y\beta} - 2\mathrm{d})\beta - 2\mathrm{d}\psi + C_L\phi = 0 \qquad (11\text{–}34a)$$

$$\mu C_{l\beta}\beta + \frac{C_{l_r}}{2}\,\mathrm{d}\psi + \left(\frac{C_{l_p}}{2}\,\mathrm{d} - J_x\mathrm{d}^2\right)\phi = 0 \qquad (11\text{–}34b)$$

$$\mu C_{n\beta}\beta + \left(\frac{C_{n_r}}{2} - J_z\mathrm{d}\right)\mathrm{d}\psi + \frac{C_{n_p}}{2}\,\mathrm{d}\phi = 0 \qquad (11\text{–}34c)$$

where
$$J_x = 2\left(\frac{k_x}{b}\right)^2$$

$$J_z = 2\left(\frac{k_z}{b}\right)^2$$

The characteristic modes of lateral motion for the airplane with controls locked will be studied in the same manner that the longitudinal modes were investigated. This method is to assume the solution of these equations as

$$\beta = \beta_1 e^{\lambda t/\tau}, \quad \mathrm{d}\psi = \mathrm{d}\psi_1 e^{\lambda t/\tau}, \text{ etc.}$$

If these assumed solutions are substituted into (11–34a, b, c) and $e^{\lambda t/\tau}$ divided out, the equations reduce to three simultaneous equations in λ. To determine the value of λ the determinant of the coefficients of λ is equated to zero and expanded. This determinant becomes:

$$
\begin{vmatrix}
C_{y\beta} - 2\lambda & -2 & C_L \\[2ex]
\mu C_{l\beta} & \dfrac{C_{l_r}}{2} & \left(\dfrac{C_{l_p}}{2}\lambda - J_x\lambda^2\right) \\[2ex]
\mu C_{n\beta} & \left(\dfrac{C_{n_r}}{2} - J_z\lambda\right) & \dfrac{C_{n_p}\lambda}{2}
\end{vmatrix} = 0 \quad (11\text{--}35)
$$

Expanding this determinant, a quartic in λ is obtained:

$$A\lambda^4 + B\lambda^3 + C\lambda^2 + D\lambda + E = 0 \quad (11\text{--}36)$$

where

$$A = 1$$

$$B = -\frac{1}{2}\left(C_{y\beta} + C_{nr}/J_z + C_{lp}/J_x\right)$$

$$C = \frac{1}{4J_xJ_z}(C_{l_p}C_{nr} - C_{l_r}C_{np}) + \frac{C_{y\beta}}{4}\left(\frac{C_{nr}}{J_z} + \frac{C_{lp}}{J_x}\right) + \frac{\mu C_{n\beta}}{J_z} \quad (11\text{--}37)$$

$$D = -\frac{\mu}{2J_xJ_z}(C_{n\beta}C_{lp} - C_{l\beta}C_{np}) - \frac{\mu}{2J_x}C_LC_{l\beta}$$

$$\qquad - \frac{C_{y\beta}}{8J_xJ_z}(C_{l_p}C_{nr} - C_{l_r}C_{np})$$

$$E = \frac{\mu C_L}{4J_xJ_z}(C_{l\beta}C_{nr} - C_{n\beta}C_{l_r})$$

The coefficients of the quartic in λ given above (11–36) are constants depending for their values on the inertia terms, the airplane relative density parameter, μ, and the partial or stability derivatives. Before it is possible to go further with this solution, it is necessary to determine the value of these derivatives.

11-3   Evaluation of Stability Derivatives

The values of the derivatives $C_{y\beta}$, $C_{n\beta}$, and $C_{l\beta}$ are the static stability derivatives generally referred to as the side force derivative, directional stability, and dihedral effect, respectively. These derivatives can be readily obtained from the wind tunnel, while $C_{n\beta}$ and $C_{l\beta}$ may also be estimated by the methods outlined in Chapters 8 and 9, respectively. It should be noted that these slopes are per radian and in terms of sideslip, β.

The derivatives C_{l_p} and C_{n_r} are the damping derivatives in roll and yaw, respectively. The damping in roll, C_{l_p}, was discussed in Chapter 9, and curves for its evaluation are given in Figure 9–14. The damping in yaw, C_{n_r}, has not been discussed before but is quite similar to the airplane's damping in pitch, $C_{m_{\dot{\theta}}}$, in that most of the damping arises because of the increase in angle of attack of the tail surface due to the angular velocity of the airplane. If the airplane rotates with a yawing velocity, r (radians per second), the angle of attack of the vertical tail will be increased by:

$$\Delta \alpha_t = -\frac{rl_v}{V} \text{ radians} \tag{11–38}$$

The yawing moment coefficient due to the vertical tail is simply

$$C_n = -C_{L_v} \frac{S_v}{S_w} \frac{l_v}{b} \eta_t \tag{11–39}$$

or

$$C_n = -a_v \frac{S_v}{S_w} \frac{l_v^2}{b} \frac{r}{V} \eta_t \tag{11–40}$$

and

$$C_{n_r} = \frac{dC_n}{d\left(\dfrac{rb}{2V}\right)} = -2a_v \frac{S_v}{S_w} \frac{l_v^2}{b^2} \eta_t \tag{11–41}$$

A contribution of the wing to the damping in yaw is usually taken as $-C_{D_w}/4$, giving finally,

$$C_{n_r} = -\frac{C_{D_w}}{4} - 2a_v \frac{S_v}{S_w} \left(\frac{l_v}{b}\right)^2 \eta_t \tag{11–42}$$

The derivatives C_{l_r} and C_{n_p} are usually referred to as the rotary or cross derivatives. They are the rolling moment due to the yawing velocity and the yawing moment due to the rolling velocity. The value of these derivatives has never been very carefully established, as wind-tunnel tests for them are very difficult. They have been evaluated analytically by several investigators and have been found to be functions of the lift coefficient, C_L.

The rolling moment due to yawing arises because of the increased velocity of the outside wing over the inside wing, thereby introducing a rolling moment tending to raise the outside wing. A positive yawing velocity will therefore introduce a positive rolling moment.

The yawing moment due to rolling arises from the increase in angle of attack of the down-going wing and the decrease in the angle of attack of the up-going wing. As the lift vector tilts forward with respect to the wing chord with angle of attack, the down-going wing will have a forward component from the lift vector, while the up-going wing will have an aft component from the lift vector. The result of this is to create a negative yawing moment for a positive rolling velocity. By simple strip integration, and assuming an elliptic lift distribution,

$$C_l = \frac{C_L}{4} \frac{rb}{2V} \qquad (11\text{--}43)$$

and

$$C_n = -\frac{C_L}{8} \frac{pb}{2V} \qquad (11\text{--}44)$$

which gives

$$C_{l_r} = \frac{C_L}{4} \qquad (11\text{--}45)$$

$$C_{n_p} = -\frac{C_L}{8} \qquad (11\text{--}46)$$

With the stability derivative information in hand, it is now possible to evaluate the coefficients of the quartic (11–36) and to study the types of lateral motion that exist. An inspection of the signs of the terms making up the constants of the quartic in λ shows that, for all positive lift coefficients and for airplanes that are stable directionally with positive or stable dihedral, the terms A, B, C, D must always be positive, as all the terms included are positive. Whether the constant E of the quartic is positive or negative depends on the sign of the parentheses $(C_{l_\beta} C_{n_r} - C_{n_\beta} C_{l_r})$. If the product $C_{n_\beta} C_{l_r} > C_{l_\beta} C_{n_r}$, then the constant E will be negative. It will be shown later that in the normal airplane design this constant E is usually negative, and from the sign rule discussed in Chapter 10, this means that there will be a positive real root, or that one of the roots will indicate a divergence. In the normal lateral case with controls fixed there are usually two real roots, one positive and one negative, and a complex pair. The normal lateral motions then will include a pure divergence, a pure convergence, and an oscillation.

In order to demonstrate these roots, a typical airplane will be assumed and its modes of motion determined. The airplane will be assumed flying at a $C_L = 1.0$, with a density factor $\mu = 10$. The stability derivatives assumed are typical and are as follows:

$$C_{y\beta} = -.28$$

$$C_{n\beta} = .09$$

$$C_{l\beta} = -.04$$

$$C_{l_p} = -.45$$

$$C_{n_r} = -.12 \tag{11-47}$$

$$C_{l_r} = .25$$

$$C_{n_p} = -.125$$

$$J_x = .02$$

$$J_z = .03$$

If the terms given above (11-47) are substituted into (11-37), the stability quartic becomes:

$$\lambda^4 + 13.4\lambda^3 + 67.4\lambda^2 + 394\lambda - 73.8 = 0 \tag{11-48}$$

If the real roots are extracted by Horner's method, they are found to be:

$$\lambda_1 = .1815$$

$$\lambda_2 = -10.61$$

These roots indicate a mild divergence and a very heavy convergence. If the original quartic is reduced by these two roots, the remaining quadratic will be:

$$\lambda^2 + 2.97\lambda + 38.36 = 0 \tag{11-49}$$

which on solution gives

$$\lambda_{3,4} = -1.48 \pm i\,6.01$$

an oscillation whose period is 1.05τ seconds and which requires $.468\tau$ seconds to damp to one-half amplitude. The value of τ for a given μ depends on the airplane's span and velocity.

$$\tau = \frac{b}{V}\mu \tag{11-50}$$

For this example, if the span is 40 ft and the velocity 200 ft/sec, τ will equal 2.0 and the period will be 2.10 seconds. The damping is good as the oscillation is damped to half amplitude in .445 cycle.

These characteristics are typical of modern airplanes. The divergent mode of motion is known as the spiral divergence and can easily be demonstrated in flight by disturbing the airplane from a trimmed flight condition by a rudder or aileron kick and watching the ensuing motion. In the usual case, the airplane will start a slow spiral in the direction of the disturbance which will tighten up if uncorrected until a steep high-speed spiral dive develops.

The heavy convergent mode is not easily recognized in flight as it is damped out so rapidly. The lateral short period oscillation, however, is quite noticeable in flight and has been objectionable in some cases because of light damping. This oscillation is usually referred to as the Dutch roll. In most airplane designs the short period oscillation, controls locked, is not objectionable as the damping is normally heavy. We will see in a later section, however, that with controls free the Dutch roll, under certain conditions, can have very weak damping, which is quite objectionable, especially in transport- and bombardment-type aircraft.

The stability boundaries for neutral spiral divergence and neutrally damped Dutch roll can be investigated by equating the constant term E and Routh's discriminant to zero and plotting the stability boundaries in the $\mu C_{n\beta}$, $\mu C_{l\beta}$ plane.

$$E = 0 \quad \text{(spiral boundary)} \qquad (11\text{--}51)$$

$$BCD - AD^2 - B^2E = 0 \quad \text{(oscillation boundary)} \quad (11\text{--}52)$$

These boundaries are shown in Figure 11-3 for the case of the airplane assumed (11-47).

From these boundaries it can be seen that there is very little chance of ever encountering an unstable lateral oscillation with the controls locked. This is due to the fact that $C_{n\beta}$ is almost always made larger than $C_{l\beta}$, keeping the airplane far from the oscillatory boundary. This also means that the airplane will nearly always be spirally unstable at lift coefficients as high as $C_L = 1.0$. It can be seen that, to obtain spiral stability, the dihedral should be increased to increase $\mu C_{l\beta}$; however, this will bring the airplane closer to the oscillatory boundary, meaning that the damping of the lateral oscillations will be reduced. Increasing the dihedral will also tend to impair the aileron operation and to give objectionable rolling at high speed, and at low speed asymmetric power conditions for multiengined aircraft. For these reasons the rather mild spiral divergence is permitted in order to obtain the more important flying qualities that go with high $C_{n\beta}$ and low $C_{l\beta}$.

An inspection of the constant term of the quartic, E, shows that the spiral divergence will become more severe as the lift coefficient is increased. This is due to the dependence of the rotary derivative, C_{l_r}, on the lift coefficient. The condition for the constant E equalling zero or the divergence boundary may be written:

$$C_{l\beta}C_{n_r} - C_{n\beta}\frac{C_L}{4} = 0 \qquad (11\text{-}53)$$

As the lift coefficient is increased, the value of (11-53) will become more negative, while as the lift coefficient is reduced, the constant E

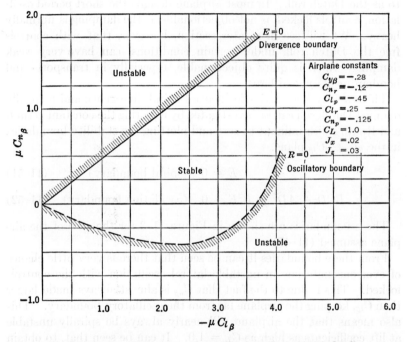

FIGURE 11-3. Typical lateral stability boundaries (stick-fixed).

will become more positive and the spiral divergence very mild and possibly even stable. For the case of the airplane assumed in the previous example, the divergence boundary becomes:

$$.0048 - .0225C_L = 0 \qquad (11\text{-}54)$$

The airplane will then be spirally unstable at all lift coefficients higher than, and spirally stable for all lift coefficients lower than, $C_L = .213$.

A study of the aerodynamic parameters involved in the short period lateral oscillations indicates the following trends: Increasing the directional stability, $C_{n\beta}$, tends to reduce the period of the oscillation but has little effect on the damping.   An increase in dihedral effect reduces the damping and shortens the period.   An increase in the damping in yaw derivative, C_{n_r}, increases the period and improves the damping.   An increase in side force parameter, $C_{y\beta}$, has little effect on the period but improves the damping of the oscillations, while an increase in airplane inertia will lengthen the period but reduce the damping a great deal.

The lateral oscillations of the airplane with controls locked have been studied at great length in the literature.*   In most airplane designs to date, this mode of motion has not been very important, as it is usually heavily damped.   In a few cases airplanes having excessive dihedral have had such poor damping of this mode that corrective action had to be taken after the airplane had made its first flight.   In these cases the damping was poor, requiring more than two cycles to damp the oscillation to one-half amplitude, and the period was so short that the pilot's reactions couldn't keep ahead of the disturbance to damp the motion by proper control action.

In most cases, however, inadequate damping of the lateral oscillations is tied in with the floating characteristics of the rudder, which, under certain conditions of balance, can introduce a poorly damped lateral oscillation known as "snaking."   This type of motion will be discussed in a later section.

11-4   Response to Aileron Control (One Degree of Freedom)

The response of the airplane to its lateral or aileron control is of great importance when analyzing its handling characteristics.   In the following sections, the response of the airplane to a step-function aileron disturbance will be studied from the point of view of the airplane, considered, first, as a single-degree-of-freedom system in roll; second, as a two-degree-of-freedom system in roll and sideslip; and finally, as an unrestricted three-degree-of-freedom system.

The disturbance or forcing function is considered, for simplicity, to be a step-function aileron deflection defined as an abrupt deflection from zero or the trim angle to some arbitrary deflection, δ_a, instantaneously when $t = 0$.   See Figure 11-4.

The equation of motion for the airplane restricted to pure roll, β and $d\psi = 0$, can be very simply written down from equations (11-33)

* NACA TR 589.

with the aileron deflection term, $\mu C_{l\delta_a}\delta_a$, taken on the right-hand side as the forcing action.

$$\left(\frac{C_{l_p}}{2}\,\mathrm{d} - J_x\mathrm{d}^2\right)\phi = -\mu C_{l\delta_a}\delta_a \tag{11-55}$$

or

$$\mathrm{d}^2\phi - \frac{C_{l_p}}{2J_x}\,\mathrm{d}\phi = \frac{\mu C_{l\delta_a}\delta_a}{J_x} \tag{11-56}$$

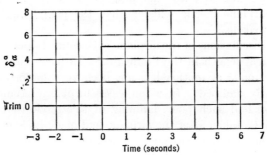

FIGURE 11-4. The aileron step-function.

If equation (11-56) is solved for $\mathrm{d}\phi$ by classical means, the transient solution is obtained from the reduced equation:

$$\mathrm{d}^2\phi - \frac{C_{l_p}}{2J_x}\,\mathrm{d}\phi = 0 \tag{11-57}$$

as

$$\mathrm{d}\phi_{\text{transient}} = \mathrm{d}\phi_0 e^{(C_{l_p}/2J_x)\,t/\tau} \tag{11-58}$$

The particular integral for the step-function disturbance is simply:

$$\mathrm{d}\phi_{ss} = A$$

The complete solution $\quad\mathrm{d}\phi = \mathrm{d}\phi_t + \mathrm{d}\phi_{ss}$
becomes

$$\mathrm{d}\phi = \mathrm{d}\phi_0 e^{(C_{l_p}/2J_x)\,t/\tau} + A \tag{11-59}$$

The two constants $\mathrm{d}\phi_0$ and A can be determined from the end conditions,

when $\qquad \dfrac{t}{\tau} = 0 \quad \mathrm{d}\phi = 0 \quad \therefore\ \mathrm{d}\phi_0 = -A$

when $\qquad \dfrac{t}{\tau} = 0 \quad \mathrm{d}^2\phi = \dfrac{\mu C_{l\delta_a}\delta_a}{J_x}$

giving $\qquad A = \dfrac{-2\mu C_{l\delta_a}\delta_a}{C_{l_p}}$

finally

$$d\phi = \frac{2\mu C_{l\delta a}\delta_a}{C_{l_p}} e^{(C_{l_p}/2J_x)t/\tau} - \frac{2\mu C_{l\delta a}\delta_a}{C_{l_p}} \qquad (11\text{-}60)$$

or

$$d\phi = \frac{2\mu C_{l\delta a}\delta_a}{C_{l_p}} [e^{(C_{l_p}/2J_x)t/\tau} - 1] \qquad (11\text{-}61)$$

The units of $d\phi$ are radians per t/τ seconds.   To convert to radians per second, multiply by $1/\tau$, which is equal to $V/b\mu$, giving:

$$p = \frac{2V C_{l\delta a}\delta_a}{C_{l_p}b} [e^{(C_{l_p}/2J_x)t/\tau} - 1] \qquad (11\text{-}62)$$

Equation (11-62) can be expressed in terms of the rolling criterion, $pb/2V$, as follows:

$$\frac{pb}{2V} = \frac{C_{l\delta a}\delta_a}{C_{l_p}} [e^{(C_{l_p}/2J_x)t/\tau} - 1] \qquad (11\text{-}63)$$

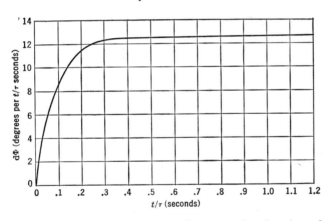

FIGURE 11-5.   Response of airplane to aileron step-function (one degree of freedom).

The rate of roll, or $pb/2V$, builds up exponentially with time, depending on the magnitude of the factor $C_{l_p}/2J_x$. As C_{l_p} is always negative, below the stall [for the case of the airplane whose characteristics were assumed in (11-47) $C_{l_p}/2J_x = -11.25$], the airplane's rolling velocity accelerates very rapidly to its steady-state value, after which the rolling velocity will remain constant.   The response of this typical airplane with normal aerodynamic parameters to a rolling disturbance due to an instantaneous five-degree total deflection of the ailerons giving ($C_l = .005$) is shown in Figure 11-5.

The solution to the equation of motion for a single-degree-of-freedom roll is quite simple and straightforward. In the more complicated equations to follow, it is usually inconvenient to do the work necessary to find the transient and steady-state solution and substitute end conditions. A short-hand method of operators is often a much simpler approach and will be described for the simple dynamic system in order that it may be more readily understood.

If the equation of motion (11–56) is rewritten with $d\phi = p'$, $C_{l_p}/2J_x = L_p$, and $\mu C_{l\delta a}\delta a/J_x = L_0$, it simply becomes:

$$dp' - L_p p' = L_0 \qquad (11\text{–}64)$$

which, when put in operator form, becomes:

$$(d - L_p)p' = L_0$$
$$p' = \frac{L_0}{(d - L_p)} \qquad (11\text{–}65)$$

or

Now $1/d - L_p = (d - L_p)^{-1}$, which, upon expanding by the binomial theorem, becomes:

$$(d - L_p)^{-1} = d^{-1} + d^{-2}L_p + d^{-3}L_p^2 + \cdots, \text{ etc.} \qquad (11\text{–}66)$$

but

$$d^{-1} = \int d\left(\frac{t}{\tau}\right) = \frac{t}{\tau}$$

$$d^{-2} = \int\int d\left(\frac{t}{\tau}\right)d\left(\frac{t}{\tau}\right) = \frac{(t/\tau)^2}{!2}$$

$$d^{-n} = \frac{(t/\tau)^n}{!n}$$

Therefore

$$(d - L_p)^{-1} = \frac{t}{\tau} + \frac{(t/\tau)^2}{!2}L_p + \frac{(t/\tau)^3}{!3}L_p^2 + \cdots, \text{ etc.} \qquad (11\text{–}67)$$

$$L_p(d - L_p)^{-1} = \frac{t}{\tau}L_p + \frac{(t/\tau)^2}{!2}L_p^2 + \frac{(t/\tau)^3}{!3}L_p^3 + \cdots, \text{ etc.} \qquad (11\text{–}68)$$

or

$$L_p(d - L_p)^{-1} = e^{L_p t/\tau} - 1 \qquad (11\text{–}69)$$

$$(d - L_p)^{-1} = \frac{e^{L_p t/\tau} - 1}{L_p} \qquad (11\text{–}70)$$

or

$$p' = \frac{L_0}{L_p}(e^{L_p t/\tau} - 1) \qquad (11\text{–}71)$$

If the original expression for p', L_p and L_0 are substituted in (11–71), the original expression for $d\phi$ is obtained

$$d\phi = \frac{2\mu C_{l\delta a}\delta_a}{C_{l_p}} \left(e^{(C_{l_p}/2J_x)\,t/\tau} - 1\right) \tag{11-72}$$

A systematic method for this procedure is known as Heaviside's expansion theorem, which can be developed as follows. The solution in operator form is arranged as $p' = f(\mathrm{d})/F(\mathrm{d})$, where $f(\mathrm{d})$ and $F(\mathrm{d})$ are polynomials in (d), but with $F(\mathrm{d})$ of higher order than $f(\mathrm{d})$. In the example just given, $f(\mathrm{d}) = L_0$ and $F(\mathrm{d}) = (\mathrm{d} - L_p)$.

The polynomial $F(\mathrm{d})$ may be broken up as follows:

$$F(\mathrm{d}) = (\mathrm{d} - \lambda_1)\ (\mathrm{d} - \lambda_2)\ (\mathrm{d} - \lambda_3) \cdots, \text{etc.}$$

where λ_1, λ_2, λ_3, etc., are roots of $F(\mathrm{d})$. Then by partial fractions

$$\frac{f(\mathrm{d})}{F(\mathrm{d})} = \frac{A}{\mathrm{d} - \lambda_1} + \frac{B}{\mathrm{d} - \lambda_2} + \frac{C}{\mathrm{d} - \lambda_3} + \cdots, \text{etc.} \tag{11-73}$$

$$\frac{f(\mathrm{d})}{F(\mathrm{d})} = (\mathrm{d} - \lambda_1)^{-1}A + (\mathrm{d} - \lambda_2)^{-1}B$$
$$+ (\mathrm{d} - \lambda_3)^{-1}C + \cdots, \text{etc.} \tag{11-74}$$

as before $(\mathrm{d} - \lambda_1)^{-1} = \dfrac{\dfrac{t}{\tau}\lambda_1 + \dfrac{\lambda_1{}^2(t/\tau)^2}{!2} + \dfrac{\lambda_1{}^3(t/\tau)^3}{!3} + \cdots}{\lambda_1}$ \hfill (11-75)

$$(\mathrm{d} - \lambda_1)^{-1} = \frac{e^{\lambda_1 t/\tau} - 1}{\lambda_1} \tag{11-76}$$

also $(\mathrm{d} - \lambda_2)^{-1} = \dfrac{e^{\lambda_2 t/\tau} - 1}{\lambda_2}$ etc. \hfill (11-77)

Therefore

$$\frac{f(\mathrm{d})}{F(\mathrm{d})} = \frac{A}{\lambda_1}\left(e^{\lambda_1 t/\tau} - 1\right) + \frac{B}{\lambda_2}\left(e^{\lambda_2 t/\tau} - 1\right) + \cdots, \text{etc.} \tag{11-78}$$

and

$$\frac{f(\mathrm{d})}{F(\mathrm{d})} = \left[-\frac{A}{\lambda_1} - \frac{B}{\lambda_2} - \cdots + \frac{A}{\lambda_1}e^{\lambda_1 t/\tau} + \frac{B}{\lambda_2}e^{\lambda_2 t/\tau} + \cdots \right] \tag{11-79}$$

$$\frac{f(\mathrm{d})}{F(\mathrm{d})} = \left[\frac{f(0)}{F(0)} + \frac{Ae^{\lambda_1 t/\tau}}{\lambda_1} + \frac{Be^{\lambda_2 t/\tau}}{\lambda_2} + \cdots \right] \tag{11-80}$$

To determine the coefficients A, B, C, etc., multiply (11–73) first by $(d - \lambda_1)$, giving:

$$\frac{(d - \lambda_1)f(d)}{F(d)} = A + \frac{B(d - \lambda_1)}{(d - \lambda_2)}$$

$$+ \frac{C(d - \lambda_1)}{(d - \lambda_3)} + \cdots \quad (11\text{–}81)$$

or

$$\frac{f(d)}{(d - \lambda_2)(d - \lambda_3) \cdots} = A + \frac{B(d - \lambda_1)}{(d - \lambda_2)}$$

$$+ \frac{C(d - \lambda_1)}{(d - \lambda_3)} + \cdots \quad (11\text{–}82)$$

therefore

$$\frac{f(\lambda_1)}{F'(\lambda_1)} = A$$

likewise

$$\frac{f(\lambda_2)}{F'(\lambda_2)} = B \text{ etc.}$$

giving finally

$$\frac{f(d)}{F(d)} = \left[\frac{f(0)}{F(0)} + \sum_{\lambda=1}^{\lambda=n} \frac{f(\lambda)}{\lambda F'(\lambda)} e^{\lambda t/\tau} \right] \quad (11\text{–}83)$$

which is Heaviside's expansion theorem. This expansion falls down for the condition of equal roots, but as this condition is almost never encountered in the dynamic studies of the airplane, this special case will not be dealt with.

The solution of the single-degree-of-freedom roll by use of Heaviside's expansion theorem proceeds as follows:

$$p' = \frac{L_0}{(d - L_p)} = \frac{f(d)}{F(d)} \quad (11\text{–}84)$$

where

$$f(d) = L_0$$
$$F(d) = d - L_p$$
$$F'(d) = 1$$
$$\lambda_1 = L_p$$
$$f(0) = L_0$$
$$F(0) = -L_p$$
$$f(\lambda_1) = L_0$$
$$F'(\lambda_1) = 1$$

The solution obtained by direct substitution into (11–83) becomes

$$p' = \left[\frac{L_0}{-L_p} + \frac{L_0}{L_p} e^{L_p\, t/\tau}\right] \tag{11-85}$$

or

$$p' = \frac{L_0}{L_p}[e^{L_p\, t/\tau} - 1] \tag{11-86}$$

as before.

The use of Heaviside's expansion theorem is very helpful in more complicated equations. This is especially true if the roots of $F(d)$ are real. If there are real and complex roots, it is sometimes convenient to use the Heaviside's expansion theorem to obtain part of the solution and substitute end conditions to obtain the rest. This will be demonstrated in subsequent sections.

11-5   Response to Aileron Control (Two Degrees of Freedom)

The response of the airplane to a step-function aileron deflection, with the airplane allowed to roll and sideslip but not yaw, is more or less an academic study, as this condition actually cannot be realized in flight without severe demands on the pilot's use of the rudder control. It does afford a logical second step, however, in the solution of the equation of motion in response to a control disturbance and brings out at the same time an interesting characteristic of the airplane's lateral motion.

The equations of motion for the airplane free to roll and sideslip, but not yaw, are as follows. It is assumed that the yawing moment equation is satisfied at all times by the application of rudder control by the pilot.

$$(C_{y\beta} - 2d)\beta + C_L\phi = 0 \tag{11-87a}$$

$$\mu C_{l\beta}\beta + \left(\frac{C_{lp}}{2}d - J_x d^2\right)\phi = -\mu C_{l\delta a}\delta_a \tag{11-87b}$$

The solution of these equations for angle of bank, ϕ, in radians is as follows:

$$\phi = \frac{\begin{vmatrix} C_{y\beta} - 2d & 0 \\ \mu C_{l\beta} & -\mu C_{l\delta a}\delta_a \end{vmatrix}}{\begin{vmatrix} C_{y\beta} - 2d & C_L \\ \mu C_{l\beta} & (C_{lp}/2)d - J_x d^2 \end{vmatrix}} = \frac{f(d)}{F(d)} \tag{11-88}$$

expanding

$$f(\mathrm{d}) = -\mu C_{l_{\delta_a}} \delta_a C_{y\beta} + 2\mu C_{l_{\delta_a}} \delta_a \mathrm{d} \qquad (11\text{–}89)$$

$$F(\mathrm{d}) = 2J_x \mathrm{d}^3 - (C_{l_p} + J_x C_{y\beta})\mathrm{d}^2 + \frac{C_{l_p}}{2} C_{y\beta} \mathrm{d} - \mu C_L C_{l_\beta} \qquad (11\text{–}90)$$

The solution for the response of the typical airplane assumed in
(11–47) to a step-function aileron deflection $C_{l_{\delta_a}} \delta_a = .005$ is as follows:
Substitution of airplane parameters gives:

$$f(\mathrm{d}) = .014 + .10\mathrm{d}$$

$$F(\mathrm{d}) = .04\mathrm{d}^3 + .456\mathrm{d}^2 + .063\mathrm{d} + .40$$

$$F'(\mathrm{d}) = .12\mathrm{d}^2 + .912\mathrm{d} + .063$$

$$f(0) = .014$$

$$F(0) = .40$$

Roots of $F(\mathrm{d})$ $\lambda_1 = -11.34$ $\lambda_{2,3} = -.03 \pm .945\,i$

Heaviside's expansion theorem can be used quite readily for any
constant term and for the components of the solution involving real
roots. Substitution of a complex root into the expansion equation
involves a rather complex shifting of the form of the complex numbers
and careful consideration of phase angles. Unless one has great facility
in the handling of complex numbers, it is wiser to restrict the use of
the expansion theorem to those parts involving only the real roots, and
to express the parts of the solution involving the complex roots as a
damped sine function with an amplitude A and a phase lag B, which
can be readily determined from end conditions.

As an example of this approach, the solution for the angle of bank
as a function of time can be written down as follows:

$$\phi = \frac{f(0)}{F(0)} + \frac{f(\lambda_1)}{\lambda_1 F'(\lambda_1)}\, e^{\lambda_1 t/\tau} + \frac{f(\lambda_2)}{\lambda_2 F'(\lambda_2)}\, e^{\lambda_2 t/\tau} + \frac{f(\lambda_3)}{\lambda_3 F'(\lambda_3)}\, e^{\lambda_3 t/\tau} \qquad (11\text{–}91)$$

The terms $f(0)/F(0)$ and $f(\lambda_1)/\lambda_1 F'(\lambda_1)$ can be obtained by substitution of the proper terms given above. The substitution of the
complex roots (λ_2, λ_3), however, is quite troublesome. To avoid this
substitution, use can be made of the knowledge that this complex
pair of roots indicates a damped oscillation, which can be written in
general form as:

$$A e^{-.03\, t/\tau} \sin\left(.945\, \frac{t}{\tau} + B\right) \qquad (11\text{–}92)$$

where A and B are the amplitude and phase of the oscillation which must be determined from the end conditions of the problem. If the Heaviside's expansion theorem is used on the first two terms, as suggested, the solution for ϕ becomes

$$\phi = .035 + .0191\ e^{-11.34t/\tau} + A\ e^{-.03t/\tau} \sin\left(.945\ \frac{t}{\tau} + B\right) \quad (11\text{–}93)$$

If the end conditions, that ϕ and $d\phi$ are zero when $t/\tau = 0$, are substituted into (11–93), the constants A and B are readily determined, giving

$$A = .231$$
$$B = 6.05$$

The final equation for the angle of bank in radians versus t/τ becomes

$$\phi = .035 + .0191\ e^{-11.34t/\tau}$$
$$+ .231\ e^{-.03t/\tau} \sin\left(.945\ \frac{t}{\tau} + 6.05\right) \quad (11\text{–}94)$$

From similar developments, the solution for $d\phi$ and β can be obtained. These are as follows:

$$d\phi = -.217\ e^{-11.34t/\tau}$$
$$+ .221\ e^{-.03t/\tau} \sin\left(.945\ \frac{t}{\tau} + 1.37\right) \quad (11\text{–}95)$$

and

$$\beta = .125 - .0009\ e^{-11.34t/\tau}$$
$$- .125\ e^{-.03t/\tau} \sin\left(.945\ \frac{t}{\tau} + 1.455\right) \quad (11\text{–}96)$$

The first few seconds of this response are shown in Figure 11–6, plotted in degrees versus t/τ.

It is interesting to note that restricting the yawing motion of the airplane practically destroys the damping of the lateral motion and increases the length of the period. From the rear, the airplane's motion would appear as sketched in Figure 11–7, a rather slow, sliding, rolling oscillation practically undamped. This maneuver can be demonstrated in flight if the pilot gives the aileron an abrupt deflection and attempts to hold zero yaw by means of the rudder.

The rudder deflection required of the pilot to permit this restricted motion can be readily obtained by making use of the yawing equation of motion and requiring it to be satisfied at all times by use of the rudder.

$$\mu C_{n_\beta}\beta + \frac{C_{n_p}}{2}\,d\phi = -\mu C_{n_{\delta_r}}\delta_r \quad (11\text{–}97)$$

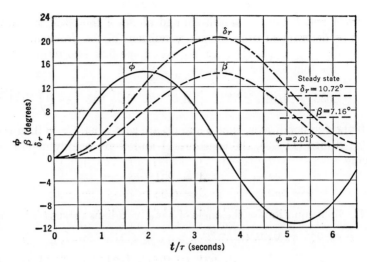

FIGURE 11-6.   Response of airplane to aileron step-function, $\dot{\psi} = 0$.

where β and $d\phi$ are given by equations (11–95) and (11–96).

$$\mu C_{n\beta} \left[.125 - .0009\, e^{-11.34\, t/\tau} - .125\, e^{-.03\, t/\tau} \right.$$
$$\left. \sin\left(.945\,\frac{t}{\tau} + 1.455\right) \right]$$

$$+ \frac{C_{np}}{2} \left[-.217\, e^{-11.34 t/\tau} + .221\, e^{-.03 t/\tau} \right.$$
$$\left. \sin\left(.945\,\frac{t}{\tau} + 1.37\right) \right] = -\mu C_{n\delta r}\delta_r$$

(11–98)

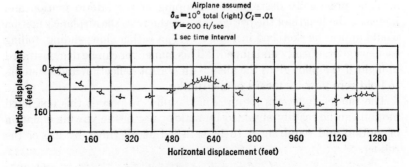

FIGURE 11-7.   Airplane response to abrupt aileron deflection. $\psi = 0$.

If the rudder power, $C_{n\delta r}$, is taken as $-.06$ per radian, the rudder angle required to maintain $\psi = 0$ during the maneuver becomes

$$\delta_r = .187 + .0213\, e^{-11.34t/\tau}$$
$$- .210\, e^{-.03t/\tau} \sin\left(.945\,\frac{t}{\tau} + 1.45\right) \quad (11\text{–}99)$$

The variation of δ_r with t/τ is shown in Figure 11–6. It should be noted that, if this motion is allowed to proceed until the oscillations are completely damped out, the steady-state condition is a straight slip with 7.16 degrees of sideslip, 10.72 degrees of left rudder, 2.01 degrees of right bank, and the 5 degrees of total right aileron.

Another type of two-degree-of-freedom response to an aileron step-function disturbance can be obtained by restricting the airplane in sideslip and allowing it freedom in roll and yaw. This is a more interesting case than the last, as it will be shown later that this is the condition for a perfectly coordinated turn. Again, rather vigorous rudder action is required of the pilot to keep the sideslip zero.

The equations of motion for this case are obtained from the rolling and side force equations, with the yawing equation used to determine the rudder action required.

$$-2\mathrm{d}\psi + C_L\phi = 0$$
$$\frac{C_{lr}}{2}\,\mathrm{d}\psi + \left(\frac{C_{lp}}{2}\,\mathrm{d} - J_x\mathrm{d}^2\right)\phi = -\,\mu C_{l\delta a}\delta_a \quad (11\text{–}100)$$

If these equations are solved for the case of the typical airplane assumed in the previous work, the angle of bank, yawing velocity, and rate of roll are as follows. The aileron disturbance, as before, is taken as 5 degrees total aileron ($C_l = .005$).

$$\phi = .018\, e^{-11.53\,t/\tau} + .782\, e^{.27\,t/\tau} - .80 \quad (11\text{–}101)$$
$$\mathrm{d}\phi = -.209\, e^{-11.53\,t/\tau} + .209\, e^{.27\,t/\tau} \quad (11\text{–}102)$$
$$\mathrm{d}\psi = .009\, e^{-11.53\,t/\tau} + .390\, e^{.27\,t/\tau} - .399 \quad (11\text{–}103)$$

The plot of these variables in degrees as functions of t/τ is given in Figure 11–8. It should be noted that for $\beta = 0$ there is no oscillatory mode, the normal heavily damped rolling mode and the spiral divergence mode coming through unaffected in magnitude.

The rudder angle required to maintain $\beta = 0$ during this maneuver can be solved by again making use of the yawing equation and substituting the values obtained above for $\mathrm{d}\psi$ and $\mathrm{d}\phi$ [equations (11–102) and (11–103) with $\beta = 0$].

$$\left(\frac{C_{nr}}{2} - J_z d\right) d\psi + \frac{C_{np}}{2} d\phi = -\mu C_{n\delta_r}\delta_r \qquad (11\text{--}104)$$

The rudder angle required for a rudder whose power is $C_{n\delta_r} = -.06$ becomes

$$\delta_r = .0399 + .0263\, e^{-11.53\, t/\tau} - .0662\, e^{.27\, t/\tau} \qquad (11\text{--}105)$$

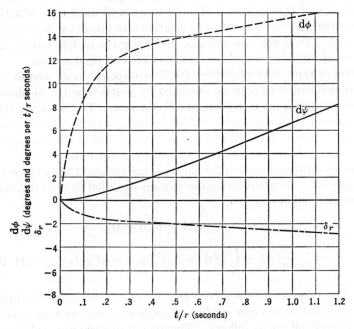

FIGURE 11–8. Response to aileron step-function, $\beta = 0$.

This angle is also plotted in Figure 11–8 in degrees as a function of t/τ. This is the case of the perfectly coordinated roll into a turn. The very rapid application of rudder required to maintain perfect coordination, when responding to an abrupt aileron deflection, is apparent.

11–6   Response to Aileron Control (Three Degrees of Freedom)

In the previous sections, the response of the airplane to an abrupt aileron deflection has been studied with one or more of its degrees of freedom restricted. The next step is to study the response of the airplane to the step-function aileron deflection with all three degrees of freedom unrestricted. This is similar to the normal flight test pro-

cedure for measurement of aileron effectiveness in which an abrupt
aileron deflection is made with rudder locked.

It will first be assumed that the aileron control produces no adverse
yawing moments.   The equations of motion for this maneuver become:

$$(C_{y\beta} - 2\mathrm{d})\beta - 2\mathrm{d}\psi + C_L\phi = 0$$

$$\mu C_{l\beta}\beta + \frac{C_{lr}}{2}\mathrm{d}\psi + \left(\frac{C_{lp}}{2}\mathrm{d} - J_x\mathrm{d}^2\right)\phi = -\mu C_{l\delta a}\delta_a \quad (11\text{-}106)$$

$$\mu C_{n\beta}\beta + \left(\frac{C_{nr}}{2} - J_z\mathrm{d}\right)\mathrm{d}\psi + \frac{C_{np}}{2}\mathrm{d}\phi = 0$$

The solution of these equations for the variables ϕ, $\mathrm{d}\phi$, β, and $\mathrm{d}\psi$
as functions of t/τ proceeds as before.   By the method of determinants,

$$\phi = \frac{\begin{vmatrix} (C_{y\beta} - 2\mathrm{d}) & -2 & 0 \\ \mu C_{l\beta} & \dfrac{C_{lr}}{2} & -\mu C_{l\delta a}\delta_a \\ \mu C_{n\beta} & \left(\dfrac{C_{nr}}{2} - J_z\mathrm{d}\right) & 0 \end{vmatrix}}{\begin{vmatrix} (C_{y\beta} - 2\mathrm{d}) & -2 & C_L \\ \mu C_{l\beta} & \dfrac{C_{lr}}{2} & \left(\dfrac{C_{lp}}{2}\mathrm{d} - J_x\mathrm{d}^2\right) \\ \mu C_{n\beta} & \left(\dfrac{C_{nr}}{2} - J_z\mathrm{d}\right) & \dfrac{C_{np}}{2}\mathrm{d} \end{vmatrix}} \quad (11\text{-}107)$$

If the stability derivatives are substituted into (11-107) and the
determinants expanded, the expression for ϕ becomes

$$\phi = \frac{f(\mathrm{d})}{F(\mathrm{d})} \quad (11\text{-}108)$$

the familiar form for operation by Heaviside's expansion theorem.
It should be noted that the polynomial $F(\mathrm{d})$ is just the stability
quartic discussed in Section 11-3.   If the values for the typical airplane assumed in the previous sections are substituted for a 5-degree
deflection of the ailerons:

$$\phi = \frac{2.5\mathrm{d}^2 + 5.35\mathrm{d} + 75.7}{\mathrm{d}^4 + 13.4\mathrm{d}^3 + 67.4\mathrm{d}^2 + 394\mathrm{d} - 73.8} \quad (11\text{-}109)$$

If the solution is expanded by Heaviside's expansion theorem, the following result is obtained:

$$\phi = -1.025 + 1.01\ e^{.1815\ t/\tau} + .022\ e^{-10.61\ t/\tau}$$
$$- .0096\ e^{-1.48\ t/\tau} \sin\left(6.01\ \frac{t}{\tau} + 2.325\right) \quad (11\text{-}110)$$

By similar processes:

$$d\phi = .183\ e^{.1815\ t/\tau} - .233\ e^{-10.61\ t/\tau}$$
$$+ .0513\ e^{-1.48\ t/\tau} \sin\left(6.01\ \frac{t}{\tau} + 1.345\right) \quad (11\text{-}111)$$

$$\beta = - .0339 + .048\ e^{.1815\ t/\tau} - .0048\ e^{-10.61\ t/\tau}$$
$$- .0155\ e^{-1.48\ t/\tau} \sin\left(6.01\ \frac{t}{\tau} + .66\right) \quad (11\text{-}112)$$

$$d\psi = -.508 + .489\ e^{.1815\ t/\tau} - .04\ e^{-10.61\ t/\tau}$$
$$+ .092\ e^{-1.48\ t/\tau} \sin\left(6.01\ \frac{t}{\tau} + 2.45\right) \quad (11\text{-}113)$$

A build-up of $d\phi$ versus t/τ is shown in Figure 11–9 as a function of its component modes. It should be noted that, as the assumed airplane is spirally unstable, the rate of roll will build up indefinitely as time goes to infinity.

Plots of $d\phi$, $d\psi$, and β as functions of t/τ are shown in Figure 11–10. It can be seen that, even for the case of an aileron control giving no adverse yaw, the airplane will start to yaw in the wrong direction under the pressure of the yawing moment due to the rolling velocity, C_{np}. The sideslip builds up rather slowly, and for rolls up to 45-degree banks never becomes excessively large. This is due to the fact that the assumed airplane had rather high directional stability, $C_{n\beta}$. The low sideslip and low dihedral effect, $C_{l\beta}$, for this airplane prevented any large drop in rolling velocity, $d\phi$.

Similar time histories of the response of the assumed airplane altered by a large decrease in vertical tail area, thereby reducing $C_{n\beta}$ and C_{nr}, and a large increase in the wing dihedral angle, thereby increasing the dihedral effect, $C_{l\beta}$, are shown in Figure 11–11. The heavy fall-off in rolling velocity due to the large dihedral can be noted, as well as the increase in adverse yawing and angle of sideslip.

11-7 Response to Aileron with Adverse Yaw

The results given in the previous section were for the airplane responding to an aileron that introduced no yawing moments, adverse or

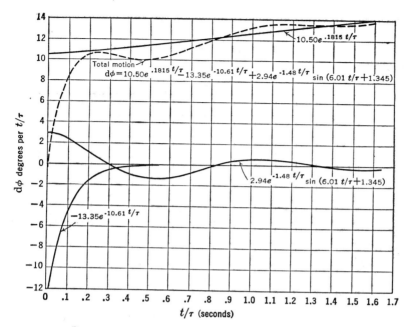

FIGURE 11–9. Build-up of $d\phi$ from component modes.

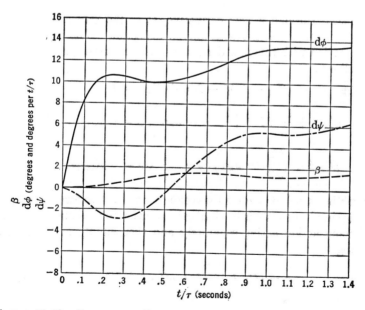

FIGURE 11–10. Response to aileron step-function (three degrees of freedom).

otherwise. In nearly all modern airplanes, the yaw, due to the aileron, is adverse (right aileron giving yawing moment to the left), and this yawing moment has large effects on the nature of the airplane's response.

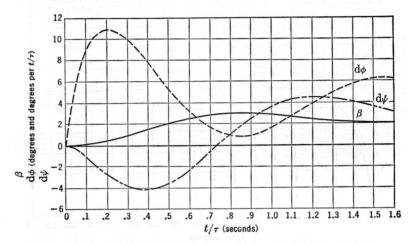

FIGURE 11-11. Response to aileron step-function, spirally stable airplane.

The introduction of aileron yaw into the equations of motion is accomplished through a new derivative, $C_{n\delta a}$, the rate of change of yawing moment coefficient with aileron deflection. The three equations of motion with this new derivative included in the yaw equation are as follows:

$$(C_{y\beta} - 2d)\beta - 2d\psi + C_L\phi = 0$$

$$\mu C_{l\beta}\beta + \frac{C_{lr}}{2}d\psi + \left(\frac{C_{lp}}{2}d - J_xd^2\right)\phi = -\mu C_{l\delta a}\delta_a \quad (11\text{-}114)$$

$$\mu C_{n\beta}\beta + \left(\frac{C_{nr}}{2} - J_zd\right)d\psi + \frac{C_{np}}{2}d\phi = -\mu C_{n\delta a}\delta_a$$

The method used for solving these equations with two forcing actions occurring simultaneously is to obtain the solution to the equations with each forcing action acting separately and then simply add the two results together.

$$\phi = \mu C_{l\delta a}\delta_a \frac{f(d)_1}{F(d)} + \mu C_{n\delta a}\delta_a \frac{f(d)_2}{F(d)} \quad (11\text{-}115)$$

As an example of this operation, consider that the aileron of the airplane being investigated introduces an adverse yawing moment that is just 40 per cent of the rolling moment. For the airplane assumed in the previous section, this would give $\mu C_{l_{\delta a}}\delta_a = .050$ and $\mu C_{n_{\delta a}}\delta_a = -.020$. The first part of the solution for ϕ (11–115) has already been obtained in the previous section.

$$\mu C_{l_{\delta a}}\delta_a \frac{f(d)_1}{F(d)} = -1.025 + 1.01\, e^{.1815\, t/\tau} + .022\, e^{-10.61\, t/\tau}$$
$$- .0096\, e^{-1.48\, t/\tau} \sin\left(6.01\frac{t}{\tau} + 2.325\right) \quad (11\text{–}116)$$

The solution for ϕ in response to the adverse yawing moment is obtained by a similar process, assuming that the rolling moment due to the aileron is zero.

$$\phi = \frac{\begin{vmatrix} (C_{y\beta} - 2d) & -2 & 0 \\[2mm] \mu C_{l\beta} & \dfrac{C_{l_r}}{2} & 0 \\[2mm] \mu C_{n\beta} & \left(\dfrac{C_{n_r}}{2} - J_z d\right) & -\mu C_{n_{\delta a}}\delta_a \end{vmatrix}}{F(d)}$$
$$= \mu C_{n_{\delta a}}\delta_a \frac{f(d)_2}{F(d)} \quad (11\text{–}117)$$

Expansion of these determinants gives the solution

$$\mu C_{n_{\delta a}}\delta_a \frac{f(d)_2}{F(d)} = .189 - .193\, e^{.1815\, t/\tau} + .0022\, e^{-10.61\, t/\tau}$$
$$+ .0103 e^{-1.48\, t/\tau} \sin\left(6.01\frac{t}{\tau} + .176\right) \quad (11\text{–}118)$$

The total response for ϕ will then simply be the addition of equations (11–116) and (11–118), giving

$$\phi = -.836 + .817 e^{.1815\, t/\tau} + .0242\, e^{-10.61\, t/\tau}$$
$$- .0175 e^{-1.48\, t/\tau} \sin\left(6.01\frac{t}{\tau} + 2.84\right) \quad (11\text{–}119)$$

By similar processes, the variation of $d\phi$, $d\psi$, and β may be obtained for the case of the aileron with adverse yaw. These variations are given in Figure 11–12 for the assumed airplane as functions of t/τ.

The detrimental effects of the large adverse yaw due to the ailerons can be seen in Figure 11–12. The airplane will start to yaw in the wrong direction for the turn, much more severely than for the case of zero adverse aileron yaw already studied. The sideslip builds up

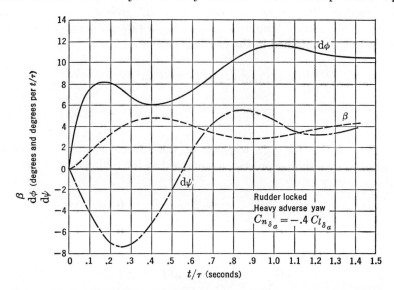

FIGURE 11–12. Response to aileron step-function (three degrees of freedom).

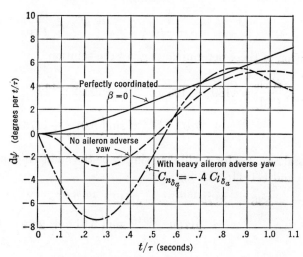

FIGURE 11–13. Comparison of adverse yawing for different type control configurations.

much faster and thereby cuts down the peak rolling velocity obtained. The transient response of the airplane to a rapid roll into a turn can be made much more favorable if the ailerons could be made to give favorable yawing moments, or if the rudder were geared to the ailerons to give favorable yaw. In Figure 11–13 the yawing velocities versus time for the case of a perfect turn, $\beta = 0$, and for the cases of zero and adverse aileron yaw are compared.

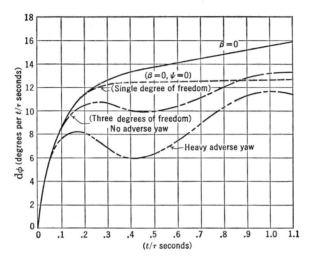

FIGURE 11–14.   Comparison of transient rolling motion in response to aileron step-function.

The comparison between the rolling velocities as functions of time for the single-degree-of-freedom response ($\beta = 0$, $\psi = 0$), the two-degree-of-freedom response ($\beta = 0$), and the three-degree-of-freedom response is given in Figure 11–14.

11–8   Steady-state Turns

In analyzing the lateral controls of the airplane, it is often useful to study their operation in steady-state turning maneuvers. In such a maneuver, the derivatives $d\phi$, $d^2\phi$, $d^2\psi$, $d\delta$, and $d\beta$ are all considered to be equal to zero, reducing the equations of motion to the following for the case of zero adverse aileron yaw.

$$C_{y\beta}\beta - 2d\psi + C_L\phi = 0 \qquad (11\text{–}120a)$$

$$\mu C_{l\beta}\beta + \frac{C_{l_r}}{2}\,d\psi + \mu C_{l\delta a}\delta_a = 0 \qquad (11\text{–}120b)$$

$$\mu C_{n\beta}\beta + \frac{C_{n_r}}{2}\,d\psi + \mu C_{n\delta r}\delta_r = 0 \qquad (11\text{–}120c)$$

These equations can be solved for the steady-state values of the variables under specified conditions of the turn. One of the most important conditions for any turn is the so-called perfect turn, in which the resultant acceleration on the pilot is in the plane of symmetry or, in pilot's parlance, through the seat of the pants.

It can be shown from a simple resolution of forces that a perfect turn is possible only if the sideslip of the airplane during the turn is held zero. This is always true except for the special case where the derivative $C_{y\beta}$ is zero. All-wing tailless airplanes approach this condition in very pure designs, and therefore the seat of the pants is a very poor indicator of the nicety of the turn for these aircraft. It can be said that the higher the side force derivative, $C_{y\beta}$, the easier it will be for the pilot to feel any error in the turn.

In the perfect turn with $\beta = 0$, the unbalanced side force due to gravity is totally absorbed in yawing the airplane. From $(11-120a)$ it can be seen that for this condition

$$d\psi = \frac{C_L}{2}\phi$$

If this value for $d\psi$ is substituted into equations $(11-120b$ and $c)$, the requirements on aileron and rudder in a perfect turn are obtained at once:

$$\delta_a = -C_L\phi \frac{C_{l_r}}{4\mu C_{l\delta a}} \qquad (11-121)$$

$$\delta_r = -C_L\phi \frac{C_{n_r}}{4\mu C_{n\delta r}} \qquad (11-122)$$

In the perfect turn, then, the aileron must always be held against the turn to balance the rolling moment due to the yawing velocity, while the rudder must always be held into the turn to balance out the damping in yaw. These solutions are correct if the deflections of the ailerons do not introduce yawing moments (adverse yaw). If the ailerons do introduce yawing moments, the rudder angle required will be altered. If the ailerons introduce adverse yaw, then the rudder angle into the turn will be reduced.

A more interesting study of the steady-state turn is the investigation of the so-called two-control turn, in which either the aileron or rudder is used independently. Two-control airplanes having elevator and aileron, or elevator and rudder controls only, have not only been studied academically but have also been built for production on the theory that the amateur pilot is less likely to get into trouble if a simplified control system could be devised.

Insight into the ability of the aileron or rudder alone to establish equilibrium in a steady banked turn can be obtained by studying the steady-state equation for the control angle and the sideslip essential to this maneuver. A measure of the perfection of the turn is the sideslip angle required in this two-control turn.

If the two controls are elevator and aileron, and if the aileron deflection introduces no adverse yawing moments, the equations of motion may be written down as follows, with the angle of bank considered the forcing action:

$$C_{y\beta}\beta - 2\mathrm{d}\psi + 0 = -C_L\phi \qquad (11\text{--}123a)$$

$$\mu C_{l\beta}\beta + \frac{C_{l_r}}{2}\,\mathrm{d}\psi + \mu C_{l\delta_a}\delta_a = 0 \qquad (11\text{--}123b)$$

$$\mu C_{n\beta}\beta + \frac{C_{n_r}}{2}\,\mathrm{d}\psi + 0 = 0 \qquad (11\text{--}123c)$$

which, if solved for the aileron angle required, becomes

$$\delta_a = \frac{\begin{vmatrix} C_{y\beta} & -2 & -C_L\phi \\[1ex] \mu C_{l\beta} & \dfrac{C_{l_r}}{2} & 0 \\[1ex] \mu C_{n\beta} & \dfrac{C_{n_r}}{2} & 0 \end{vmatrix}}{\begin{vmatrix} C_{y\beta} & -2 & 0 \\[1ex] \mu C_{l\beta} & \dfrac{C_{l_r}}{2} & \mu C_{l\delta_a} \\[1ex] \mu C_{n\beta} & \dfrac{C_{n_r}}{2} & 0 \end{vmatrix}} \qquad (11\text{--}124)$$

Expanding these determinants gives

$$\delta_a = \frac{C_L\phi}{C_{l\delta_a}}\left[\frac{C_{l\beta}C_{n_r} - C_{n\beta}C_{l_r}}{C_{n_r}C_{y\beta} + 4\mu C_{n\beta}}\right] \qquad (11\text{--}125)$$

It should be noted that the sign of the aileron deflection required depends on the spiral stability of the airplane. If spirally stable ($C_{l\beta}C_{n_r} > C_{n\beta}C_{l_r}$), the ailerons must be deflected into the turn.

The sideslip angle for equilibrium in this steady turn may be solved from the equations of motion (11–123) by a similar process, giving

$$\beta = -\frac{C_L\phi}{C_{y\beta} + 4\mu\dfrac{C_{n\beta}}{C_{n_r}}} \qquad (11\text{--}126)$$

The sideslip during an aileron-only, two-control turn is therefore a function of C_L, and the angle of bank, ϕ. The sideslip will be small if the directional stability is high and if the airplane has a large relative density, μ.

The equilibrium values of rudder angle and sideslip for a steady two-control (rudder-elevator) turn can be obtained from the following equations of motion:

$$C_{y\beta}\beta - 2d\psi + 0 = -C_L\phi \qquad (11\text{--}127a)$$

$$\mu C_{l\beta}\beta + \frac{C_{lr}}{2}\,d\psi + 0 = 0 \qquad (11\text{--}127b)$$

$$\mu C_{n\beta}\beta + \frac{C_{nr}}{2}\,d\psi + \mu C_{n\delta r}\delta_r = 0 \qquad (11\text{--}127c)$$

The solutions of these equations for the rudder angle and sideslip required are as follows:

$$\delta_r = -\frac{C_L\phi}{C_{n\delta r}}\left[\frac{C_{l\beta}C_{nr} - C_{n\beta}C_{lr}}{C_{y\beta}C_{lr} + 4\mu C_{l\beta}}\right] \qquad (11\text{--}128)$$

$$\beta = \frac{-C_L\phi}{C_{y\beta} + 4\mu\dfrac{C_{l\beta}}{C_{lr}}} \qquad (11\text{--}129)$$

The solution for the rudder angle shows that, for a spirally stable airplane, the rudder must be held with the turn, while for a spirally unstable airplane the rudder must be held against the turn. The sideslip in this type of turn is a function of C_L and angle of bank, ϕ, and the term $\mu(C_{l\beta}/C_{lr})$. The expression for the sideslip in an aileron-only turn was very similar, being a function of the term $\mu(C_{n\beta}/C_{nr})$. In almost all airplane designs, $C_{n\beta}/C_{nr}$ is greater than $C_{l\beta}/C_{lr}$, indicating that the aileron-elevator two-control airplane is a great deal better in this respect than the rudder-elevator operation, especially at the higher lift coefficients. The sideslip in a rudder-only turn becomes much worse at high C_L because of the fact that the derivative C_{lr} is a function of C_L.

Comparative sideslip angles in two-control turns are shown in Figure 11–15 for the aileron-elevator and rudder-elevator operations of a typical airplane as a function of the lift coefficient.

In the study of the two-control aileron turn just completed, the aileron was assumed not to give any adverse yawing moments. It is of interest to investigate the case where aileron deflection can introduce yawing moments, either due to the adverse yawing moments of the aileron or due to gearing the ailerons to the rudder. For this study, a yawing moment due to the aileron deflection, $C_{n\delta a}$, will be assumed

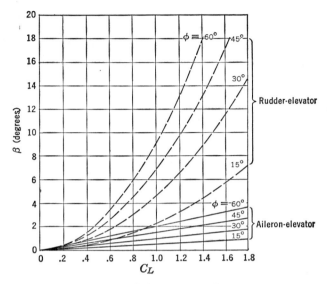

FIGURE 11-15. Comparison of sideslip in steady state, two-control turns.

to be some proportion of the rolling moment due to aileron deflection.

$$C_{n\delta a} = kC_{l\delta a} \tag{11-130}$$

The equations of motion for the steady-state turn for this case become

$$C_{y\beta}\beta - 2\mathrm{d}\psi + 0 = -C_L\phi$$

$$\mu C_{l\beta}\beta + \frac{C_{lr}}{2}\mathrm{d}\psi + \mu C_{l\delta a}\delta_a = 0 \tag{11-131}$$

$$\mu C_{n\beta}\beta + \frac{C_{nr}}{2}\mathrm{d}\psi + \mu k C_{l\delta a}\delta_a = 0$$

Solution of these equations for the aileron angle and sideslip gives:

$$\delta_a = \frac{-C_L\phi}{C_{l\delta a}}\left[\frac{C_{l\beta}C_{nr} - C_{n\beta}C_{lr}}{(C_{y\beta}C_{lr} + 4\mu C_{l\beta})k - (C_{nr}C_{y\beta} + 4\mu C_{n\beta})}\right] \tag{11-132}$$

$$\beta = C_L\phi\left[\frac{C_{nr} - kC_{lr}}{(C_{y\beta}C_{lr} + 4\mu C_{l\beta})k - (C_{nr}C_{y\beta} + 4\mu C_{n\beta})}\right] \tag{11-133}$$

The requirement on k for a perfect turn, $\beta = 0$, is obtained readily from (11–133). For zero sideslip, the value of k becomes:

$$k = \frac{C_{n_r}}{C_{l_r}} \tag{11–134}$$

a condition which could have been developed directly from equations (11–121) and (11–122). As C_{l_r} is a function of the lift coefficient, there is no single value for the gearing factor, k, that will satisfy the requirements for zero sideslip throughout the speed range. For the perfect turn, it can be seen again that k must be negative, or the coupled yawing moment must be adverse.

If equation (11–132) is solved for ϕ, the following expression is obtained:

$$\phi = -\frac{C_{l_{\delta a}}\delta_a}{C_L} \left[\frac{(C_{y\beta}C_{l_r} + 4\mu C_{l\beta})k - (C_{n_r}C_{y\beta} + 4\mu C_{n\beta})}{C_{l\beta}C_{n_r} - C_{n\beta}C_{l_r}} \right] \tag{11–135}$$

If the gearing ratio is adjusted so that:

$$k = \frac{C_{n_r}C_{y\beta} + 4\mu C_{n\beta}}{C_{l_r}C_{y\beta} + 4\mu C_{l\beta}} \tag{11–136}$$

then the airplane will not bank in response to a slow application of the ailerons. In this condition, the heavy yawing moments due to the aileron deflection will create enough sideslip for the dihedral effect to just offset the aileron rolling moment. In response to the slow application of the aileron, such an airplane would start to skid around in a flat turn under the pressure of the side force derivative, $C_{y\beta}$.

It can also be shown that if the factor k be given the value:

$$k = \frac{C_{n\beta}}{C_{l\beta}} \tag{11–137}$$

the airplane in response to the aileron will not yaw, making only a straight slip when in equilibrium with the aileron deflected. It is obvious that values of k close to those given in (11–136) and (11–137) would result in an impossible two-control airplane design.

The equilibrium aileron and sideslip angles for a 30-degree steady banked turn are shown in Figure 11–16 for the typical airplane assumed earlier in this chapter. This airplane has good directional stability and low dihedral effect. At a lift coefficient of unity, it can be seen that the sideslip would be quite small, independent of the value of k. A similar set of curves is also given in Figure 11–16 for an airplane made spirally stable at the same lift coefficient by reducing the vertical tail area and increasing the wing dihedral. For values of k slightly less

than zero, the aileron angles and sideslip go to infinity. This is due to
the fact that the conditions given by equations (11–136) and (11–137)
are reached and the ailerons will no longer control the airplane.

The value of k that should give the best all-around two-control
performance is a debatable point. From the point of view of the

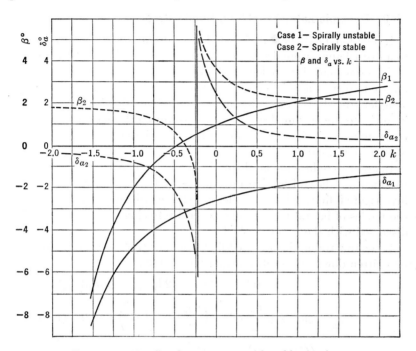

FIGURE 11–16. Steady-state turns with rudder gearing.

transient motion at the start of a turn, the airplane will require a positive or favorable k in order to avoid the very unpleasant tendency of
the airplane to swing in the wrong direction in response to the lateral
controls. It was shown in Figure 11–13 that even for an airplane
with $k = 0$ there is adverse yawing. For k, negative, the adverse
yawing is severe, while if k is made positive, the adverse yawing can
be minimized.

When equilibrium has been established in the steady turn, the value
of k required for a perfect turn is negative and can satisfy conditions at
only one value of the lift coefficient. If the impossibility of a perfect
turn is accepted by using a positive k, then there will always be sideslip, but it can be kept acceptably small by proper design.

A good start in the design of a two-control airplane can be made by making the factor k just positive enough to make the entry into the turn satisfactory. Beyond this, it is advisable to have high directional stability and low dihedral to increase the damping of the transient motion. With care in design, excellent two-control airplanes are quite possible.

11–9   Dynamic Lateral Stability, Rudder-free

In Section 11–3 the natural modes of airplane motion, in response to a disturbance, were discussed under the assumption that the rudder and aileron controls were locked. The resultant analysis was therefore referred to as the stick-fixed or control-fixed motion. In order to complete the study of the airplane's lateral dynamics, it is necessary to investigate the new modes of motion introduced through the freeing of the rudder and ailerons to float under the influence of their hinge moment parameters.

In order to reduce the mathematical complexity of this investigation, the modes of lateral motion introduced by freeing the rudder will be investigated first, with the ailerons still considered fixed. Second, the modes of lateral motion introduced by freeing the ailerons will be discussed with the rudder locked.

If the free rudder is discussed first, the four equations of motion can be written down at once from equations (11–33).

$$(C_{y\beta} - 2\mathrm{d})\beta - 2\mathrm{d}\psi + C_L\phi = 0 \quad (11\text{–}138a)$$

$$\mu C_{l\beta}\beta + \frac{C_{lr}}{2}\,\mathrm{d}\psi + \left(\frac{C_{lp}}{2}\,\mathrm{d} - J_x\mathrm{d}^2\right)\phi = 0 \quad (11\text{–}138b)$$

$$\mu C_{n\beta}\beta + \left(\frac{C_{nr}}{2} - J_z\mathrm{d}\right)\mathrm{d}\psi + \frac{C_{np}}{2}\,\mathrm{d}\phi$$
$$+ \mu(C_{n\delta r} + C_{nd\delta r}\mathrm{d})\delta_r = 0 \quad (11\text{–}138c)$$

$$C_{hr\beta}\beta + (C_{hrd\psi} + C_{hrd^2\psi}\mathrm{d})\mathrm{d}\psi + C_{hrd^2\phi}\mathrm{d}^2\phi$$
$$+ (C_{hr\delta r} + C_{hrd\delta r}\mathrm{d} - h_r\mathrm{d}^2)\delta_r = 0 \quad (11\text{–}138d)$$

If the equations (11–138) are solved in detail by substitution of the solution in the form $\beta = \beta_0 e^{\lambda\,t/\tau}$ the resulting determinant of the coefficients will yield a sextic in λ. If the six roots of this equation are extracted, they will be found to include two real roots, one indicating a heavily damped convergence, easily recognized as the normal rolling convergence encountered in the stick-fixed case, while the second is the spiral mode, usually such a slow divergence that it is of no real design importance. The other four roots combine to form two complex pairs. One of these oscillations is of very short period and is heavily damped. It can be thought of as the rapid return of the rudder to its proper floating angle after a disturbance. The second oscillation has a period somewhat similar to the oscillatory mode with controls fixed, discussed before, but the damping can be greatly reduced and even made unstable under certain rudder balance conditions. This mode is of design importance and is usually referred to as the "snaking" mode. It is the only one of all these modes that is of interest to the aerodynamicist and, therefore, any reductions to the equations of motion that will eliminate any of the unimportant modes and still leave the major characteristics of the snaking mode unchanged would be a welcome simplification.

It has been proven by English investigators[*] and substantiated later by the NACA,[†] that, if the coupling of the lateral motion with the degrees of freedom in roll and sidewise motion is neglected, the rolling convergence and spiral modes are eliminated, while the major characteristics of the "snaking" mode are only slightly affected. Under these assumptions the airplane can be thought of as oscillating in yaw, while in flight along a straight flight path (Figure 11-17). In this condition, the angle of yaw, ψ, is just equal to the negative of the angle of sideslip, β, and the terms in β and ψ in the equations of motion may be grouped together.

With these simplifying assumptions the equations of motion become

$$\left(-\mu C_{n\beta} + \frac{C_{nr}}{2}\mathrm{d} - J_z\mathrm{d}^2\right)\psi + \mu(C_{n\delta r} + C_{n\mathrm{d}\delta r}\mathrm{d})\delta_r = 0 \quad (11\text{-}139)$$

$$(-C_{h\beta} + C_{h\mathrm{d}\psi}\mathrm{d} + C_{h\mathrm{d}^2\psi}\mathrm{d}^2)\psi + (C_{h\delta} + C_{h\mathrm{d}\delta}\mathrm{d} - h_r\mathrm{d}^2)\delta_r = 0$$

* British AD 3164.
† NACA WR L-394.

Straight flight path

FIGURE 11-17.
Simplified lateral motion.

The solution of equations (11–139) for the characteristic modes results in a quartic equation in λ, the four roots of which form two complex pairs, indicating oscillations that fit the "snaking" and rudder flapping modes of the unreduced equation very closely. Careful analysis of these two oscillations shows that only the rudder flapping mode has a frequency rapid enough to be affected by the rudder inertia, and therefore when studying the "snaking" mode it is possible to assume the inertia of the rudder negligible. Under the assumption of zero rudder inertia ($h_r = 0$), the equations yield a cubic in λ, the one real root indicating a pure and rapid convergence of the rudder to its floating angle. The "snaking" mode again is only slightly affected. Finally, the derivative $C_{hd^2\psi}$, which arises as a result of any mass unbalance of the rudder, can be neglected as nearly all modern airplanes have mass balanced rudders to avoid flutter difficulties. Under these further assumptions equations (11–139) reduce as follows:

$$\left(-\mu C_{n\beta} + \frac{C_{nr}}{2}\,\mathrm{d} - J_z\mathrm{d}^2\right)\psi + \mu(C_{n\delta r} + C_{nd\delta r}\mathrm{d})\delta_r = 0$$

$$(-C_{h\beta} + C_{hd\psi}\mathrm{d})\psi + (C_{h\delta} + C_{hd\delta}\mathrm{d})\delta_r = 0$$

(11–140)

If the familiar substitution of the solution in the form $\psi = \psi_0 e^{\lambda\, t/\tau}$ etc., is made, and the resulting determinant of the coefficients expanded, a cubic in λ is obtained:

$$A\lambda^3 + B\lambda^2 + C\lambda + D = 0 \qquad (11\text{–}141)$$

where

$$A = -J_z C_{hd\delta} \qquad\qquad (11\text{–}142)$$

$$B = C_{hd\delta}\frac{C_{nr}}{2} - J_z C_{h\delta} - \mu C_{nd\delta r}C_{hd\psi} \qquad (11\text{–}143)$$

$$C = \frac{C_{nr}}{2}C_{h\delta} - \mu C_{n\beta}C_{hd\delta} + \mu C_{nd\delta}C_{h\beta} - \mu C_{n\delta}C_{hd\psi} \qquad (11\text{–}144)$$

$$D = \mu(C_{n\delta}C_{h\beta} - C_{n\beta}C_{h\delta}) \qquad (11\text{–}145)$$

Before it is possible to investigate the nature of the roots of (11–141), it is necessary to evaluate the stability derivatives involved.

The derivatives $C_{n\beta}$ and C_{nr} were discussed in the sections on the control-fixed lateral motions. $C_{n\beta}$ is the static directional stability, rudder fixed, and C_{nr} is the damping in yaw. The derivative $C_{n\delta r}$ is the rudder power discussed in detail in Chapter 8. It should be noted again that all derivatives are taken per radian.

The derivative $C_{h\delta}$ is the rate of change of rudder hinge moment with rudder angle, or the rudder's so-called restoring tendency. $C_{h\beta}$ is the rudder's floating tendency. Under the assumption of left rudder as a positive deflection and the angle of sideslip taken as positive with the wind coming in from the right, $C_{h\beta}$ must be positive for an unbalanced rudder. In terms of the normal hinge moment parameter

$$C_{h\beta} = -C_{h\alpha} \qquad (11\text{--}146)$$

The derivative $C_{hd\psi}$ arises as a result of the increase in vertical tail angle of attack due to the yawing velocity, r. If this angle of attack of the vertical tail is taken as

$$\Delta\alpha_v = \frac{rl_v}{V} \qquad (11\text{--}147)$$

the hinge moment becomes

$$C_h = C_{h\alpha}\frac{rl_v}{V} \qquad (11\text{--}148)$$

and

$$\frac{\partial C_h}{\partial r} = C_{h\alpha}\frac{l_v}{V} \qquad (11\text{--}149)$$

If (11–149) is divided on each side by $1/\tau$, the derivative sought for is obtained:

$$C_{hd\psi} = C_{h\alpha}\frac{l_v}{\mu b} \qquad (11\text{--}150)$$

The derivatives $C_{nd\delta r}$ and $C_{hd\delta r}$ arise from the alteration in pressure distribution over the vertical tail due to a rate of deflection of the rudder. This effect was discussed in Chapter 10 for the case of the elevator, and was found to arise as the result of the acceleration of the potential flow with rate of flap deflection and effective change in the tail's mean camber.

The derivatives in terms of the partial derivatives $\partial C_L/\partial d\delta_r$ given as A, B, C, and D in Figure 10–10 can be expressed as follows:

$$C_{nd\delta r} = -\frac{c_v}{2\mu b}[A + a_vB]\frac{S_v}{S_w}\frac{l_v}{b} \qquad (11\text{--}151)$$

and

$$C_{hd\delta r} = -\frac{c_v}{2\mu b}[C + a_vD] \qquad (11\text{--}152)$$

where c_v is the mean chord of the vertical tail, and a_v is the slope of the lift curve of the vertical tail per radian.

In order to investigate the nature of the roots of (11–141), typical derivatives will be developed based on the airplanes studied earlier in this chapter.

Typical derivatives are as follows:

$$C_{n\beta} = \quad .09$$

$$C_{n_r} = \quad -.12$$

$$C_{n\delta r} = \quad -.060$$

$$C_{nd\delta r} = \quad -.001$$

$$C_{h\alpha} = \quad -.24$$

$$C_{h\delta} = \quad -.48$$

$$C_{hd\psi} = \quad -.012$$

$$C_{hd\delta} = \quad -.01$$

$$\mu = \quad 10$$

$$J_z = \quad .03$$

If these derivatives are substituted into (11–142) to (145) the stability cubic will be as follows:

$$\lambda^3 + 49.6\lambda^2 + 94\lambda + 960 = 0 \qquad (11\text{--}153)$$

The roots of this cubic are as follows:

$$\lambda_1 = -48.1$$

$$\lambda_{2,3} = -.75 \pm 4.6\,i$$

These roots indicate an extremely heavy convergence, which can be thought of as the motion of the rudder snapping back to its normal floating angle, and a damped oscillation, quite similar to the controls-fixed oscillation but with lighter damping. The damping of the oscillatory mode is a very sensitive function of the hinge moment parameters, $C_{h\alpha}$ and $C_{h\delta}$. To illustrate this, assume the airplane derivatives all held constant except for the terms dependent on $C_{h\alpha}$, and assume aerodynamic balance giving $C_{h\alpha} = +.24$ and $C_{h\delta} = -.05$, an extremely close balance for a rudder.

With these new hinge moment parameters, the cubic in λ becomes:

$$\lambda^3 + 7.4\lambda^2 + 45\lambda + 630 = 0 \qquad (11\text{--}154)$$

the roots becoming

$$\lambda_1 = -9.6$$

$$\lambda_{2,3} = 1.1 \pm 8.05\,i$$

These roots indicate that balancing out the rudder reduces the heavy convergence, because of the reduction of $C_{h\delta}$, and the damping of the oscillatory mode goes negative.   In other words an airplane with very satisfactory stick-fixed lateral dynamics may encounter a serious instability if the balance of the rudder is carried too far.

As the hinge moment parameters, $C_{h\alpha}$ and $C_{h\delta}$, are the sensitive variables for the oscillatory or "snaking" mode, it is convenient to plot the boundary between stability and instability in the $C_{h\alpha}$, $C_{h\delta}$ plane. This boundary is defined by Routh's discriminant, which, for a cubic equation, $A\lambda^3 + B\lambda^2 + C\lambda + D = 0$, is simply:

$$BC - AD = 0 \qquad (11\text{-}155)$$

The plot of the stability boundary for the assumed airplane is given in Figure 11–18.   It is sometimes enlightening not only to solve for the conditions of zero damping, but to obtain the variation of the damping and the frequency of the oscillations throughout the $C_{h\alpha}$, $C_{h\delta}$ plane.

Lines of constant damping are obtained by assuming the root λ equal to $\lambda_1 - a$, where a is any damping constant whatever:

$$\lambda = -a + \lambda_1 \qquad (11\text{-}156)$$

If this is substituted into the original cubic in λ, a new cubic involving λ_1 and a will be obtained.   Application of Routh's discriminant to this new equation will determine the boundary or line of constant damping in the $C_{h\alpha}$, $C_{h\delta}$ plane.   Constant damping lines are shown for a typical airplane in Figure 11–18.

Lines of constant frequency can also be drawn in the same plane by making use of the condition obtained from the development of Routh's discriminant that, for zero damping with $\lambda = \pm i\omega$, the frequency, ω, in radians per t/τ seconds becomes

$$\omega = \sqrt{\frac{C}{A}} = \sqrt{\frac{D}{B}} \qquad (11\text{-}157)$$

As the coefficients of the cubic A, B, C, D are functions of $C_{h\alpha}$ and $C_{h\delta}$, the value of the frequency for any point on any of the boundaries may be obtained and lines of constant frequency plotted.   Constant-frequency lines are shown in Figure 11–18 for the assumed airplane.

The points A and B spotted on Figure 11–18 are the particular solutions developed before.   It should be noted that the modern trend in aerodynamic balance is carrying the airplane at least in the direction of low damping of the snaking mode.

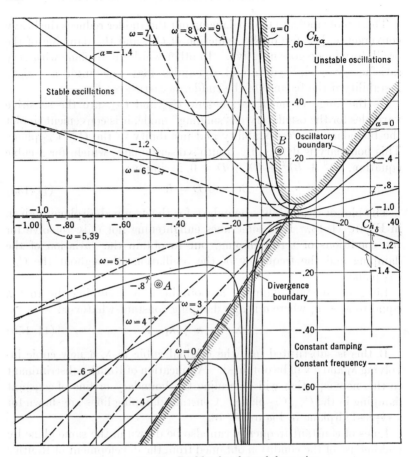

FIGURE 11–18.   Rudder-free lateral dynamics.

Although the variables C_{h_α} and C_{h_δ} are the most sensitive ones in determining the damping of the snaking mode of lateral oscillations, other airplane parameters also can have considerable influence. Exhaustive studies of the effect of all airplane parameters on the damping of this mode are reported in the literature,* and the reader is referred to them for the complete treatment. Suffice it to say that large increases in the rudder damping derivative, $C_{h_{d\delta}}$, through aerodynamic hinge moments or the effect of viscous friction in the control system have detrimental effects on the damping, while large changes in rudder inertia have a relatively small influence. Large rudder mass unbalance also reduces the damping, but this is not usually a serious

* NACA WR L-394.

problem as all modern airplanes have rather close mass balance to limit the possibility of rudder flutter.

One control characteristic that may introduce serious difficulty is the effect of solid friction. It was first shown by Boeing engineers that the effect of solid friction and a positive floating tendency can combine to permit steady lateral oscillations, the amplitude of which are a function of the magnitude of the friction. It is important, therefore, to keep the friction level very low if a positive C_{h_α} is used to implement the airplane's stick-free directional stability.

11–10   Dynamic Lateral Stability (Ailerons Free)

The equations of lateral motion with the ailerons free, but with rudder locked, can be written down from equations (11–33).

$$(C_{y_\beta}\beta - 2\mathrm{d})\beta - 2\mathrm{d}\psi + C_L\phi = 0 \quad (11\text{–}158a)$$

$$\mu C_{l_\beta}\beta + \frac{C_{l_r}}{2}\,\mathrm{d}\psi + \left(\frac{C_{l_p}}{2}\,\mathrm{d} - J_x\mathrm{d}^2\right)\phi$$
$$+ \mu(C_{l_{\delta a}} + C_{l_{\mathrm{d}\delta a}}\mathrm{d})\delta_a = 0 \quad (11\text{–}158b)$$

$$\mu C_{n_\beta}\beta + \left(\frac{C_{n_r}}{2} - J_z\mathrm{d}\right)\mathrm{d}\psi + \frac{C_{n_p}}{2}\,\mathrm{d}\phi = 0 \quad (11\text{–}158c)$$

$$C_{h_a\beta}\beta + (C_{h_{ad\phi}}\mathrm{d} + C_{h_{ad^2\phi}}\mathrm{d}^2)\,\phi$$
$$+ (C_{h_{a\delta a}} + C_{h_{ad\delta a}}\mathrm{d} - h_a\mathrm{d}^2)\delta_a = 0 \quad (11\text{–}158d)$$

If these equations are solved for typical values of the stability derivatives, it will be found that for normal designs they define four real roots and one complex pair. Only one of the real roots is divergent and is the now familiar spiral mode. One of the real roots is equivalent to the rolling convergence mode with controls locked, a function mainly of the damping in roll C_{l_p}; the other two real roots are usually heavy convergences that may, under certain conditions of aileron hinge moments, combine to define a normally heavily damped oscillation. Finally, the complex pair is the control-fixed lateral oscillation mode.

As the spiral mode and the control-fixed lateral or "Dutch roll" mode are not of particular interest to the present investigation, it is possible to restrict the equations for simpler mathematical treatment. It has been demonstrated* that, if the yawing and the sideslipping motion is neglected, the spiral mode and normal oscillatory lateral mode will be eliminated, with only small influence on the rolling modes under investigation.

* NACA WR L-361.

Under this assumption, and with the additional assumption that the aileron's inertia can be neglected and that the ailerons are mass balanced, the general equations (11–158) reduce to the following pair:

$$\left(\frac{C_{l_p}}{2} - J_x \mathrm{d}\right) \mathrm{d}\phi + \mu(C_{l\delta a} + C_{l\delta a}\mathrm{d})\delta_a = 0$$

$$C_{h\mathrm{d}\phi}\mathrm{d}\phi + (C_{h\delta} + C_{h\delta}\mathrm{d})\delta_a = 0$$

(11–159)

If the usual procedure for solution of these equations is followed, the characteristic motion will be defined by a quadratic in λ, the roots of which indicate the nature of the modes. Before it is possible to study these characteristic modes, it is necessary to evaluate the stability derivatives involved.

In setting up these derivatives, it is convenient to simplify the two-control aileron system by the following assumptions. A positive aileron deflection is one tending to roll the airplane to the right; this means that the right aileron will be up and the left aileron will be down. The aileron deflection (δ_a) will be considered the total aileron deflection, and the hinge moment is considered positive if it tends to deflect the aileron positively.

The derivatives C_{l_p}, $C_{l\delta a}$, and $C_{h\delta}$ have been discussed elsewhere in this book and will not be treated here. The derivative $C_{h\mathrm{d}\phi}$ is the change in aileron hinge moment due to the rolling velocity, p, developed as follows. Under the influence of the rolling velocity, the angle of attack of each aileron is changed on the average by the amount

$$\Delta\alpha = \frac{p\bar{y}}{V}$$

(11–160)

where p is the rolling velocity in radians per second and $\bar{y}$ the distance from the airplane centerline to the centroid of the wing area covered by the aileron. The hinge moment coefficient due to this increment in angle of attack under this new sign convention becomes:

$$C_h = -C_{h\alpha}\frac{p\bar{y}}{V}$$

(11–161)

or

$$\frac{\partial C_h}{\partial p} = -C_{h\alpha}\frac{\bar{y}}{V}$$

(11–162)

If each side of (11–162) is divided by $1/\tau$, the equation reduces to the derivative sought:

$$C_{h\mathrm{d}\phi} = -C_{h\alpha}\frac{\bar{y}}{b\mu}$$

(11–163)

The derivatives $C_{hd\delta a}$ and $C_{ld\delta a}$ arise as the result of the alteration in the chordwise pressure distribution due to the rate of flap deflection. This phenomenon has been discussed in previous sections. The derivative $C_{hd\delta a}$ can be obtained from the following formula, where the terms C and D may be obtained from the curves given in Figure 10–10:

$$C_{hd\delta a} = -\frac{1}{2\mu A_w}[C + a_w D] \qquad (11\text{–}164)$$

where A_w is the aspect ratio of the wing and a_w the slope of the wing's lift curve per radian.

The derivative $C_{ld\delta a}$ is the change in rolling moment due to the rate of flap deflection and can be developed in terms of the rolling moment due to flap deflection $C_{l\delta a}$ as follows. The factors A and B can be obtained from the data given in Figure 10–10.

$$C_{ld\delta a} = \frac{1}{2\mu A_w}[A + a_w B]\frac{C_{l\delta a}}{a_w \tau} \qquad (11\text{–}165)$$

Equations (11–159), if expanded in terms of the parameter λ as before, yield a quadratic of the general form

$$A\lambda^2 + B\lambda + C = 0 \qquad (11\text{–}166)$$

where

$$A = -C_{hd\delta}J_x \qquad (11\text{–}167)$$

$$B = \left(C_{hd\delta}\frac{C_{lp}}{2} - J_x C_{h\delta} - \mu C_{ld\delta a}C_{hd\phi}\right) \qquad (11\text{–}168)$$

$$C = \left(\frac{C_{h\delta}C_{lp}}{2} - \mu C_{l\delta a}C_{hd\phi}\right) \qquad (11\text{–}169)$$

Typical values of the derivatives based on the airplane assumed in other sections of this chapter give:

$$C_{lp} = -.45$$

$$C_{l\delta a} = .15$$

$$C_{hd\delta a} = -.014$$

$$C_{ld\delta a} = .0015$$

$$C_{h\delta} = -.48$$

$$C_{h\alpha} = -.24$$

$$C_{hd\phi} = .0096$$

If these derivatives are substituted in (11–167) to (169), the quadratic in λ becomes

$$\lambda^2 + 45.2\lambda + 336 = 0 \qquad (11\text{–}170)$$

the roots of which are

$$\lambda_1 = -9.8$$

$$\lambda_2 = -35.8$$

These roots indicate two very heavy convergences, which can be thought of as the heavy rolling convergence, a function of C_{l_p}, and the snapping of the aileron back to its normal floating angle under the influence of the hinge moment parameter, C_{h_δ}.

If extreme aerodynamic balance is resorted to, as was done for the case of the free rudder, the motion is changed considerably. If all the derivatives assumed above are held constant except C_{h_α} and C_{h_δ}, with $C_{h_\alpha} = +.24$ and $C_{h_\delta} = -.05$, the quadratic in λ becomes

$$\lambda^2 + 15.3\lambda + 91.6 = 0 \qquad (11\text{–}171)$$

the roots of which become

$$\lambda_{1,2} = -7.65 \pm i\, 5.75$$

a heavily damped oscillation of rather short period. A careful analysis of these equations indicates that the most complete aerodynamic balance is required to give any oscillation at all, and as a matter of fact, throughout the entire useful aileron hinge moment range, the possibility of any undamped oscillation is remote.

In order to demonstrate the fact that even the closest aerodynamic balance of the ailerons will not introduce any dynamic difficulties, for normal mass balanced ailerons, the boundary between regions where the roots indicate oscillations and where they indicate a convergence or divergence is plotted in the C_{h_α}, C_{h_δ} plane in Figure 11–19 for the assumed airplane. This boundary is defined through the condition of infinite period or zero frequency. The expression for the frequency for a quadratic $A\lambda^2 + B\lambda + C = 0$ is simply:

$$\omega = \sqrt{\frac{C}{A} - \left(\frac{B}{2A}\right)^2} \qquad (11\text{–}172)$$

For zero frequency, or the oscillation boundary, the relationship between the coefficients of the quadratic becomes simply

$$B^2 = 4AC \qquad (11\text{–}173)$$

Lines of constant damping are also imposed on the oscillatory region of Figure 11–19. The damping constant for a quadratic is simply

$B/2A$; the condition for zero damping or Routh's discriminant for a quadratic is simply $B = 0$.

The boundary of zero damping is shown in Figure 11–19 to be well into the positive C_{h_δ} area, a region where controls should never enter. In all normal ranges of the hinge moments, the motion is aperiodic

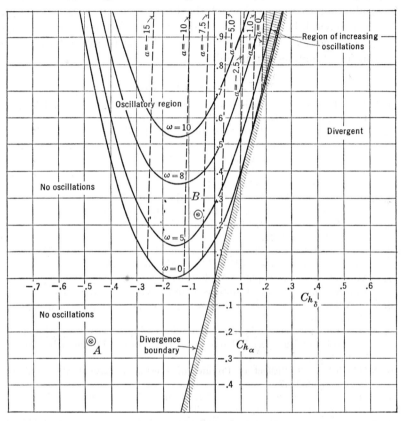

FIGURE 11–19. Aileron free dynamics.

and heavily convergent.   For close aerodynamic balance, the motion may be oscillatory, but it is still heavily damped.

The divergence boundary ($C = 0$, for a quadratic) indicates the values of the aileron hinge moments where their floating rate is just sufficient to balance out the damping of the wings in roll.   For hinge moments to the right of this boundary, the rolling divergence will be objectionably rapid and should be avoided.   Again abnormal hinge moment characteristics are required to get even close to this divergent region.   The two typical examples A and B are spotted in Figure 11–19.

The NACA* has conducted a rather exhaustive study of the aileron free modes, and the general trends can be stated as follows. For all normal mass balanced ailerons, freeing the ailerons introduces no dynamic complications. If the ailerons are not mass balanced and if the friction level is large, there may be some danger of oscillatory instability for extreme values of the hinge moment parameters C_{h_α}, C_{h_δ}. In general, because of the tendency of modern designers to mass balance all controls and especially the ailerons to avoid flutter, the aileron free dynamics can usually be neglected.

SUGGESTED READING

1. NACA TR 589, "An Analysis of Lateral Stability in Power-off Flight with Charts for Use in Design," by C. H. Zimmerman.
2. NACA WR L-394, "A Theoretical Investigation of the Lateral Oscillations of an Airplane with Free Rudder with Special Reference to the Effect of Friction," by Greenberg and Sternfield, 1943.
3. NACA TR 709, "An Analysis of the Stability of an Airplane with Free Controls," by Jones and Cohen, 1941.
4. British NPL S & C 4304, "An Investigation of the Lateral Stability of Aeroplanes with Rudder Free," by Bryant and Gandy, 1939.
5. British Report AD 3164, "Lateral Instability and Rudder-fuselage Flutter," by H. M. Lyon, 1941.
6. NACA WR L-361, "A Theoretical Investigation of the Rolling Oscillations of an Airplane with Ailerons Free," by Doris Cohen, 1944.
7. NACA TR 560, "A Simplified Application of the Method of Operators to the Calculation of Disturbed Motions of an Airplane," by R. T. Jones.
8. NACA TR 579, "A Study of the Two-control Operation of an Airplane," by R. T. Jones.
9. NACA TR 638, "The Influence of Lateral Stability on Disturbed Motions of an Airplane with Special Reference to the Motions Produced by Gusts," by R. T. Jones.
10. NACA TN 1193, "Effect of Product of Inertia on Lateral Stability," by L. Sternfield, 1947.
11. NACA TN 1370, "Correlation of Experimental and Calculated Effects of Product of Inertia on Lateral Stability."
12. Boeing Airplane Co. Report #D-3397, "Dynamic Control Surface Instability," by Schairer and Bush, 1941.

PROBLEMS

11-1. The fighter-type airplane, whose $\frac{1}{5}$ scale wind-tunnel model is shown in part *a* of the figure (p. 471), is tested in the wind tunnel and is found to have the aerodynamic characteristics shown by the wind-tunnel curves in parts *b* and *c* of the figure (pp. 472, 473). What are the stability derivatives $C_{n\beta}$, $C_{l\beta}$, $C_{y\beta}$, C_{l_r}, C_{n_r}, C_{l_p}, C_{n_p} at a $C_L = 1.0$?

11-2. If the full-scale airplane given in Problem 11-1 has the following inertia and weight, what will be the period and damping of the control-fixed lateral oscil-

* NACA WR L-361.

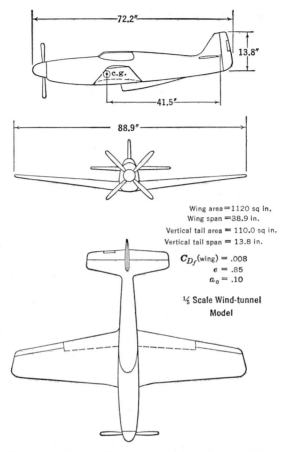

Wing area=1120 sq in.
Wing span =38.9 in.
Vertical tail area = 110.0 sq in.
Vertical tail span = 13.8 in.

C_{D_f}(wing) = .008
e = .85
a_o = .10

⅕ Scale Wind-tunnel
Model

PROBLEM 11–1a. Three views of wind-tunnel model.

lation when flying at $C_L = 1.0$ at 10,000 ft? Will the airplane be spirally stable? At what speed in miles per hour will the airplane have neutral spiral stability?

$$\text{Weight} = 8000 \text{ lb}$$
$$I_x = 5600 \text{ slug ft}^2$$
$$I_y = 8000 \text{ slug ft}^2$$
$$I_z = 8800 \text{ slug ft}^2$$

11–3. The airplane given in Problems 11–1 and 11–2 is flying at 10,000 ft at an indicated speed of 150 mph. The pilot abruptly deflects the ailerons to a total deflection of 5 degrees for a right roll. During the subsequent roll, the pilot maintains perfect coordination ($\beta = 0$) by use of the rudder. Calculate the rate of roll, $d\phi/dt$, in degrees per second, the rate of yaw, $d\psi/dt$, in degrees per second, and the angle of sideslip, β, in degrees, all as functions of time, t, in seconds. Also, calculate the variation of the rudder angle, δ_r, in degrees as a function of time for

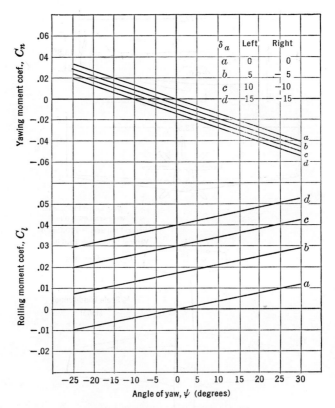

δ_a	Left	Right
a	0	0
b	5	5
c	10	10
d	15	15

PROBLEM 11–1*b*.   Wind-tunnel data.

this maneuver.   Plot all answers versus t for at least 1 cycle of the natural oscillatory frequency, with sufficient points to give a smooth curve.

11–4. If the rudder of the airplane given in Problems 11–1 and 11–2 were geared to the ailerons in such a way that 1 degree of aileron throw gave 1 degree of rudder throw (aileron deflection for right roll giving right rudder deflection), what would be the sideslip in a steady 30-degree banked right turn throughout the lift coefficient range from $C_L = .2$ to $C_L = 1.8$?   Plot results.

11–5. An airplane has the following lateral characteristics:

$C_{n\beta} = .06$	$C_{l\delta a} = .172$	$W = 10,000$ lb	$S_v = 30$ sq ft
$C_{l\beta} = -.12$	$C_{n\delta r} = -.0573$	$S_w = 250$ sq ft	$c_v = 3$ ft
$C_{n_r} = -.18$	$C_{y\beta} = -.20$	$b = 40$ ft	$b_v = 10$ ft
$C_{l_r} = .1$	$C_L = .4$	$\rho = \rho_0$	$c_a/c_w = .20$
$C_{n_p} = -.05$	$J_x = .03$	$l_v = 18$ ft	$c_r/c_v = .30$
$C_{l_p} = -.45$	$J_z = .05$	$\bar{y} = 14$ ft	Frise ailerons

This airplane is disturbed by a rudder step-function of 5 degrees right rudder.
a. What will be the transient motion (β and $d\psi/dt$) after this disturbance if the

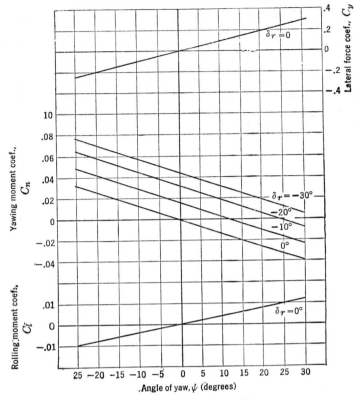

Problem 11–1c. Wind-tunnel data.

ailerons are deflected in such a way that the airplane always maintains zero bank ($\phi = 0$). The ailerons introduce no yawing moments. Plot all answers in degrees and degrees per second versus time in seconds.

b. What will be the steady-state values of β, $d\psi/dt$, δ_r, and δ_a in degrees?

c. What will be the transient motion of δ_a required to maintain zero ϕ during this maneuver?

11–6. If the airplane whose characteristics are given in Problem 11–5 has a rudder and ailerons whose hinge moments are as follows, calculate the stability derivatives, $C_{ha d\phi}$, $C_{ha d\delta_a}$, $C_{l d\delta_a}$, $C_{h d\delta_r}$, $C_{h rd\psi}$, and $C_{n d\delta_r}$

$$C_{h a \alpha} = -.20$$
$$C_{h a \delta_a} = -.40$$
$$C_{h r \alpha} = .08$$
$$C_{h r \delta_r} = -.04$$

11–7. If the rudder is completely mass balanced and the inertia is considered negligible, will the "snaking" mode of lateral oscillation with rudder free and ailerons locked be stable or unstable?

11–8. If the ailerons are mass balanced and their inertias considered negligible, what will be the nature of the ailerons-free rolling mode, with rudder locked?

11–9. In the rudder $C_{h\alpha}$, $C_{h\delta}$ plane plot the boundary describing zero damping of the snaking mode for the airplane given in Problem 11–6. Also plot the boundary describing oscillations whose frequency is 3.8 radian per second.

11–10. In the aileron $C_{h\alpha}$, $C_{h\delta}$ plane plot the boundary between the regions of oscillations and no oscillation for the ailerons-free rudder-locked oscillatory mode. On the same graph, plot curves of constant frequency ($w = 5$ and 10 radians per second) and lines of constant damping, including the zero damping boundary.

APPENDIX

NATURE OF THE ATMOSPHERE

The most important fluid with which the aerodynamicist is concerned is the air which constitutes the earth's atmosphere. The atmosphere at sea level is composed approximately of 78 per cent by volume of nitrogen, 21 per cent of oxygen, and small amounts of water vapor, carbon dioxide, hydrogen, and other rare gases. Up to altitudes of approximately 40,000 ft, fluctuating winds in all directions within the troposphere keep the air commingled in nearly the same proportions. Above the tropopause, which is the dividing point between the troposphere and the stratosphere, the different gases begin to settle or separate out according to their respective densities. The motion of air within the stratosphere is mainly horizontal. The stratosphere extends some 50 miles above the earth's surface. Very little is known about the characteristics of the upper stratosphere, as direct measurements of pressure and wind velocity have been made only to an altitude of about 25 miles. Thermometers installed in rockets of the German V-2 type disclose a torrid zone with temperature of 170°F between 30 and 40 miles in altitude, indicating an ozone layer which has the property of absorbing and holding more heat than the air below it. From 40 to 50 miles the temperature may drop as low as 150°F below zero. Above 50 miles altitude, in the ionosphere, recorded temperatures as high as 250°F indicate another torrid zone. Even less is known about this third blanket of air except that within this layer many electrical phenomena are known to exist. It is believed that the accumulation of cosmic dust particles is responsible for the heat absorption in this layer.

With the flight of the German V-2 rocket to altitudes approaching 100 miles and the future design of man-carrying rocket-propelled aircraft to reach even higher altitudes, it is hoped that scientists will soon be able to supply aeronautical engineers with the necessary knowledge of the upper atmosphere, so that the realm of flight may be extended upward until flight to other planets will be accomplished.

The air near the surface of the earth is compressed by the weight of the air above it. The highest pressure on a unit area is at the earth's surface and is caused by the weight of a column of air extending to the top of the atmosphere. Since the density of air decreases with altitude, the rate of decrease of pressure with an increase in altitude is greater than linear. The change in density of the air follows basic thermodynamic laws and is caused by changes in temperature as influenced by the sun's heat cycle. The sun's rays pass through the atmosphere with little heating effect and transmit heat energy to the relatively dense surface of the earth. The air adjacent to the earth's surface is heated by conduction, rises, and expands as it reaches regions of lesser pressure. The expansion is usually adiabatic because of the rapid movement of the air, so that all the heat energy is used in internal work of expansion until a state of equilibrium is reached at the tropopause. Since the air is heated to a greater extent at the equator than at the poles, it ascends to a higher altitude before dissipating its heat energy. Because the pressure at the higher altitude is also smaller, the air expands to a greater degree and hence has a lower temperature. At the tropics the height of the troposphere is approximately 56,000 ft and corresponding temperature $-110°F$; at the poles 28,000 ft and $-50°F$; and at 40° latitude 35,000 ft and $-67°F$. This shows that the temperature lapse rate or rate of change of temperature with altitude varies with latitude. Since the physical properties of the air are dependent upon temperature, and the performance of aircraft is dependent upon the air density, pressure, and temperature, correlation of performance data is dependent upon some assumed standard lapse rate. For convenience, an International Standard Atmosphere has been adopted based on an average linear lapse rate at 40° north latitude which has been empirically chosen after a study of average lapse rates observed throughout the world.

The equation of the linear variation of temperature with altitude within the troposphere is given by

$$T = T_0 - \alpha h \tag{1}$$

where

$$T_0 = 518.4°F \text{ and } \alpha \simeq .003566°F/ft$$

To determine the pressure and density within the standard atmosphere, the following additional assumptions are made:

1. The air is dry.
2. The air behaves as a perfect gas as given by the equation of state:

$$p = \rho g R T$$

or

$$\frac{p}{p_0} = \frac{\rho}{\rho_0} \frac{T}{T_0} \tag{2}$$

3. The gravity field is constant.

The variation of pressure with altitude is determined from the summation of vertical forces acting on an infinitesimal volume of air within the atmosphere (Figure 1).

$$\Sigma \text{ Vertical forces} = 0$$

and assuming a positive direction is upward,

$$p - p - dp - w\,dh = 0$$

$$dp = -\rho g\,dh \qquad (3)$$

Substituting for ρ its value from the equation of state and rearranging,

$$\frac{-dp}{p} = \frac{1}{R}\frac{dh}{T} \qquad (4)$$

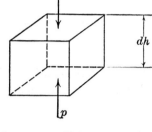

FIGURE 1.   Unit volume of air in atmosphere.

and from equation (1), substituting the value of T within the troposphere,

$$\frac{-dp}{p} = \frac{1}{R}\frac{dh}{T_0 - \alpha h}$$

$$-\int_{p_0}^{p}\frac{dp}{p} = \frac{1}{R}\int_{0}^{h}\frac{dh}{T_0 - \alpha h}$$

$$-\log_e p]_{p_0}^{p} = \frac{-1}{\alpha R}\log_e (T_0 - \alpha h)]_{0}^{h}$$

$$\log_e \frac{p}{p_0} = \frac{1}{\alpha R}\log_e \frac{T_0 - \alpha h}{T_0}$$

Since

$$\alpha = \frac{T_0 - T}{h}$$

$$\frac{p}{p_0} = \left(\frac{T}{T_0}\right)^{1/\alpha R} \qquad (5)$$

or

$$\frac{p}{p_0} = \left(1 - \frac{\alpha h}{T_0}\right)^{1/\alpha R} \qquad (6)$$

which gives the pressure ratio variation with altitude within the troposphere. From equations (2) and (5)

$$\frac{\rho}{\rho_0}\frac{T}{T_0} = \left(\frac{T}{T_0}\right)^{1/\alpha R}$$

$$\frac{\rho}{\rho_0} = \sigma = \left(\frac{T}{T_0}\right)^{(1/\alpha R)-1}$$

$$\sigma = \left(1 - \frac{\alpha h}{T_0}\right)^{(1/\alpha R)-1} \tag{7}$$

which gives the density ratio variation with altitude within the troposphere. Also from equations (2) and (5)

$$\frac{p}{p_0} = \frac{\rho}{\rho_0}\left(\frac{p}{p_0}\right)^{\alpha R}$$

$$\sigma = \left(\frac{p}{p_0}\right)^{1-\alpha R} \tag{8}$$

which gives the density ratio variation with pressure ratio within the troposphere.

Since the temperature is constant in the stratosphere to an altitude of approximately 20 miles, the pressure ratio above the tropopause may be obtained by direct integration of equation (4), using conditions at the tropopause for the lower limit of integration. The corresponding density ratio is obtained by use of equation (2). Table 1 lists the most important physical properties of the air for standard altitudes up to 65,000 ft.

A density- and pressure-altitude conversion chart (Figure 2) is a convenient graphical representation of pressure-temperature-density relationships computed from the equation of state. Constant-density lines are labeled "density altitude, h_d, in feet." Cross-plotted are lines of constant pressure in feet as computed from the equation of state and labeled "pressure altitude, h_p." These latter altitudes correspond to pressure readings of an airplane altimeter calibrated to read pressure in feet of air. The calibration is chosen so that the altimeter in the standard atmosphere would read the actual altitude. On this chart is drawn the standard temperature variation with density altitude based on the standard lapse rate. Pressure and density altitudes are identical only along the standard temperature line. Auxiliary scales near the bottom of Figure 2 are given for the pressure ratio, p/p_0, at standard temperature, and density ratio, σ, corresponding to the density altitude. When flight tests of airplanes are made, atmospheric conditions of p and T are rarely if ever standard. Performance data are usually

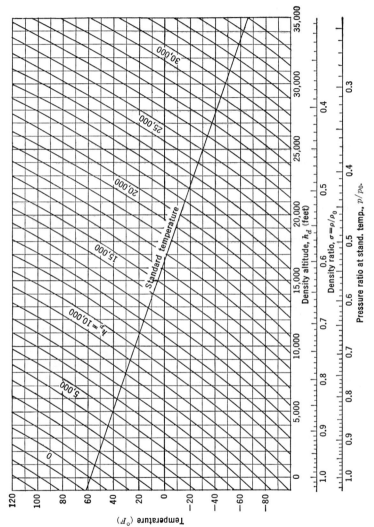

FIGURE 2. Density and pressure altitude conversion chart.

presented for the density altitude obtained from Figure 2 corresponding to these atmospheric conditions and not for the pressure altitude measured by the altimeter. In this way a common basis for the correlation of many tests flown at non-standard conditions is possible.

To construct the lines of constant pressure or pressure altitude the following form of the equation of state is used:

$$\rho T = \frac{p}{gR}$$

where

$$R = \frac{p_0}{\rho_0 g T_0} = \text{Constant}$$

For example, to determine the line of 10,000 ft pressure altitude, select from Table 1 the standard value of p at 10,000 ft such that

$$\rho T = \text{Constant}$$

By selecting several values of ρ corresponding to several density altitudes, values of T can be computed. A plot of density altitude versus T then determines the 10,000 ft pressure-altitude line.

TABLE 1

STANDARD ATMOSPHERE TABLE

Altitude (ft)	Pressure (in. Hg)	Pressure ratio, p/p_0	Temperature (°C)	Temperature (°F)	Density, ρ, (slugs/ft³)	Specific weight, $w = \rho g$ (lb/ft³)	Density ratio, σ	$\sigma^{1/2}$	Speed of sound, a (mph)	Kinematic viscosity, $\nu = \mu/\rho$ (ft²/sec ×10⁴)
−1,000	31.02	1.0367	17.0	62.6	.002448	.07878	1.0296	1.0147	763.5	1.527
− 500	30.47	1.0182	16.0	60.8	.002413	.07764	1.0147	1.0073	762.2	1.545
0	29.92	1.0000	15.0	59.0	.002378	.07651	1.0000	1.0000	760.9	1.564
500	29.38	.9821	14.0	57.2	.002343	.07540	.9854	.9927	759.6	1.583
1,000	28.86	.9644	13.0	55.4	.002309	.07430	.9710	.9854	758.2	1.602
1,500	28.33	.9469	12.0	53.7	.002275	.07321	.9568	.9782	756.9	1.621
2,000	27.82	.9298	11.0	51.9	.002242	.07213	.9427	.9709	755.6	1.641
2,500	27.31	.9129	10.0	50.1	.002209	.07106	.9288	.9638	754.3	1.661
3,000	26.82	.8962	9.1	48.3	.002176	.07001	.9151	.9566	753.0	1.681
3,500	26.32	.8798	8.1	46.5	.002144	.06897	.9015	.9495	751.6	1.702
4,000	25.84	.8636	7.1	44.7	.002112	.06794	.8880	.9424	750.3	1.723
4,500	25.36	.8477	6.1	43.0	.002080	.06693	.8747	.9353	749.0	1.744
5,000	24.89	.8320	5.1	41.2	.002049	.06592	.8616	.9282	747.7	1.766
5,500	24.43	.8165	4.1	43.4	.002018	.06493	.8486	.9212	746.3	1.788
6,000	23.98	.8013	3.1	37.6	.001988	.06395	.8358	.9142	745.0	1.810
6,500	23.53	.7863	2.1	35.8	.001957	.06298	.8231	.9072	743.7	1.833
7,000	23.09	.7715	1.1	34.0	.001928	.06202	.8106	.9003	742.3	1.856
7,500	22.65	.7570	+0.1	32.3	.001898	.06107	.7982	.8934	741.0	1.879
8,000	22.22	.7427	−0.9	30.5	.001869	.06013	.7859	.8865	739.6	1.903
8,500	21.80	.7286	−1.8	28.7	.001840	.05920	.7738	.8797	738.3	1.927
9,000	21.38	.7147	−2.8	26.9	.001812	.05829	.7618	.8728	736.9	1.951
9,500	20.98	.7010	−3.8	25.1	.001784	.05739	.7500	.8660	735.6	1.977
10,000	20.57	.6876	−4.8	23.3	.001756	.05649	.7384	.8593	734.2	2.002
10,500	20.18	.6743	−5.8	21.6	.001728	.05561	.7268	.8525	732.9	2.028
11,000	19.79	.6613	−6.8	19.8	.001701	.05474	.7154	.8458	731.5	2.054
11,500	19.40	.6484	−7.8	18.0	.001675	.05388	.7042	.8391	730.1	2.081
12,000	19.02	.6358	−8.8	16.2	.001648	.05302	.6930	.8325	728.8	2.107
12,500	18.65	.6234	−9.8	14.4	.001622	.05218	.6820	.8259	727.4	2.135
13,000	18.29	.6112	−10.8	12.6	.001596	.05135	.6712	.8193	726.0	2.163
13,500	17.93	.5991	−11.7	10.9	.001571	.05053	.6605	.8127	724.7	2.192
14,000	17.57	.5873	−12.7	9.1	.001545	.04972	.6499	.8061	723.3	2.220
14,500	17.22	.5756	−13.7	7.3	.001521	.04892	.6394	.7996	721.9	2.250
15,000	16.88	.5642	−14.7	5.5	.001496	.04813	.6291	.7931	720.5	2.280
15,500	16.54	.5529	−15.7	3.7	.001472	.04736	.6189	.7867	719.2	2.310
16,000	16.21	.5418	−16.7	1.9	.001448	.04658	.6088	.7802	717.8	2.341
16,500	15.88	.5309	−17.7	+0.2	.001424	.04582	.5988	.7738	716.4	2.373
17,000	15.56	.5201	−18.7	−1.6	.001401	.04507	.5890	.7675	715.0	2.404
17,500	15.25	.5096	−19.7	−3.4	.001378	.04432	.5793	.7611	713.6	2.438
18,000	14.94	.4992	−20.7	−5.2	.001355	.04359	.5697	.7548	712.2	2.470
18,500	14.63	.4890	−21.7	−7.0	.001332	.04287	.5603	.7485	710.8	2.504
19,000	14.33	.4789	−22.6	−8.8	.001310	.04215	.5509	.7422	709.4	2.539
19,500	14.03	.4690	−23.6	−10.5	.001288	.04145	.5417	.7360	708.0	2.573
20,000	13.74	.4593	−24.6	−12.3	.001267	.04075	.5326	.7298	706.6	2.608
20,500	13.46	.4498	−25.6	−14.1	.001245	.04006	.5236	.7236	705.2	2.645
21,000	13.18	.4404	−26.6	−15.9	.001224	.03939	.5148	.7175	703.8	2.682
21,500	12.90	.4312	−27.6	−17.7	.001203	.03872	.5060	.7113	702.3	2.719
22,000	12.63	.4221	−28.6	−19.5	.001183	.03805	.4974	.7053	700.9	2.757
22,500	12.36	.4132	−29.6	−21.2	.001163	.03740	.4889	.6992	699.5	2.795
23,000	12.10	.4044	−30.6	−23.0	.001143	.03676	.4804	.6931	698.1	2.836
23,500	11.84	.3958	−31.6	−24.8	.001123	.03612	.4721	.6871	696.7	2.876
24,000	11.59	.3874	−32.5	−26.6	.001103	.03550	.4640	.6811	695.2	2.917
24,500	11.34	.3790	−33.5	−28.4	.001084	.03488	.4559	.6752	693.8	2.958
25,000	11.10	.3709	−34.5	−30.2	.001065	.03427	.4479	.6693	692.4	3.000

STANDARD ATMOSPHERE TABLE—*Continued*

Altitude (ft)	Pressure (in. Hg)	Pressure ratio, p/p_0	Temperature (°C)	Temperature (°F)	Density ρ, (slugs/ft³)	Specific weight, $w = \rho g$ (lb/ft³)	Density ratio σ	$\sigma^{\frac12}$	Speed of sound, a (mph)	Kinematic viscosity, $\nu = \mu/\rho$ (ft²/sec ×10⁴)
25,000	11.10	.3709	−34.5	−30.2	.001065	.03427	.4479	.6693	692.4	3.000
25,500	10.86	.3628	−35.5	−31.9	.001046	.03367	.4400	.6634	690.9	3.044
26,000	10.62	.3550	−36.5	−33.7	.001028	.03307	.4323	.6575	689.5	3.087
26,500	10.39	.3472	−37.5	−35.5	.001010	.03249	.4246	.6516	688.0	3.132
27,000	10.16	.3396	−38.5	−37.3	.000992	.03191	.4171	.6458	686.6	3.178
27,500	9.94	.3321	−39.5	−39.1	.000974	.03134	.4096	.6400	685.1	3.224
28,000	9.72	.3248	−40.5	−40.9	.000957	.03078	.4023	.6343	683.7	3.272
28,500	9.50	.3176	−41.5	−42.6	.000939	.03022	.3950	.6285	682.2	3.320
29,000	9.29	.3105	−42.5	−44.4	.000922	.02968	.3879	.6228	680.8	3.369
29,500	9.08	.3036	−43.4	−46.2	.000906	.02914	.3809	.6171	679.3	3.420
30,000	8.88	.2967	−44.4	−48.0	.000889	.02861	.3739	.6115	677.8	3.470
30,500	8.68	.2900	−45.4	−49.8	.000873	.02808	.3671	.6059	676.4	3.523
31,000	8.48	.2835	−46.4	−51.6	.000857	.02757	.3603	.6003	674.9	3.575
31,500	8.29	.2770	−47.4	−53.3	.000841	.02706	.3537	.5947	673.4	3.629
32,000	8.10	.2707	−48.4	−55.1	.000825	.02656	.3471	.5891	671.9	3.684
32,500	7.91	.2645	−49.4	−56.9	.000810	.02606	.3406	.5836	670.4	3.741
33,000	7.73	.2584	−50.4	−58.7	.000795	.02557	.3342	.5781	668.9	3.797
33,500	7.55	.2524	−51.4	−60.5	.000780	.02509	.3280	.5727	667.5	3.856
34,000	7.38	.2465	−52.4	−62.3	.000765	.02462	.3218	.5672	666.0	3.915
34,500	7.20	.2407	−53.4	−64.0	.000751	.02415	.3157	.5618	664.5	3.976
35,000	7.03	.2351	−54.3	−65.8	.000736	.02369	.3096	.5565	663.0	4.037
35,332	6.92	.2314	−55.0	−67.0	.000727	.02339	.3057	.5529	662.0	4.080
36,000	6.71	.2242	−55.0	−67.0	.000704	.02266	.2961	.5442	662.0	4.212
37,000	6.39	.2137	−55.0	−67.0	.000671	.02160	.2823	.5313	662.0	4.418
38,000	6.10	.2037	−55.0	−67.0	.000640	.02059	.2692	.5188	662.0	4.633
39,000	5.81	.1942	−55.0	−67.0	.000610	.01963	.2566	.5066	662.0	4.860
40,000	5.54	.1852	−55.0	−67.0	.000582	.01872	.2446	.4946	662.0	5.098
41,000	5.28	.1765	−55.0	−67.0	.000555	.01784	.2332	.4829	662.0	5.347
42,000	5.04	.1683	−55.0	−67.0	.000529	.01701	.2224	.4715	662.0	5.609
43,000	4.80	.1605	−55.0	−67.0	.000504	.01622	.2120	.4604	662.0	5.883
44,000	4.58	.1530	−55.0	−67.0	.000481	.01546	.2021	.4495	662.0	6.171
45,000	4.36	.1458	−55.0	−67.0	.000458	.01474	.1927	.4389	662.0	6.473
46,000	4.16	.1390	−55.0	−67.0	.000437	.01405	.1837	.4286	662.0	6.790
47,000	3.97	.1326	−55.0	−67.0	.000416	.01340	.1751	.4185	662.0	7.122
48,000	3.78	.1264	−55.0	−67.0	.000397	.01277	.1669	.4086	662.0	7.470
49,000	3.60	.1205	−55.0	−67.0	.000379	.01218	.1592	.3990	662.0	7.836
50,000	3.44	.1149	−55.0	−67.0	.000361	.01161	.1517	.3895	662.0	8.219
51,000	3.28	.1095	−55.0	−67.0	.000344	.01107	.1447	.3803	662.0	8.621
52,000	3.12	.1044	−55.0	−67.0	.000328	.01055	.1379	.3714	662.0	9.042
53,000	2.98	.0995	−55.0	−67.0	.000313	.01006	.1315	.3626	662.0	9.485
54,000	2.84	.0949	−55.0	−67.0	.000298	.00959	.1253	.3540	662.0	9.949
55,000	2.71	.0905	−55.0	−67.0	.000284	.00914	.1195	.3457	662.0	10.436
56,000	2.58	.0862	−55.0	−67.0	.000271	.00872	.1139	.3375	662.0	10.947
57,000	2.46	.0822	−55.0	−67.0	.000258	.00831	.1086	.3296	662.0	11.482
58,000	2.35	.0784	−55.0	−67.0	.000246	.00792	.1035	.3218	662.0	12.044
59,000	2.24	.0747	−55.0	−67.0	.000235	.00755	.0987	.3142	662.0	12.633
60,000	2.13	.0712	−55.0	−67.0	.000224	.00720	.0941	.3068	662.0	13.251
61,000	2.03	.0679	−55.0	−67.0	.000213	.00687	.0897	.2995	662.0	13.899
62,000	1.94	.0647	−55.0	−67.0	.000203	.00655	.0855	.2925	662.0	14.579
63,000	1.85	.0617	−55.0	−67.0	.000194	.00624	.0816	.2856	662.0	15.293
64,000	1.76	.0589	−55.0	−67.0	.000185	.00595	.0777	.2788	662.0	16.041
65,000	1.68	.0561	−55.0	−67.0	.000176	.00567	.0741	.2723	662.0	16.825

QUANTITIES USED IN DERIVING VALUES FOR STANDARD DAY

Temperature at sea level: $t_0 = 15°C = 59°F$
Absolute temperature at sea level: $T_0 = 288°C = 518.4°F$
Pressure at sea level: $p_0 = 29.92117$ in. of Hg $= 2116.229$ lb /ft^2
Acceleration of gravity: $g = 32.1740$ ft /sec^2
Density at sea level: $\rho_0 = .002378$ lb sec^2 /ft^4 or slugs /ft^3
Temperature gradient: $\alpha = .0065°C$ /meter $= .00356617°F$ /ft
Isothermal temperature: $t_i = -55°C = -67°F$
Ratio of specific heats: $\gamma = 1.4000$

Absolute viscosity: $\mu = \dfrac{3.059 \times 10^{-8} T^{3/2}}{T + 114}$ (T in °C absolute)

NACA AIRFOIL SECTION DATA

TABLE 2

Low Drag Airfoils

Section	$c_{l_{max}}$	α_{0L}	a_0	c_{l_i}	$c_{d_{min}}$	c_d			$c_{m_{ac}}$	a.c.	M_{cr}
						$c_l = 0$	$c_l = .4$	$c_l = .8$			
64_1–006	.83	0	.104	0	.0038	.0038	.0057		0	.256	.836
64_1–009	1.17	0	.108	0	.0040	.0040	.0061	.0082	0	.262	.785
64_1–012	1.44	0	.110	0	.0042	.0042	.0062	.0081	0	.262	.744
64_1–206	1.03	−1.2	.104	.18	.0038	.0050	.0057	.0062	−.040	.253	.793
64_1–209	1.40	−1.4	.104	.20	.0040	.0053	.0060	.0075	−.040	.261	.760
64_1–212	1.55	−1.2	.108	.19	.0042	.0043	.0050	.0077	−.028	.262	.728
64_2–215	1.57	−1.5	.111	.22	.0045	.0046	.0048	.0081	−.030	.265	.700
64_1–412	1.67	−2.6	.112	.36	.0044	.0058	.0046	.0076	−.070	.267	.700
64_2–415	1.66	−2.8	.114	.40	.0046	.0060	.0049	.0080	−.070	.264	.678
64_3–418	1.56	−3.0	.117	.40	.0050	.0060	.0050	.0062	−.065	.273	.655
64_3–421	1.48	−2.8	.119	.40	.0050	.0056	.0050	.0058	−.066	.276	.642
65_1–006	.93	0	.105	0	.0035	.0035	.0058		0	.258	.838
65_1–009	1.08	0	.106	0	.0041	.0041	.0060	.0081	0	.264	.790
65_1–012	1.36	0	.106	0	.0038	.0038	.0061	.0081	0	.261	.750
65_1–206	1.06	−1.2	.106	.20	.0036	.0050	.0055	.0071	−.030	.257	.791
65_1–209	1.30	−1.2	.108	.20	.0038	.0052	.0058	.0075	−.032	.259	.755
65_1–212	1.48	−1.2	.108	.20	.0040	.0047	.0057	.0082	−.032	.261	.726
65_1–410	1.52	−2.6	.110	.36	.0037	.0058	.0038	.0072	−.068	.262	.714
65_1–412	1.64	−3.0	.109	.38	.0038	.0056	.0038	.0075	−.070	.265	.697
65_2–415	1.62	−2.6	.107	.38	.0042	.0060	.0042	.0084	−.060	.268	.675
65_3–418	1.55	−2.6	.106	.40	.0043	.0062	.0043	.0072	−.060	.265	.656
65_4–421	1.55	−3.2	.111	.42	.0044	.0048	.0044	.0053	−.065	.272	.637
0006	.92	0	.106			.0052	.0058	.0090	0	.250	.805
0009	1.33	0	.110			.0056	.0060	.0084	0	.250	.766
1412	1.57	−1.2	.103			.0058	.0061	.0076	−.023	.252	.720
2412	1.67	−2.0	.104			.0060	.0060	.0072	−.047	.247	.690
2415	1.65	−2.0	.106			.0064	.0064	.0077	−.045	.246	.677
2418	1.47	−2.1	.099			.0068	.0071	.0081	−.045	.241	.650
2421	1.47	−1.9	.099			.0071	.0072	.0088	−.040	.241	.630
2424	1.30	−1.9	.093			.0074	.0078	.0099	−.040	.231	.606
4412	1.66	−4.0	.106			.0062	.0060	.0064	−.092	.247	.647
4415	1.64	−4.0	.106			.0066	.0064	.0070	−.093	.245	.635
4418	1.54	−4.0	.100			.0070	.0066	.0076	−.088	.242	.620
4421	1.46	−4.0	.096			.0075	.0073	.0081	−.086	.238	.602
23012	1.78	−1.2	.104			.0068	.0060	.0068	−.015	.247	.672
23015	1.72	−1.0	.103			.0072	.0064	.0072	−.008	.243	.663
23018	1.62	−1.2	.104			.0069	.0068	.0078	−.005	.243	.655
23021	1.51	−1.2	.100			.0070	.0073	.0089	−.004	.238	.623

INDEX